Green Energy and Technology

Climate change, environmental impact and the limited natural resources urge scientific research and novel technical solutions. The monograph series Green Energy and Technology serves as a publishing platform for scientific and technological approaches to “green”—i.e. environmentally friendly and sustainable—technologies. While a focus lies on energy and power supply, it also covers “green” solutions in industrial engineering and engineering design. Green Energy and Technology addresses researchers, advanced students, technical consultants as well as decision makers in industries and politics. Hence, the level of presentation spans from instructional to highly technical. **Indexed in Scopus**.

More information about this series at http://www.springer.com/series/8059

Zujian Huang

Application of Bamboo in Building Envelope

Springer

Zujian Huang
School of Architecture
South China University of Technology
Guangzhou, China

ISSN 1865-3529 ISSN 1865-3537 (electronic)
Green Energy and Technology
ISBN 978-3-030-12031-3 ISBN 978-3-030-12032-0 (eBook)
https://doi.org/10.1007/978-3-030-12032-0

This Springer imprint is published by the registered company Springer Nature Switzerland AG.
The registered company address is: Gewerbestrasse 11, 6330 Cham, Switzerland

Foreword I

Widely distributed in mid- and low-latitude regions worldwide, bamboo is a large fast-growing herb. For a quite long period, this easily available and affordable material was called 'the poor people's wood' and played an importable role in the daily life of peoples in Asia Pacific and the Americas.

Globally, most bamboo forests see extensive growth. Since the 1950s, a few countries including China and Japan began the intensive farming of artificial bamboo forests, gradually modernizing the bamboo-related industries. Since the 1980s, with reference to wood material technology, the forestry and artificial board experts made breakthroughs in bamboo processing, making bamboo industry to embark on the track of modern material technology and standardized utilization.

The sustainable architecture prompts the sourcing and R&D of eco-building materials. The existing forest resource, material technology, and market demand draw new attention to bamboo, the traditional building material. The strengths of bamboo include low even negative carbon footprint during the whole life cycle and close to or even better comprehensive mechanical properties than the wood product of the same kind. The modified bamboo trumps the raw bamboo in size, durability, and standardization. However, without a systemic study, the replacement of wood with bamboo in architecture has not been well justified yet.

As an architect, Dr. Zujian Huang concerns more about how it functions including construction durability, indoor hygrothermal comfort, and energy consumption than the traditional perspective that how it constructs or how it looks. At the level of material and architecture, the book discusses industrialized utilization of bamboo and bamboo building envelope, explores how to optimize the physical performance of bamboo components and enclosed space, and studies the construction method adaptive to bamboo properties and typical climatic characteristics. On this basis, it looks further into bamboo industry with a focus on high-performance bamboo suitable for building envelope and its production. The well-grounded material parameters and construction method are expected to lay a solid foundation for promoting bamboo application in building industry, R&D of high-quality products of bamboo and bamboo envelope and developing standardized bamboo building system.

The research conducted by Dr. Zujian Huang in Southern China with abundant bamboo resources is of great pragmatic value. I believe the book will greatly promote the research and application of bamboo in building industry.

Guangzhou, China

Yimin Sun
Changjiang Scholar Distinguished Professor
Dean School of Architecture, South China University of Technology

Foreword II

Bamboo is a material that architects like. This is mainly due to its form, which is seen by most as the ideal natural formalization of a tube, consisting of a tubular stem with perpendicular reinforcing nodes. It is often cited in the context of bionic architecture. But besides its great structural effectiveness, bamboo has other qualities: It is a fast-growing natural resource which thrives in areas of the world that are currently showing the fastest economic growth. So the idea of putting the accent not on the structural form but on the material properties and especially its fibers is not as far-fetched as it may at first seem.

Materials such as wood have experienced a similar development and are now more important than ever. Thus, the process of de- and re-composition of natural materials is of great currency and contributes to their application in sectors where consistent material qualities are of utmost importance. They contribute to a more environment-friendly economy and open up a new spectrum of industrial applications of bamboo-based materials. This book addresses the subject with the necessary depth and establishes basic knowledge about the use of bamboo and bamboo-based building products in the context of the building envelope. It outlines typical constructions, investigates their hygrothermal properties, and further compares the results to already better-known wood applications.

Thus, it is revealed as a necessary step in the further promotion of environment-friendly materials in the building sector, opening up strategic options for bamboo-producing (mostly hot and humid) areas. For ecological reasons, energy and transport will and will have to become more expensive. As a result, it is in the best interest of bamboo-producing countries to develop the full potential of this more sustainable and locally available natural building material for its fast-growing markets. Industrially produced modified bamboo products pose no danger to the wide range of current, mostly structural uses of bamboo, except perhaps for increasing demand and thus raising bamboo prices. They are, on the contrary, adding a new range of more ecological and in many ways better materials for the fast-developing markets in the regions in which they grow.

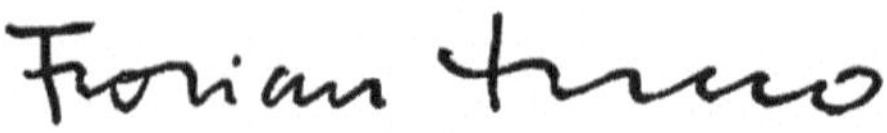

Munich, Germany

Florian Musso
Professor, Chair of Building
Construction and Material Science
Faculty of Architecture
Technical University of Munich

Preface

The book is based on the author's doctoral research project completed from 2013 to 2018, 'Study on the Application of Bamboo in Building Envelope,' which is funded by the China Scholarship Council (CSC) as a 'Joint-Training Ph.D. Program' between the South China University of Technology (SCUT, Guangzhou, China) and the Technical University of Munich (TUM, Munich, Germany). During the program, an academic paper related to the study on bamboo properties was published in the Journal of Construction and Building Materials, which caused the attention of Springer-Verlag, who then invited the author to write a book on this topic, thus started up the publication of this book.

The reason for the author and his supervisors decided to carry out this project was that the application of bamboo in building industry possesses multiple potentials such as high-value utilization of bamboo forest resources and improvement of building sustainability, which is of especial significance for those regions that are rich in bamboo resources and short of wood forest. Since 1980s, learning from the timber processing technology, forestry scholars, and wood-based panel experts has made progress in the field of bamboo processing and successfully developed bamboo-based panels, bamboo fiber, bamboo charcoal, bamboo vinegar, and other series products, resulting from the policy of 'full utilization on bamboo,' which gradually moved the bamboo industry toward the track of industrial utilization. Similarly, the mature timber construction systems from Europe and Americas have also been introduced into the research of 'building with bamboo' in other countries, for which the differences in material properties and climate adaptability have yet to be resolved. Due to the bottlenecks of lacking material property parameters, construction design methods, and technical standards, etc., a bamboo construction system for promotable application has not been formed in the latest progress and research status.

The research project defined typical industrial bamboo products, including raw bamboo, bamboo-based panels, natural bamboo fiber, and bamboo charcoal, as the research objects on material level and building envelope, respectively, on building level. Since 2014, the author has completed research on manufacturers, universities, and laboratories related to bamboo production, research, and development in China. Then, in the following years, from 2015 to 2017, the author completed the establishment of a research model for bamboo building envelope, the property study on bamboo, and the performance study on bamboo building component and enclosed space in Germany. During this period, the German timber construction system, and the research on building materials and their construction systems of the Fraunhofer IBP Hygrothermics Department, played an important role as a reference of research method to this project. In March 2018, the author finished his dissertation *Application Fundamental Research on Bamboo in Building Envelope*, which summarized the achievements of the study on material properties, building component, and enclosed space performance evaluation and optimization. Then, the dissertation is integrated into this book, which is expected to contribute material parameter database and construction design method, as well as a basis for the promotion of bamboo application in the building industry, the development of high-quality bamboo and bamboo building envelope products, as well as a modular and standardized bamboo construction system.

For the completion of this book, the author must first thank his two doctoral supervisors, Prof. Yimin Sun (School of Architecture, SCUT) and Prof. Florian Musso (Chair of Building Construction and Material Science, TUM), who jointly provided the best platform and effective guidance for the research project. It was a greatest honor of the author for being guided by the two professors in this stage as the beginning of his scientific career. Then the author should thank Prof. Thomas Herzog and Dr. Lingyun Zhang, who established contact between Prof. Musso and the author, and Markus Heinsdorff who gave a wealth of professional information and inspiration during the frequent exchanges and cooperation. The author would also like to thank Prof. Qinglin Meng (SCUT), Prof. Ya Feng (China Southwest Architecture Design and Research Institute Corp. Ltd), Prof. Yufeng Zhang (SCUT), Prof. Jing Wang (SCUT), and Prof. Nianping Li (Hunan University), who gave suggestions for modifying the dissertation; Prof. Yan Xiao (Nanjing Tech University), Mr. Hai Lin (Dasso Industrial Group Co, Ltd), Prof. Wenbiao Zhang (Zhejiang A&F University), Dr. Bo Shan (Hunan University), Dr. Chi Feng (KU Leuven), who provided support for the material test; Dr. Cornelia Fitz and Mr. Matthias Pazold (Fraunhofer IBP) and Dr. Simon Schmidt (TUM), who provided help on the operation of WUFI; as well as the colleagues, Stefan Giers, Ursula Schurmann, Lavinia Herzog, and Vesna Pungercar (Chair of Building Construction and Material Science, TUM), who offered friendly help to the author during his stay in Germany.

The author is well aware that a large number of basic research and application practices are needed to form a complete modern bamboo building envelope construction system. This book hopes to give a preliminary outline and start a discussion in this field. Owing to the limited knowledge of the author, there must be mistakes and errors in the book. Thus, welcome all readers to offer kind criticism and correction.

Guangzhou, China
December 2018

Zujian Huang

Acknowledgements

The research project is funded by: the China Postdoctoral Science Foundation (2018M640782), the Fundamental Research Funds for the Central Universities (D2181640), and the State Scholarship Fund by the Chinese Scholarship Council (201506150017).

There are several sections, including some figures and texts that are reorganized from the author's previous paper, which are:

Sections 3.2.3 and 3.2.4 of Chap. 3, as well as Figs. 3.3, 3.4, 3.14–3.16, 3.18, 3.20–3.23, and Table 3.5, reprinted from Construction and Building Materials, 155, Huang Z., Sun Y., Musso F., *Experimental study on bamboo hygrothermal properties and the impact of bamboo-based panel process*, 1112–1125, 2017, with permission from Elsevier;

Section 4.2.3 of Chap. 4, as well as Table 3.3 (Chap. 3), Table 3.4 (Chap. 3) and Table 4.1, reprinted from Construction and Building Materials, 156, Huang Z., Sun Y., Musso F., *Assessment of bamboo application in building envelope by comparison with reference timber*, 844–860, 2017, with permission from Elsevier;

Sections 3.3.3 and 3.3.4 of Chap. 3, as well as Figs. 3.25, 3.26, 3.31–3.34, reprinted from Construction and Building Materials, 177, Huang Z., Sun Y., Musso F., *Hygrothermal performance of natural bamboo fiber and bamboo charcoal as local construction infills in building envelope*, 342–357, 2018, with permission from Elsevier;

Sections 5.2, 5.3.1, 5.3.2 and 5.3.3 of Chap. 5, as well as Figs. 5.2–5.25, and Table 5.3, reprinted from Construction and Building Materials, 202, Huang Z., Sun Y., Musso F., *Hygrothermal Performance Optimization on Bamboo Building Envelope in Hot-Humid Climate Region*, 223–245, 2019, with permission from Elsevier;

Sections 4.1.2 and 4.1.3 of Chap. 4, as well as Figs. 4.2–4.5, and Tables 4.2–4.4, reprinted from IOP Conference Series: Materials Science and Engineering, 264, Zujian Huang, Yimin Sun, Florian Musso, *Assessment on bamboo scrimber as a substitute for timber in building envelope in tropical and humid subtropical climate zones—part 1 hygrothermal properties' test*, 1–9, 2017, with permission from IOP Publishing Ltd;

Other data and photographs quoted from others have been clarified by a citation description following each table or figure.

Summary

The application of bamboo in building industry possesses multiple potentials such as high-value utilization of bamboo forest resources and improvement of building sustainability. The research project defines typical industrial bamboo products and building envelope, respectively, as the research objects on material and building levels. Based on the Künzel equations of coupled heat and moisture process, the study mainly completes measurement on materials and simulation on building component and enclosed space and carried out discussion on the study model for bamboo building envelope.

Property study on material. Based on the material parameter requirement of HAM model, test items are systematically carried out, including bulk density test and vacuum saturation test for basic properties; sorption test for moisture storage properties; capillary absorption test, water vapor transmission test, and drying test for moisture transport properties; thermal analysis for heat storage properties; thermal conductivity test, surface light and thermal properties' test for heat transport properties. The test results of partition boards, including raw bamboo (FB) and five typical bamboo-based panels (BBPs), show that hygrothermal properties have a stronger correlation with open porosity Φ than with bulk density ρ; comparison between BBPs and FB shows that bamboo-based panel processes improve the homogeneity, broaden the material property spectrum, and strengthen the distinction with reference timber; the transport properties change greater than the storage properties, and the hygric properties change greater than the thermal properties; comparison among bamboo and the corresponding reference timber (RT) shows that bamboo has lower moisture storage and transport properties and higher heat storage and transport properties than RT; a dynamic test in wind tunnel with a summer typical meteorological day condition is performed and shows that bamboo scrimber has lower moisture adsorption and desorption rate than the reference hardwood; the significant difference between the static and dynamic test results shows the necessity of a comprehensive evaluation approach that can take typical meteorological conditions into consideration. The test results for construction infills, including natural bamboo fiber (BF) and bamboo charcoal (BC), show that the relation, respectively, between the hygrothermal properties of BF and the ambient air relative humidity can generally be defined with exponential function; compared with the uncarbonized bamboo panels, the moisture-control capacity and rate of BC are improved.

Performance study on building component and enclosed space. With ten groups of meteorological data from nine North American typical climate zones as external conditions, timber and timber units are set as reference model, accordingly bamboo and bamboo units of the same construction and space size as evaluation model in computer program WUFI Plus, by which performance on levels of material, building component, and space unit is compared with consideration to external climate, internal heat and moisture load, construction type, and HVAC conditions. Simulation results show that bamboo units have better heat storage but worse heat transport performance and surpass timber units in lightweight construction in hot regions. Furthermore, with four groups of meteorological data from North American Hot-Humid climate regions as external conditions, a large sample of exterior walls are constructed in WUFI Plus and the coupled heat and moisture process simulation is performed. Annual exterior walls hygrothermal performance, indoor hygrothermal environment, and HVAC demand are chosen as indicators, with which factor impact analysis is carried out to investigate the effect of the partition boards, air layer, thermal insulation and moisture-control infill layers, and the facade rainfall treatment on hygrothermal performance of the layered constructions. Climate-responsive design suggestions in Hot-Humid climate regions are generated from the analysis results, in terms of material and construction parameter optimization.

Discussion on bamboo building envelope study model. Correlation analyses are carried out between the simulation results and the meteorological parameters, showing that solar radiation, air temperature, relative humidity, and driving rain have complementary significant correlations with the corresponding results; liquid water-related material parameters and variation in heat and moisture transport properties caused by the change of moisture content have a significant influence on the simulation results; the annual dynamic simulation shows different laws with the static assessment, which means a long-period dynamic evaluation on building envelope under practical conditions is of necessity.

As an application fundamental research on bamboo in building envelope, the study on material properties, building component, and enclosed space performance evaluation and optimization contributes material parameter database and construction design reference to the application of bamboo in building envelope.

Keywords Bamboo · Building envelope · Hygrothermal property · Hygrothermal performance

Contents

Terminologies and Abbreviations

Terminologies

Roman-Letter Notations

A	Test area of specimen (m^2)
A_{cap}	Water absorption coefficient ($kg/(m^2\ s^{0.5})$)
a_w	Moisture-related thermal conductivity supplement ((W/m K)/u(−))
c_o/c_d	Specific heat capacity of dry building material (kJ/(kg K))
d	Specimen thickness (m)
D_l/D_w	Liquid water transport coefficient (m^2/s)
G	Water vapor flow rate through specimen (kg/s)
g_v	Density of water vapor flow rate ($kg/(m^2\ s)$)
H	Total enthalpy (J/m^3)
h_{air}	Water vapor diffusion-equivalent air layer thickness (m)
h_v	Evaporation heat of water (J/kg)
m_d	Mass of a dry building material (kg)
$m_{moisture}$	Mass of moisture in a building material (kg)
m_{vac}	Mass of a building material of vacuum saturation (kg)
m_w	Mass of a moist building material (kg)
p	Partial pressure of water vapor (Pa)
p_o	Standard barometric pressure = 101,325 (Pa)
Δp_v	Water vapor pressure difference across specimen (Pa)
R_o	Gas constant for water vapor = 461.5 (N m/(kg K))
$R_{v,total}$	Total water vapor resistance (m^2 s Pa/kg)
R_v	Specimen water vapor resistance (m^2 s Pa/kg)
S	Heat storage coefficient ($W/(m^2\ K)$)

S_{24h}	24h heat storage coefficient (W/(m^2 K))
T	Thermodynamic temperature (absolute temperature) (K)
t	Time (s)
U	Drying rate/Moisture adsorption and desorption rate (kg/(m^2 s))
u	Water content in mass rate (kg/kg)
V	Bulk volume (sum of framework volume and porosity volume) (m^3)
V_{moisture}	Volume of moisture (calculated in liquid) (m^3)
w	Water content in mass–volume rate (kg/m^3)
w_{cap}	Free water saturation (capillary saturation) (kg/m^3)
w_{vac}	Maximum water saturation (vacuum saturation) (kg/m^3)
Z	The cycle of the temperature wave (s)

Greek-Letter Notations

α_e	Solar direct absorptivity (%)
δ	Water vapor transfer coefficient (kg/(m s Pa))
δ_{air}	Water vapor permeability of air with respect to partial vapor pressure (kg/(m s Pa))
ε	Hemispherical emissivity
θ	Water content in volume rate (m^3/m^3)
ϑ(T)	Celsius temperature (°C)
λ	Thermal conductivity of a moist building material (W/(m K))
λ_o/λ_d	Thermal conductivity of a dry building material (W/(m K))
μ	Water vapor diffusion resistance factor of a building material
ρ_d	Bulk density of a dry building material (kg/m^3)
ρ_e	Solar direct reflectivity (%)
ρ_l	Density of water in liquid form (kg/m^3)
ρ_v	Light reflectivity (%)
ρ_w	Bulk density of a moist building material (kg/m^3)
Φ	Porosity (open)
φ(RH)	Relative humidity (%)

Mathematical Symbols

d	Operator for total differential
∂	Operator for partial differential
Δ	Difference operator

Abbreviations

Material Name

BBP	Bamboo-based panel
BC	Bamboo charcoal
BF	Natural bamboo fiber
BFB	Bamboo scrimber
BMB	Plybamboo
BOSB	Bamboo oriented strand board
BPB	Bamboo particleboard
BSB	Bamboo laminated lumber
FB	Flattened bamboo panel
HW	Hardwood
RB	Raw bamboo
RT	Reference timber
SW	Softwood

Physical Quantity

w-BF	Annual moisture content of infill layer (kg/m^3)
ξ	Volumetric moisture capacity (kg/m^3)
ξ/A	Moisture capacity per unit area (kg/m^2)
s_d	Vapor diffusion thickness (m)
U_c	Heat transfer coefficient ($W/(m^2\ K)$)
S_c	Heat capacity ($kJ/(m^2\ K)$)
S_i	Thermal capacity inside ($kJ/(m^2\ K)$)
t_p	Temperature phase shift (h)
α	Temperature attenuation
M_{flow}	Annual moisture flow through exterior wall (kg)
$M_{exchange}$	Annual moisture exchange with exterior wall (kg)
H_{flow}	Annual heat flow through exterior wall (kWh)
$H_{exchange}$	Annual heat exchange with exterior wall (kWh)
$T_{i.mean}$	Annual indoor air temperature mean value (°C)
$T_{i.amp}$	Annual indoor air temperature amplitude (°C)
$T_{is.amp}$	Annual exterior wall inside surface temperature amplitude (°C)
$RH_{i.mean}$	Annual indoor relative humidity mean value (%)
$RH_{i.amp}$	Annual indoor relative humidity amplitude (%)
$T_{e.mean}$	Annual outdoor air temperature mean value (°C)
$T_{e.amp}$	Annual outdoor air temperature amplitude (°C)
$RH_{e.mean}$	Annual outdoor relative humidity mean value (%)
$RH_{e.amp}$	Annual outdoor relative humidity amplitude (%)
SR	Annual solar radiation ($kWh/(m^2\ a)$)

DR	Annual driving rain (Ltr/m^2)
PMV	Predicted mean vote
$HVAC_p$	Annual HVAC period (d)
P	Annual heating and cooling demand (kWh)
H	Annual humidification and dehumidification demand (kg)

Others

HAM	Coupled heat–air–moisture transfer
r	Correlation coefficient
sig	Statistical significance
Af	Tropical rainforest climate
Am	Tropical monsoon climate
Aw	Tropical savanna climate
Cwa	Humid subtropical climate
Cfa	Humid subtropical climate
BSh	Hot semiarid climate
BWh	Hot desert climate

List of Figures

List of Tables

Chapter 1
Introduction

1.1 Research Background

1.1.1 Bamboo Resource

1.1.1.1 The Forest

The world's bamboo forests are mainly distributed in tropical and subtropical climate zones in the Asia-Pacific (67%), Americas (30%), and Africa (3%). There is no native bamboo species in Europe [1] (Fig. 1.1; Table 1.1).

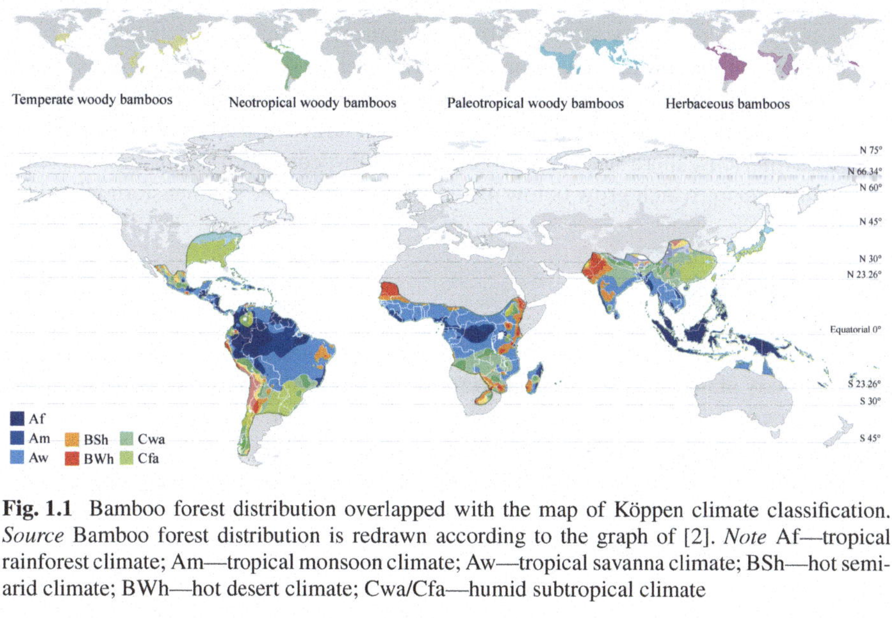

Fig. 1.1 Bamboo forest distribution overlapped with the map of Köppen climate classification. *Source* Bamboo forest distribution is redrawn according to the graph of [2]. *Note* Af—tropical rainforest climate; Am—tropical monsoon climate; Aw—tropical savanna climate; BSh—hot semi-arid climate; BWh—hot desert climate; Cwa/Cfa—humid subtropical climate

Z. Huang, *Application of Bamboo in Building Envelope*,
Green Energy and Technology, https://doi.org/10.1007/978-3-030-12032-0_1

Table 1.1 Bamboo forest distribution in the world

Asia-Pacific	More than 50 genera, 900 species	China, India, Myanmar, Thailand, Bangladesh, Cambodia, Vietnam, Japan, Indonesia, Malaysia, Philippines, South Korea, Sri Lanka, etc.
Americas	18 genera, more than 270 species	Mexico, Guatemala, Costa Rica, Nicaragua, Honduras, Colombia, Venezuela, Amazon basin of Brazil, etc.
Africa	Over a dozen of species	The northwest to the southeast oblique zone that spans the African rainforest and the evergreen and deciduous mixed forest
Europe	No native species	European countries such as Britain, France, Germany, Italy, Belgium, and Netherlands have introduced species from Asian, African, and Latin American countries

(1) Distribution of tropical bamboo. Tropical bamboo species are distributed in Asia-Pacific, the Americas, and Africa. These bamboos are not resistant to frost, so they cannot grow in either high latitudes or high altitudes. The representative bamboo species include the genus *Guadua* in the Americas, as well as the genus *Dendrocalamus* and the *Bambusa* in Asia.

(2) Distribution of temperate bamboo. Temperate bamboo is mainly distributed in China, Japan, and South Korea in Asia, where there is cold winter. These bamboos have a relatively stronger frost resistance, so they can also grow at high altitudes in the tropics. The representative bamboo species are the genus *Arundinaria* and the *Phyllostachys*, the latter including the Chinese Moso bamboo (scientific name: *Phyllostachys heterocycla* (Carr.) Mitford cv. *Pubescens*). In the Americas, there are only three native subspecies belonging to the genus *Arundinaria*, which grow below 46° north latitude in the southeastern USA [1].

1.1.1.2 The Plant

In the whole plant system, there are two major types, namely the herbaceous plants and the woody plants, and woody plants are further divided into the gymnosperms and the angiosperms. The former includes coniferous wood or 'softwood' with needle-like leaves, which, except a few species, drop almost no leaves throughout the year. The latter are further divided into the dicotyledons, which include 'hardwoods' with broad leaves that usually fall in autumn or winter, and the monocotyledons. Bamboo and palm are the most important members of the monocotyledons (Fig. 1.2).

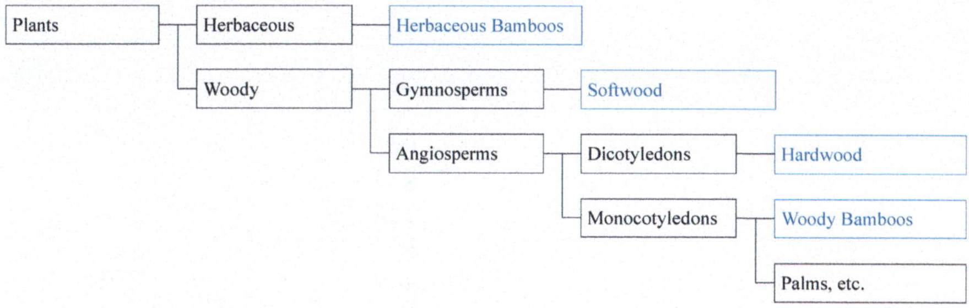

Fig. 1.2 Position of bamboo in the plant taxonomy

When classified from the perspective of bamboo itself, it belongs to the Bambusoideae of the family Poaceae, and the Bambusoideae is further divided into two major categories: herbaceous bamboos and woody bamboos. The former normally has soft bamboo culms of small diameter; the latter generally has woody and usually hollow bamboo culms separated by the membrane. There are about 1250 known bamboo species [3], of which only a few bamboo species have larger culm diameters ($\geq$6 in./15 cm) and longer bamboo culms [1].

Wood is usually divided into two groups, softwood and hardwood, the former growing in temperate and cold regions, the latter in temperate and tropical regions, and the two woods differ in the anatomy of trunks and leaves. When classifying woody bamboo species, there are also several corresponding concepts: scattered bamboo—clumping bamboo; monopodial rhizome—sympodial rhizome; leptomorph rhizome—pachymorph rhizome; temperate bamboo—tropical bamboo. Among them, the scattered bamboo corresponds to the monopodial and the leptomorph rhizomes, usually growing in the temperate region; the clumping bamboo corresponds to the sympodial and the pachymorph rhizomes, usually growing in the tropical region[1,2] (Fig. 1.3).

Rao et al. screened out 20 dominant bamboo species based on five indicators: bamboo use value, cultivation, product and processing, germplasm and genetic resources, and agricultural ecology. Except for the *Phyllostachys pubescens*, which is a scattered bamboo distributed in the temperate region, the rest are mainly clump-

[1]In 1879, the Rivieres were the first to clear the distinction between the two basic forms of the bamboo rhizomes. They used the terms 'caespitose' or 'clumping' for 'pachymorph,' and 'tracant,' or 'running' for 'leptomorph.' In 1925, when McClure was living in China, he introduced the terms 'monopodial' and 'sympodial,' and later, when he was at the Smithsonian Institution in Washington, he developed the concept of 'leptomorph' and 'paquimorph,' respectively. Today the terms 'leptomorph' and 'paquimorph' are used in the Americas by the botanists and taxonomists, to represent and the terms 'monopodial' and 'sympodial' in Asia [1].

[2]The leptomorph and pachymorph types have in common in the bamboo culm morphology, anatomical structure, and growth processes. But the two have differences in the number of chromosomes, which is $2n = 48$ in the leptomorph type, and $2n = 72$ in the pachymorph type. Bamboos with low multiple chromosomes are considered to be of the advanced type and those with high ones are of primitive type [1].

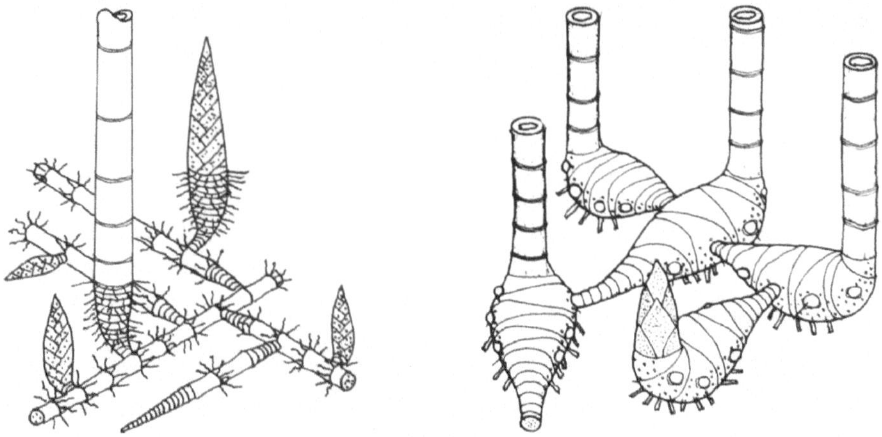

Fig. 1.3 Left—Monopodial rhizome; right—sympodial rhizome. *Source* López [1]

ing bamboo species distributed in the tropical and subtropical regions of Asia and the Americas, including seven species of *Bambusa*, four species of *Dendrocalamus*, and three species of *Gigantochloa*, and the other five species of genus *Cephalostachyum*, *Guadua*, *Melocanna*, *Ochlandra*, and *Thyrsostachys*. In general, China's *Phyllostachys edulis* and South America's *Guadua angustifolia* are the two most prominent ones [4].

1.1.1.3 The Culm

Bamboo is composed of rhizome, culm, branches, and leaves. The bamboo culm is the main part for the material utilization, which corresponds to the trunk of the wood. The bamboo culms of different bamboo species vary in size from 0.1 to 0.3 m in length and 0.2–0.3 cm in diameter to 30–35 m in length and 25–30 cm in diameter (*Dendrocalamus sinicus*), and even for the same bamboo species, affected by factors such as the climate, topography, soil, altitude, age of the plant, and the position in bamboo culm, it is impossible to obtain a uniform value[3] (Table 1.2).

(1) Internode

The internode is the main volume of bamboo culm, consisting of fiber bundles arranged almost in parallel in the longitudinal direction (Fig. 1.4, left). From the cross section, it can be divided into four types according to the relative wall thickness of the culm (Fig. 1.4, right). Types 1 and 2 are more common, including the

[3] Some scholars consider the rhizome as the main stem of bamboo and the culm as a branch. Bamboo culms scattered over brush land are connected with one or some underground stems. But in the eyes of bamboo material researchers, bamboo culm is the main object of study [11]. The object of this book is the bamboo culm, which is the main part for material utilization.

Table 1.2 Constituent unit of the bamboo culm

	Diagram	Description	Normal size
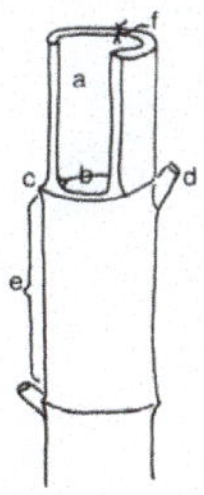	a. Cavity	The hollow space inside the culm	–
	b. Diaphragm	Transversal flaps separate the cavities from each other	Very thin
	c. Node	Corresponding to the position of the diaphragms, at the outside of the culm	3–10 mm
	d. Branch	Emerge only at the nodes	–
	e. Internode	A piece of a culm between two nodes	15–50 cm
	f. Wall	Surrounding fiber bundles, separate the cavities from the outside	10–25 mm

Source Sort out according to Janssen [5]

main large bamboo species; types 3 and 4 exist in a small amount in the bamboo species of small diameter, so that even if it is solid, its actual size is not large.

The cross-sectional shape affects the utilization way of bamboo to a certain extent. For example, the *Guadua* and *Dendrocalamus* that belong to the type 2 have large culm diameter and wall thickness, so that can play its good mechanical strength when directly applied to the load-bearing rods in the form of round bamboo. The thin-walled Moso bamboo with a wall thickness of only 8–10 mm has a cross-sectional shape belonging to the type 1 and is usually processed into bamboo-based panels.

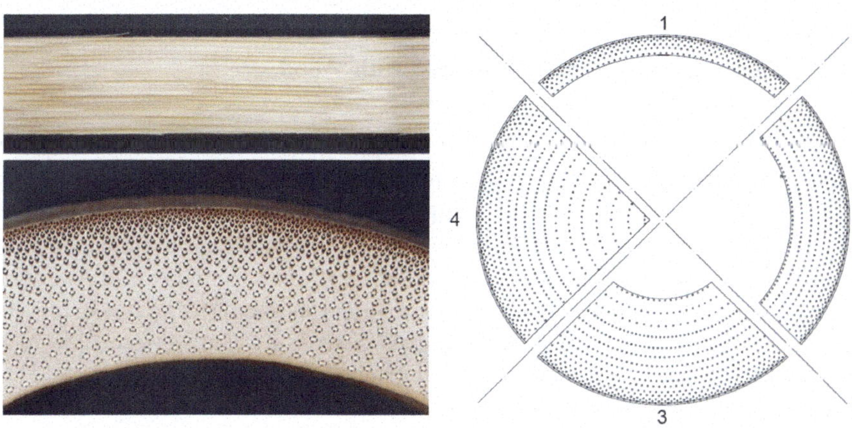

Fig. 1.4 Photographs of bamboo internodes (left: Moso bamboo) and the morphology of internodal sections (right: 1—*Guadua cebolla*, 2—*Guadua angustifolia*, 3—*Guadua amplexifolia*, 4—*Dendrocalamus strictus*). *Source* Left—De Vos [6]; right—by the author

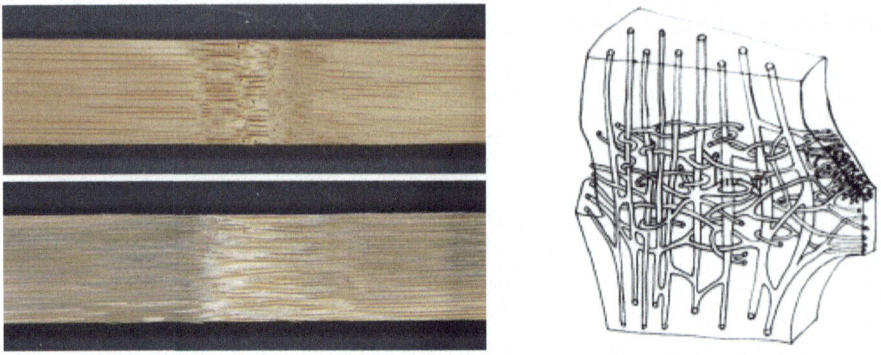

Fig. 1.5 Photographs of bamboo notes (left up: Moso bamboo; left down: *Guadua* bamboo) and the microstructure (right). *Source* left—De Vos [6]; right—Liese [7]

The prior art of technology has been able to flatten the whole bamboo to obtain a glueless panel.

(2) Node

Bamboo nodes are the core section for morphological and intersegmental growth, where the roots and branches only grow. It is the only channel for the lateral transport of water and nutrients during the early growth of bamboo culm, so that there are a small amount of radially developed fibers in the microstructure of bamboo nodes. This has an important influence on the physical properties, the mechanical strength, the drying process, and the application of the liquid preservatives of the bamboo culm (Fig. 1.5).

1.1.1.4 Characteristics When Compared with Wood

(1) Forest resources

Worldwide, bamboo forest coverage is only about 1% of the wood forest coverage, and bamboo cannot compete with wood in terms of species, quantity, management level, and market. However, in some regions, such as the Asia-Pacific region, the proportion of bamboo resources increases. For example, China's bamboo forest coverage accounts for 3% of the total forest's proportion, and in one of its largest bamboo-producing province, Hunan, 56% of the area is covered by bamboo forest. From a local perspective, bamboo resources are still competitive (Table 1.3).

In general, the management level of bamboo forest is relatively backward, mostly in a natural growth state. Partially in China, Japan, and several countries, the practice of intensive management of artificial bamboo forests shows that the quality of materials that can be harvested in bamboo forests can be higher than that of the naturally grown softwood forests (e.g., Norwegian spruce) and hardwood forests (e.g., American red oak) (Table 1.4).

Table 1.3 Forest resources comparison between bamboo and wood

Comparison item	Bamboo	Timber
Species[a]		
Total	1225	30,000
Used for commercial or engineering	20	1500–3000
Traded on international markets	5	500
Forest resources[b]		
Total (million ha)	25–30	4000
Africa (%)	3	15.6
Americas (%)	30	39.8
Asia (%)	67 (the Asia-Pacific region)	14.8
Europe (%)	–	25.4
Oceania (%)	–	4.3
Climate zone		
Tropical zone	Leptomorph rhizome	Hardwood (deciduous wood)
Temperate zone	Pachymorph rhizome	Hardwood + Softwood
Frigid zone	–	Softwood (coniferous wood)
Operation[b]		
Plantation/Natural	Mostly natural; plantation in China, Japan	Primary forest (32.0%); other naturally regenerated forest (60.8%); planted forest (7.2%)

Source
[a]Bamboo: Zhang, Industrial Utilization on Bamboo, 2001; Timber: Volz, The Material, Timber construction manual, 2004
[b]Bamboo: López, Bamboo-the gift of the gods, 2003; Timber: FAOSTAT database, Net forest conversion, 2015

(2) Plant characteristics

As the largest grass species, bamboo has different growth characteristics compared with wood. On the one hand, its height growth is extremely explosive and finishes in a short period of time; on the other hand, there is almost no increase in culm diameter and wall thickness during the height growth. Taking *Guadua* bamboo as an example, the longitudinal dimension used in the construction industry can reach 15 m, but the culm diameter is usually less than 15 cm, and the wall thickness is less than 2 cm. The cross-sectional dimension is much smaller than that of timber, making it impossible for bamboo to be used as a load-bearing member in those construction conditions with higher requirements (Table 1.5).

Table 1.4 Forest yield comparison between bamboo and wood

Comparison item	Bamboo	Timber	
Species	Moso bamboo (China)	Spruce (Norway)	Red oak (USA)
Growing environment	Intensively managed plantation	Natural managed forest	Natural managed forest
Harvest cycle (year)	3–6	60–80	80–100
Harvest method	Partial harvest to clearcutting	Selection harvesting to patch clearcutting	Selection harvesting to patch clearcutting
Average aboveground stem growth over harvest cycle—ovendry mass/green volume (mt/ha/yr)	7.0–10.0 (up to 35.0 in optimal conditions)	2.5–3.5 (2.43–3.26)[a]	2.0–3.0[b]
Regeneration method after harvest	Sprouting	Sprouting/natural regenera-tion/replanting	Sprouting/natural regenera-tion/replanting
Number of harvest cycles over an 80-year period	14–20	1–2	1–2

[a]Wood volume yield and stand structure in Norway spruce understory depending on birch shelterwood density

[b]Data are typical yields of aboveground biomass in an 80-year cycle. Yields are about one-half of this if considering only the main bole or sawlog portion

Source

Bamboo: Yiping and Henley [8]; Timber: Sander, Ontario Ministry of Natural Resources, 1998

The influence of age on the economic value of bamboo culm is much greater than that of timber. Bamboo culms reach maturity at about 3 years old and have the best mechanical strength. López's research on *Guadua* bamboo in South America shows that this species is most suitable for application in various engineering projects at the age of 6 years, as its mechanical strength starts to decline after about 7–8 years old [1].

On the other hand, bamboo, as a kind of grass, has a particularly obvious life limitation, which, unlike wood, can accumulate biomass with the increase of age over a long period of time. Most bamboo species have a maximum life span of round 10 years, and some can reach 12 years. The culms of overaged bamboo would eventually dry out, whiten, and die.

From the perspective of resource utilization, there is difficult problem of 'use it or abandon it' for bamboo, which means people cannot save the material through forest reservation like wood. Conversely, it must be harvested and stored manually.

Table 1.5 Plant characteristics comparison between bamboo and wood

Comparison item	Bamboo	Timber
Growth characteristics[a]		
Longitudinal-height	(1) Completes within 2–4 months (2) Mainly by intercalary meristems (3) Does not simultaneously begin and end in different internodes	(1) Lasts in all lifetime (2) Realized by primary meristem on apex (3) Does not take place on secondary growth tissue
Lateral culm cross-sectional dimension	(1) Does not increase after the completion of height growth (2) The diameter and wall thickness of culm increase slightly in the process of height growth from shoot to young bamboo	(1) Lasts in all lifetime (2) Realized by cambium
Dimension	*Guadua*	*Spruce*
Maximum cross-sectional dimension × height (m)	0.25 × 20	1.5 × 50
Used in building industry (m)	(0.09 – 0.13) × 15	–
Age	*Guadua*	*Spruce*
Maximum plant life (year)	10	200
Used in building industry (year)	4–6	60–120

[a]*Source* Zhang, Industrial Utilization on Bamboo, 2001

(3) Microstructure

Wood is a cylindrical solid composite of bark, sapwood, and heartwood, composed of alternating spring and summer materials of various sizes, properties, and functions, while bamboo is composed of many hollow cylinders separated by bamboo nodes. In terms of mechanics, the bamboo nodes play an important role in preventing cracking, structural deformation, and improving the rigidity of the round bamboo.

From a more microscopic perspective, bamboo has only a small amount of transverse fibers at the bamboo nodes, much less than wood, which is an important channel for the lateral transport of liquids. This, on the one hand, results in different drying characteristics of bamboo compared with wood. On the other hand, the liquid containing the preservative is more difficult to enter into the interior of bamboo when the culm is treated for corrosion prevention, so that common corrosion prevention treatment methods such as soaking, cooking, high-pressure injection are more difficult to function (Fig. 1.6; Table 1.6).

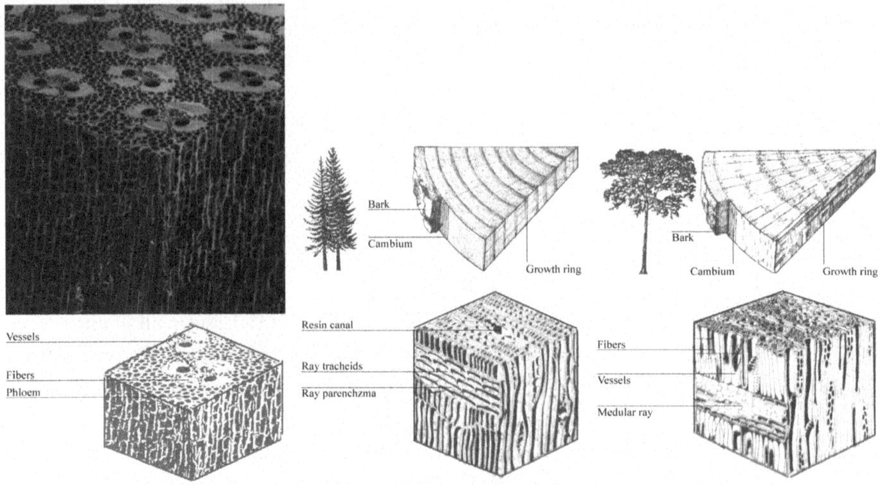

Fig. 1.6 Left up: three-dimensional view of culm tissue with vascular bundles embedded in ground parenchyma in *Oxytenanthera abyssinica* (1500×). *Source* Liese [7]; left down, middle and right: microstructure of bamboo wall, temperate trees (softwoods) and tropical trees (hardwood). *Source* López [1]

Table 1.6 Microstructure comparison between bamboo and wood

Comparison item	Bamboo	Softwood	Hardwood
Cells function	Support, conduction, storage, etc.	Support, conduction, storage, etc.	
Cells type	Parenchyma 52% Fiber 40% Conducting tissue (vessels, sieve tubes, companion cells) 8%	Has a simpler structure; consists mainly of one type of cell, which transports water and nutrients while providing support	cells are more specialized, and vessels form; position and direction of the cells and vessels, together with the growth rings, characterize its distinctive feature
Fibers longitudinal	Almost exclusively	Almost exclusively	Almost exclusively
Fibers radial	Only small amount in internode	Ray tracheids and ray parenchyma	Medular ray

Source
Bamboo: Liese and Mende [9], Grosser and Liese [10]; Timber: Volz, Timber construction manual, 2004

Table 1.7 Chemical composition comparison between bamboo and wood

Comparison item	Bamboo	Timber
Substances[c]		
Carbon (%)	50.0	≈50
Oxygen (%)	43.0	≈44
Hydrogen (%)	6.1	≈6
Other substances (%)	Nitrogen 0.04–0.26; ash 0.2–0.6	Pigments, oils, tanning agents, and resins[a]
Molecular components[d]		
Cellulose (%)	45.3	40–50
Hemicellulose (%)	25 [1]	20–35
Lignin (%)	25.5	15–35
Pentosans (the major constituent of Lignin) (%)	24.3	–
Extractive (%)[b]	2.6	<10

[a]These determine the smell, color, and degree of resistance in the sense of preservation of wood, and may account for up to 10%
[b]Extractive: For bamboo, it's Wax, Starch and Silica; For timber, it's Pigments, Oils, Tanning agents and Resins
Source
[c]Bamboo: Dunkelberg, Bamboo as a Building Material, 1985; Timber: Volz, The Material, Timber construction manual, 2004
[d]Li et al. [13]

(4) Chemical composition

The organic material content of bamboo is generally similar to timber, containing approximately 50% cellulose, 25% hemicellulose, 25% lignin (mainly pentosans) and extractives [11].

However, the difference in the content of some trace substance between bamboo and timber leads to different characteristics of bamboo from timber. The starch content is often mentioned when discussing the chemical composition of bamboo. The starch stored in parenchyma cells largely determines its susceptibility to molds, especially blue stain fungi and beetles. The starch granules in bamboo are rich in parenchyma cells that form the ground tissue and vascular bundles of bamboo, and even in fibers [12].

Due to the fact that bamboo lacks certain natural toxins that timber has, it cannot effectively resist insects and molds, making it more susceptible to biological damage (Table 1.7).

(5) Physical and mechanical properties

Since there is almost no radial fiber within bamboo, its anisotropy is stronger than timber, and the ratio between the longitudinal and lateral strength of bamboo is about 30:1, while the ratio of timber is about 20:1. The same reason causes bamboo culm to crack more easily, which is a disadvantage, and can also be a significant advantage under certain conditions, such as making the production of bamboo fiber or fiber bundle by crushing easier [14].

Takenouchi finds that for cylindrical structure, when the ratio between the wall thickness and the cylinder diameter is in the range 1/8 to 1/5, it has better bending resistance than the solid rod of the same size. *Phyllostachys bambusoides* and *Phyllostachys nigra* henonis have ratios between the wall thickness and the culm diameter of about 1/9, so they perform better to resist the damage of snow. However, this ratio of Moso bamboo is 1/11, so it is easily damaged by bending [1].

In terms of material strength, De Vos compared the Moso and *Guadua* bamboos with the Norwegian spruce and dark red Meranti, showing that for hardness, Moso bamboo was about twice as high as *Guadua*, higher than the two types of wood, and the hardness of *Guadua* was between the two woods; for flexural strength, the modulus of elasticity (MOE) of *Guadua* was higher than the two woods, while Moso was the lowest; both modular of ruptures (MOR) of the Moso and *Guadua* bamboos were significantly higher than the two woods. According to the series of test results, Moso bamboo was classified as C16, while *Guadua* bamboo as C35 (Netherlands building material strength grade) [6] (Table 1.8).

Bamboo and wood also differ in moisture content and shrinkage characteristics. Due to rapid growth, the moisture content of bamboo in green state is usually higher than that of wood, but it is lower for storage and use after harvesting and drying. The FSP (fiber saturation point) is 13–20%, which is significantly lower than the 28–30% of wood.

Bamboo shrinkage occurs at the beginning of the drying process as soon as the moisture content decreases, but it is stopped during the process when the moisture

Table 1.8 Mechanical strength comparison between bamboo and wood

Comparison item	Moso	*Guadua*	Norway spruce	Dark red Meranti
Hardness				
Internode [N]	5666.0	2685.8	1680	3570
Exterior	5859.2	2241.1	–	–
Interior	5472.7	3130.4	–	–
Modulus of elasticity (MOE)				
Internode [N/mm^2]	8261.6	14,189.6	9700	12,020
Exterior	8414.8	14,103.8	–	–
Interior	8108.4	14,275.4	–	–
Modulus of Rupture (MOR)				
Internode [N/mm^2]	113.0	127.3	63.0	87.7
Exterior	110.7	134.3	–	–
Interior	115.2	120.2	–	–

Source
Bamboo: De Vos [6]; Timber: Wiselius, Wood handbook (Centrum Hout, 9th edition), 2005

content decreases from about 70 to 40%, and then restarts. For wood, when the moisture content is above FSP (ca. 30%), the drying process has no effect on its volume and strength; when it is below FSP, the wood begins to lose moisture from the cell wall, and at this point the shrinkage begins and the strength increases. For example, when the moisture content drops from the green state to 12%, the longitudinal compressive strength increases to about two times, and when it drops to 5%, it is about three times [1].

In contrast, the strength increase of bamboo due to the drying process is much smaller than that of wood. In addition, the differences in thermal properties between bamboo and wood are always ignored, and it is generally believed that the difference between the heat capacity and thermal conductivity of the two is insignificant [1] (Table 1.9).

Table 1.9 Physical properties' comparison between bamboo and wood

Comparison item	Bamboo	Timber
Moisture content		
During growth[a]	*Phyllostachys pubescens* at the cutting age is approximately 80%	Can amount to around 70%
Cutting[b]		
Green	Bottom 48.5%, middle 38.5%, top 31.6%	Section area $\le 200\ \text{cm}^2$: $u > 30\%$ Section area $> 200\ \text{cm}^2$: $u > 35\%$ Section area $\le 200\ \text{cm}^2$: $20\% < u \le 30\%$ Section area $> 200\ \text{cm}^2$: $20\% < u < 35\%$ $u \le 20\%$
Semi-dry	–	
Dry	Bottom 15.7%, middle 15.6%, top 15.2%	
Fiber saturation point (FSP)[c]	13–20%	28–30%
The degree of swelling and shrinkage for every 1% change in the moisture content[c]	*Phyllostachys pubescens* Longitudinal: 0.024%	Longitudinal: $<0.01\%$
	Phyllostachys pubescens Tangential: 0.1822% (exterior > interior)	Tangential: 0.27–0.36%
	Phyllostachys pubescens Radial: 0.1890% (node parts: 0.2726%, internode parts: 0.1521%)	Radial: 0.15–0.19%

Source
[a]Bamboo: Zhang [11]
[b]Bamboo: Prawirohatmodjo [15], Liese [7], Sharma and Mehra [16], Kumar and Dobriyal [17]
[c]Bamboo: López, Bamboo-the gift of the gods, 2003
[a/b/c]Timber: Volz, The Material, Timber construction manual, 2004

Bamboo is 'neither grass nor wood, forming its own group.' There is a significant difference between bamboo and wood, and among different bamboo species. Therefore, specific analysis on regional resources and bamboo types is necessary in practice.

1.1.1.5 Bamboo Resource in China

(1) Possesses the main dominant temperate bamboo species. According to the research of Xue, China's bamboo forest can be divided into five major bamboo areas: the northern monopodial bamboo zone, the Jiangnan mixing bamboo zone, the southern sympodial bamboo zone, the southwest mountain bamboo zone, and the Hainan and Yunnan climbing bamboo zone. It has dominant bamboo species such as Moso (*P. edulis*), *Dendrocalamus giganteus*, and *Bambusa textilis*. Among them, Moso, as a temperate bamboo, is the main species for industrial utilization in China, and also a rare dominant bamboo species of monopodial rhizome in the world. China possesses more than 90% of the world's Moso bamboo, which is of high industrial value and suitable for producing a variety of bamboo-based panels, and can be derived to bamboo shoots, bamboo fiber, bamboo charcoal, and bamboo vinegar [18] (Fig. 1.7, left).

(2) Large-scale forestry and intensive management. In the late 1950s and early 1960s, Chinese forest ecologist and bamboo expert Xiong started the study

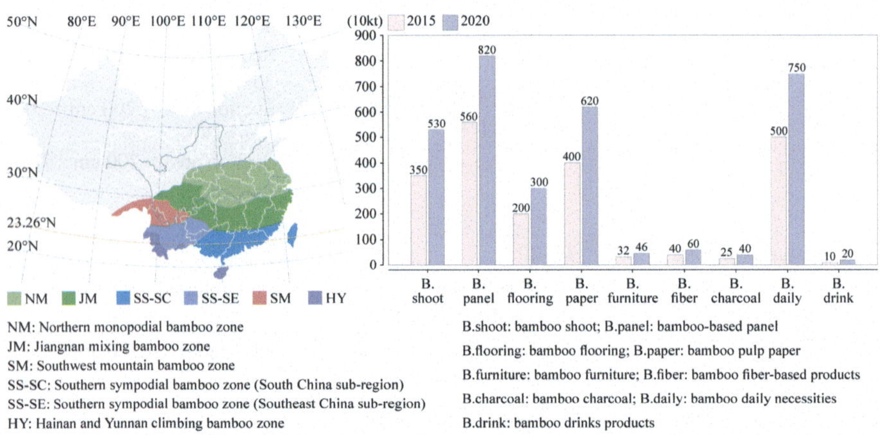

Fig. 1.7 Bamboo forest distribution in China and the market forecasting for the years 2015 and 2020. *Note* For the left one, NM—northern monopodial bamboo zone; JM—Jiangnan mixing bamboo zone; SS(SC)—southern sympodial bamboo zone (South China subregion); SS(SE)—southern sympodial bamboo zone (southeast China subregion); SM—southwest mountain bamboo zone; HY—Hainan and Yunnan climbing bamboo zone. For the right one, the graph is drawn according to the data of 'National Bamboo Industry Development Plan (2011–2020),' in which the furniture is calculated as 20 kg/piece

on expansion, high yield, and management of bamboo forests represented by Moso, which gradually modernized the bamboo forests and related industries. At present, China has a total amount of about the world's one-fourth of the bamboo forest area, one-third of the bamboo yield, and more than 85% of the bamboo-based panel production. During 2009–2013, China completed the 8th national forest resource inventory, showing that, in 2013, the area of bamboo forest in China was 6.01 million ha, the yield was 1.877 billion bamboo, and the total output value was 167.075 billion Yuan. According to the '*National Bamboo Industry Development Plan (2011–2020)*' finished by the State Forestry Administration in 2012, it is expected that the national bamboo forest area would reach 7.73 million ha in 2020, the number of people directly engaged in the bamboo industry would reach 100.175 million, the output value would reach 300 billion Yuan, and the income generated by the bamboo industry would account for 20% of the farmers' total income in the bamboo-producing area (Fig. 1.7, right).

(3) High degree of industrial utilization and rich product system. In 1980s, Chinese Academician Zhang, an expert in wood processing and wood-based panel technology, and his team made progress in the field of bamboo processing. They have successfully developed bamboo plywood, bamboo particleboard, composite bamboo laminated lumber, and other products. The bamboo industry is gradually moving toward the combination with modern material technology and industrial utilization [11]. At present, China has developed a complete product series of 'full bamboo utilization' such as raw bamboo, bamboo-based panels, bamboo fiber, bamboo charcoal bamboo vinegar, etc., owns the corresponding core patents, and takes the lead in terms of industrial utilization, mechanization, product system, and production scales.

(4) Undertakes various functions such as ecological security and wood forest protection. Due to the extremely developed underground rhizome, bamboo forest has excellent capacity of water storage and soil fixation, and China's temperate bamboo species have good cold resistance to survive on high and steep slopes. Therefore, it is used to maintain soil and water sources on the riversides [18]. Due to the shortage of wood forest in China, the '*Forest Law of the People's Republic of China*' was promulgated in 1984, and the forest protection became a national policy. In this situation, the domestic timber supplied to the building materials industry would be more constrained. In 2011, China's total timber consumption was 499.9 million m^3, of which the international dependence was as high as 44.8%. It is currently the world's largest importer of timber and timber products. The efficient utilization of fast-growing and high-yielding bamboo resources and 'substitute timber with bamboo' are regarded as one of the ideal strategies to alleviate timber shortages and protect the wood forest resources.

1.1.2 *Industrial Utilization on Bamboo and Development of the Products System*

Restricted by the regional wood forest resources, modified bamboos in standard panel or square form are considered to be an ideal substitute for timber [11]. Since the 1970s, wood processing technologies are as references for investigations on the industrial applications of bamboo in China, India, Thailand, Vietnam, Costa Rica, Malaysia, Indonesia, the Philippines, and other countries, and afterward various industrial bamboo products are successively developed, including the plybamboo in the 1980s, the bamboo particleboard, bamboo oriented strand board, and the bamboo laminated lumber in the 1990s, bamboo scrimber in the 2000s, and flattened bamboo panel in the 2010s, as well as the bamboo fiber, bamboo charcoal, bamboo vinegar, etc. [19–22], which are promoted to the concrete formworks, load-bearing components, truck and bus bottom boards, furniture and finishes, textile, daily necessities, food industries, etc. [11].

In 2012, Wang used the China Patent Database of the State Intellectual Property Office as a data source to analyze the development of bamboo industry in China from 1985 to 2011. Indicators show that the bamboo industry in China has experienced two 'development stages' of 1989–1993 and 1998–2006, and moved a 'mature stage' in 2007–2010. After 2011, it goes into a new round of technology development, in which the application ranges from traditional food, handicrafts, and furniture supplies to high-tech fields such as chemical pharmaceuticals and bamboo-based panels [23].

1.1.2.1 BBP Made of Bamboo Sliver

In 1980s, plybamboo became the earliest developed bamboo-based panel product. It is made up of bamboo sliver, or bamboo mats and curtains that are woven from the sliver units. The manufacturing process generally includes:

Raw bamboo, cross-cutting, outer node removing, splitting, inner node removing, boiling, softening, flattening, rolling, planing, drying, shaping, edge shearing, adhesive coating, assembling, pre pressing, hot pressing, storing, edge sawing, checking [11].

Due to its mature technology, low cost, and high mechanical strength, plybamboo is widely used in the truck and bus bottom, concrete formwork, packaging industries, etc. In the construction industry, the Institute of Modern Bamboo, Timber and Composite Structures (IBTCS), headed by Xiao, of Hunan University develops the 'GluBam,' a kind of plybamboo, and conducts systematic research on it. The research topics cover the production and mechanical properties of GluBam, the mechanical performance of its structural members and joints, the design and construction of GluBam roof trusses, prefabricated units, light-framed houses and bridges, and have been successfully applied in engineering practice [24] (Fig. 1.8).

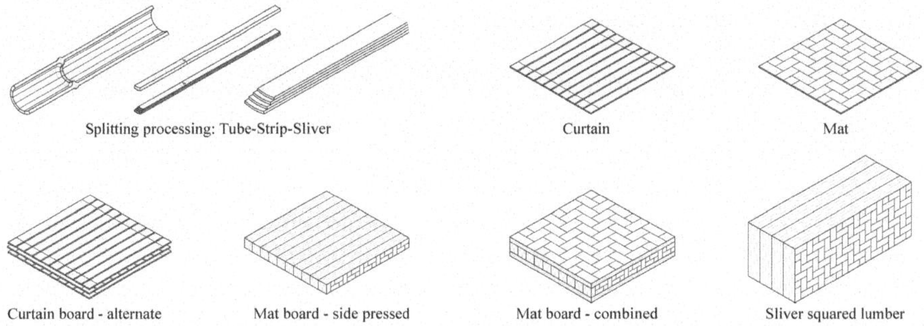

Fig. 1.8 Technical process diagram of plybamboo (constituent unit preparation, slab production, assembly method, forming, etc.). *Source* Redrawn according to the related technical patents

1.1.2.2 BBP Made of Bamboo Particle

In 1990s, bamboo particleboard and bamboo oriented strand board were successively developed. Bamboo particleboard is made up of bamboo shavings, sticks, or fibers, and to some extent, it corresponds to the wood particleboard and medium-density fiberboard (MDF). Bamboo oriented strand board is derived on the basis of bamboo particleboard, by changing the constituent units into larger bamboo shavings, and adding orientation process to the paving method. Its manufacturing process generally includes:

Raw bamboo, chip cutting, shaving, drying, dried shavings storing, adhesive using spreading, prepressing, cutting, hot pressing [11].

In the wood-based panel industry, medium-density fiberboard (MDF) and oriented strand board (OSB) are two of the three mainstream wood-based panel products (the third one is the laminated lumber). The corresponding bamboo products, namely the bamboo particleboard and bamboo oriented strand board, are already mature in manufacturing technologies, which are mainly adjusted from the technical equipments for making wood particleboard. However, the application of this bamboo products is limited in the packaging, furniture and building formwork sectors, unable to achieve large-scale marketing (Fig. 1.9).

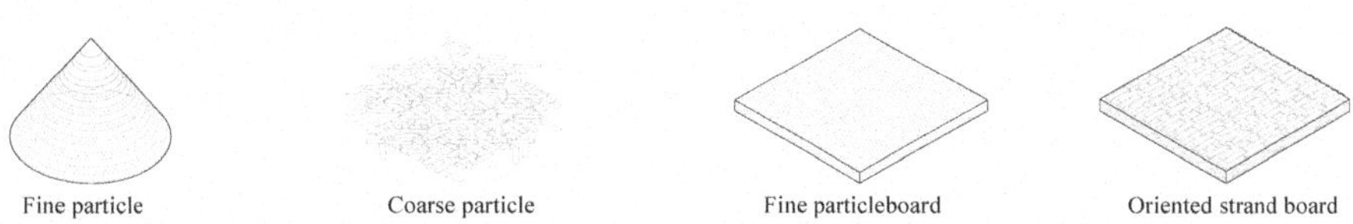

Fig. 1.9 Technical process diagram of bamboo particleboard (constituent unit and the final products). *Source* Redrawn according to the related technical patents

1.1.2.3 BBP Made of Bamboo Strip

In 1990s, bamboo laminated lumber was first developed in Taiwan, China, which is made up of bamboo strips. The manufacturing process generally includes:

Raw bamboo, strip making, rough planning, strip selection, carbonization, drying carbonized strips, fine planning, adhesive coating, assembling: pressing strips to single layer sheet, sanding the single layer sheet, adhesive coating, pressing sheets to multiple layers board, sawing, sanding, dust absorption [25].

Compared with the corresponding products of wood-based panels, the raw material used in bamboo laminated lumber is the hollow bamboo tube, so that its unit production process is quite special. The mainstream method is to divide the split bamboo strips into standard units with rectangular cross section by cutting or flat pressing. In addition to this, there are experimental attempts to explore the cross-sectional form of the isosceles trapezoid, the curvature with equal external and internal diameter, in order to improve the utilization rate of raw materials. The longitudinal shape can be either straight or curved, and the latter is mainly used in the production of furniture, handicrafts, and such products (Fig. 1.10).

There are many assembly ways to form panels or squares for bamboo laminated lumber. In general, the parallel arrangement of bamboo units is advantageous for playing the mechanical strength of the bamboo fiber in one direction, and the orthogonal arrangement can improve the anisotropy of the products. The latter must follow the 'odd layer principle' (Figs. 1.11 and 1.12).

Since bamboo laminated lumber retains the surface features of bamboo to a large extent, and its appearance is well recognized, it is widely used as a decorative panel for wall and flooring, as well as in furniture design. In 2012, China's bamboo flooring production was about 35 million m^2, of which about 20 million m^2 used bamboo laminated timber as the raw material, and the remaining 15 million m^2 was bamboo

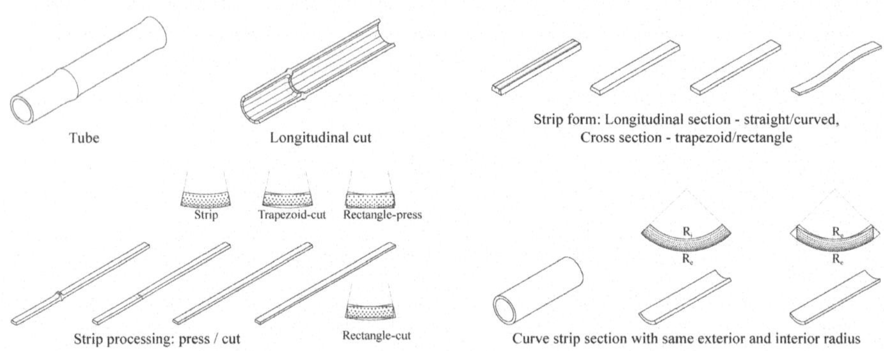

Fig. 1.10 Technical process diagram of bamboo laminated lumber (constituent unit preparation). *Source* Redrawn according to the related technical patents

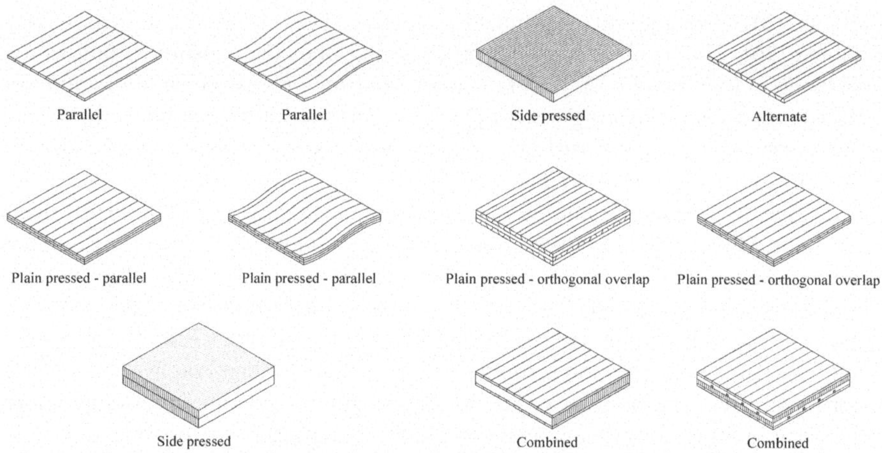

Fig. 1.11 Technical process diagram of bamboo laminated lumber (slab production, assembly method, forming, etc.). *Source* Redrawn according to the related technical patents

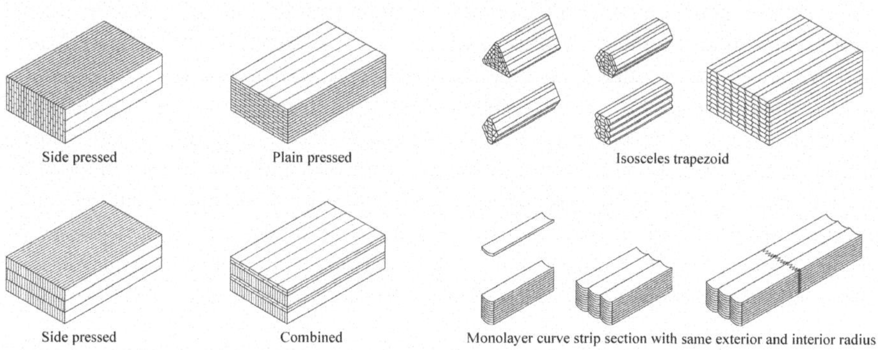

Fig. 1.12 Technical process diagram of bamboo laminated lumber (squared laminating process). *Source* Redrawn according to the related technical patents

scrimber and bamboo composites. In addition, certain bamboo laminated lumber products bonded with specific adhesives are also applied to the fields of furniture, kitchen utensils, and so on.

1.1.2.4 BBP Made of Bamboo Fiber

In 2000s, the research and development of bamboo scrimber has matured in technology in China. Microscopically, bamboo fibers are arranged in parallel and have higher strength than wood fibers [12], which prompts that bamboo fiber-based products are beneficial to play their inherent mechanical advantages. However, due to the complicated manufacturing process and low added value, bamboo medium-density fiberboard (MDF) composed of sufficiently separated fibers has not been successfully promoted [26]. The bamboo scrimber process simplifies the constituent unit as the loose bamboo fiber bundle [27], and its manufacturing process generally includes:

Raw bamboo, strip making, rough planning, strips splitting, carbonization, carbonized strips drying, strips crushing, adhesive using, strips pressing, glue activating in oven, beams sawing, beams sanding [25].

China has significant advantages of quantity and technology in bamboo scrimber. In 2014, the Chinese State Forestry Administration's Intellectual Property Research Center finished an analysis on all the wood and bamboo scrimber technology patents published worldwide in 1976–2012, which was based on the European Patent Office's Worldwide Patent Database and the Derwent World Patent Index [28]. China has 57 technology patents for bamboo scrimber and 16 patents for bamboo–wood composite scrimber. The core contents of technical research include the high-temperature decomposition of bamboo fiber and hemicellulose, manufacturing process and crack resistance of bamboo scrimber. The international distribution of Chinese patent holders is mainly in the USA, Australia, and Europe, largely because these places are its export destinations, and also areas where patent disputes are easy to occur.

Bamboo scrimber can be made from the raw materials that are derived from herbal bamboos and small-diameter bamboos, which have not been fully utilized. It can also improve the utilization rate of raw materials to 90%, have better mechanical strength, cracking resistance, and corrosion resistance, and therefore rapidly become the mainstream of bamboo fiber-based panels. Products are widely applied in load-bearing components, interior finishes, outdoor paving, furniture and wind power blades [29, 30]. The existing research on bamboo scrimber covers its manufacturing process, including the preparation of the constituent units, gluing, assembling, forming by hot and cold pressing [31, 32], as well as prevention of cracking, deformation, and corrosion [33] and flame-resistant technology of the products, etc. [34] (Fig. 1.13).

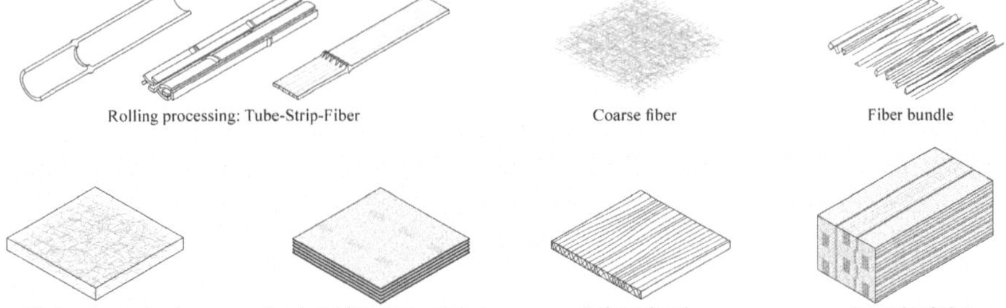

Fig. 1.13 Technical process diagram of bamboo fiber board (constituent unit preparation, assembly method, forming, etc.). *Source* Redrawn according to the related technical patents

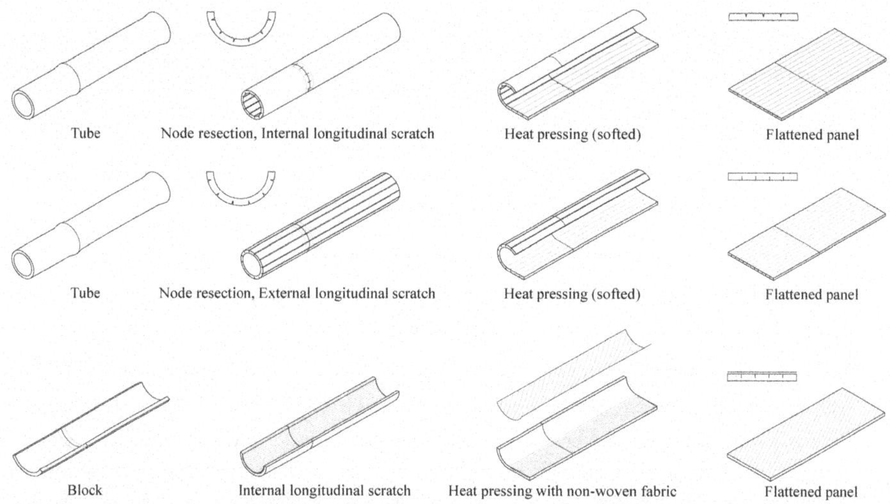

Fig. 1.14 Technical process diagram of flattened bamboo panel (whole process, for producing finishes panel). *Source* Redrawn according to the related technical patents

1.1.2.5 BBP Made of Flattened Bamboo Culm

In 2010s, by utilizing the structural characteristics of the thin-walled Moso bamboo (wall thickness 8–10 mm), the flattened bamboo culm was developed in China through softening, flattening, and shaping the complete or semi-bamboo tube. The manufacturing process generally includes:

Raw bamboo, longitudinally split into two halves (this step is usually omitted), inner node removing, outer node removing, longitudinal scratching, softening by vapor treatment, flattening, finalizing shape by pressure, surface planing, flat boards drying, cutting [25].

Since glue is not necessary in the forming process of flattened bamboo panel, the product is considered to be ecological and environmental friendly and can be used for tableware, chopping board, etc. It can also be reinforced with a non-woven fabric, or spliced and glued to obtain a panel in the size of a building component, which is the raw material for the wallboard or flooring (Fig. 1.14).

In the Americas, there are also simplified flattening methods, which mainly eliminate the step of longitudinal scratching, and the resulting product has obvious cracks on the surface, which is not suitable to be a decorative panel but can be used as a core plate for multi-layered board (Fig. 1.15).

1.1.2.6 BBP Made of Composite Materials

Bamboo composite technology is the combination among different bamboo units, or between bamboo and other materials, which make uses of their respective charac-

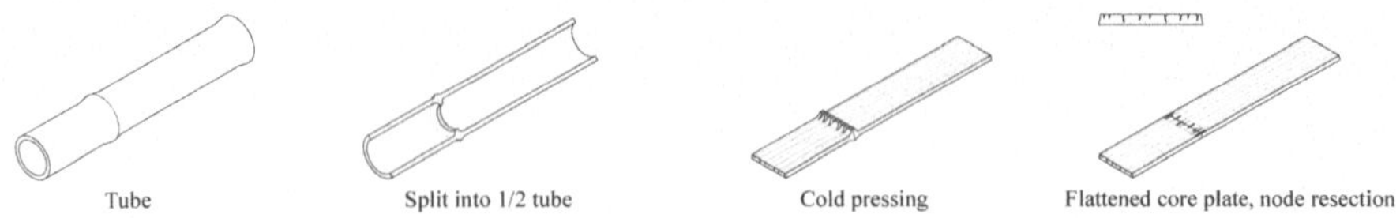

Fig. 1.15 Technical process diagram of flattened bamboo panel (simplified process, for producing core plate). *Source* Redrawn according to the related technical patents

teristics to achieve mechanical complementarity, comprehensive performance optimization, or cost control. It contains the following main categories:

(1) The compounding among different bamboo units. It takes the bamboo slivers, the bamboo curtains, and mats woven from slivers, or bamboo particleboard, bamboo oriented strand board as the inner core plate, and the bamboo veneer, fiberboard, laminated lumber as the outer layer, to form the panel products of high mechanical strength, low cost, and ideal appearance (Fig. 1.16).
(2) The compounding between bamboo and wood. Due to the similar bonding and swelling properties, bamboo and wood have a good basis to work together, and therefore the bamboo–wood composites have got a lot of attention. Similar to the principle of (1), it takes the bamboo plate as the inner core, the wood veneer/plate as the outer layer, or vice versa, the wood plate as the inner core, and the bamboo veneer/plate as the outer layer, to form an ideal panel product with ideal strength, cost, and appearance [35, 36] (Fig. 1.17).
(3) The compounding of bamboo and its complementary materials. It makes use of the high tensile strength of bamboo fiber for mechanical reinforcement of other substrates, such as the bamboo fiber-reinforced concrete, plastic, gypsum, and clay; or compensates for the weakness of bamboo brittleness by adding plastic materials, such as bamboo–plastic composite, bamboo–steel composite, bamboo–plexiglass composite [37, 38] (Fig. 1.18).

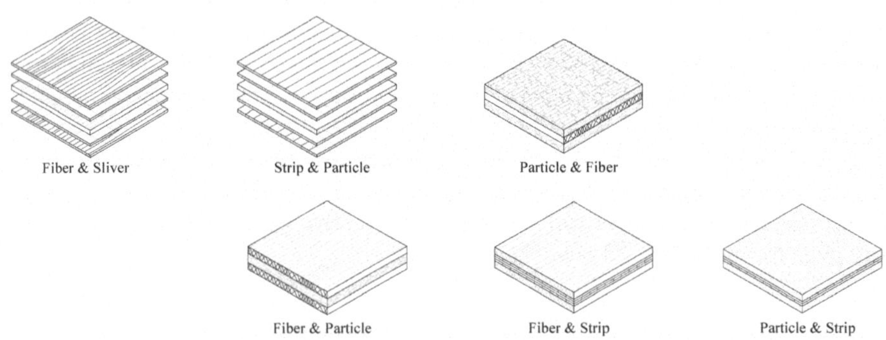

Fig. 1.16 Technical process diagram of bamboo composite panels (combination of different bamboo units). *Source* Redrawn according to the related technical patents

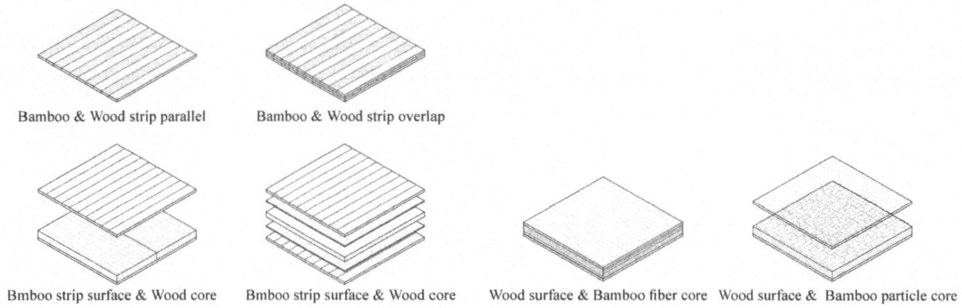

Fig. 1.17 Technical process diagram of bamboo composite panels (combination of bamboo and wood). *Source* Redrawn according to the related technical patents

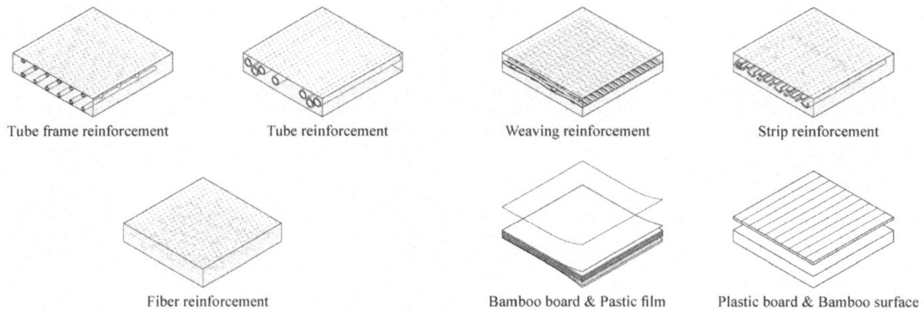

Fig. 1.18 Technical process diagram of bamboo composite panels (combination of bamboo and other complementary materials). *Source* Redrawn according to the related technical patents

1.1.2.7 Bamboo Profile

Bamboo flooring is currently the most mature bamboo profile product on the market. According to UN Comtrade database, the total import and export mass of bamboo flooring trade in 2014 was 248.06 kt, of which the total export volume in Asia accounted for 94%. The proportion of imports was in turn, 40% in Europe, 26% in Asia, 16% in the Americas, 15% in Oceania, and 3% in Africa.

The technology accumulation of bamboo flooring is also relatively rich. Apart from the researches on the cross section of the flooring, the form of the units and the joints, there are also studies on the performance improvement through the built-in heating film and the moisture-control bamboo charcoal infill (Fig. 1.19).

In addition, the wall or floor components formed by a combination of bamboo-based panels as partition boards, profiled steel as the framework, beam profiles that play the high mechanical strength of bamboo are also studied and developed (Fig. 1.20).

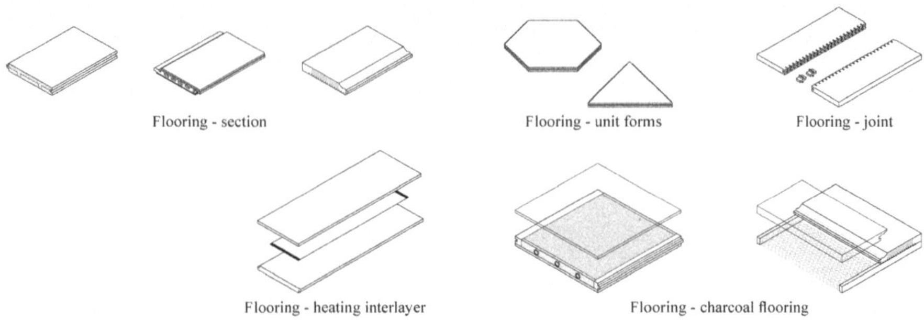

Fig. 1.19 Technical process diagram of bamboo profiles (bamboo flooring). *Source* Redrawn according to the related technical patents

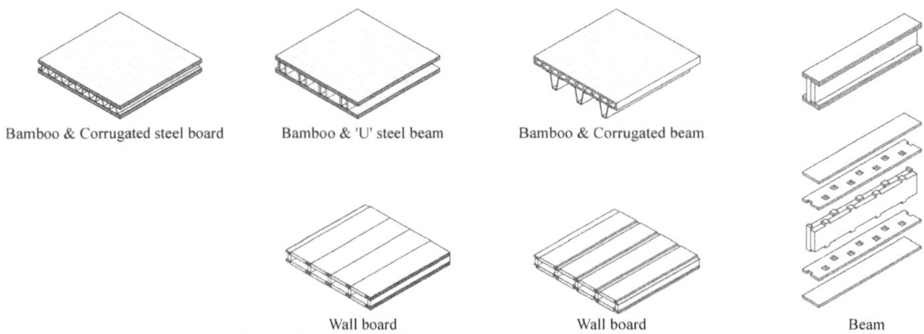

Fig. 1.20 Technical process diagram of bamboo profiles (wall, floor, beam, etc.). *Source* Redrawn according to the related technical patents

1.1.2.8 Bamboo Veneer

Bamboo veneer is usually only 0.3–0.4 mm thick and applied in veneers, furniture, hand-made articles such as the molded plates, fans, bookmarks, and screens, which belong to the higher-end products. According to the raw materials for production, it can be divided into two types, one is obtained by stripping the bamboo species of thick-walled culms, e.g., the thick-walled Moso bamboo. It is technically difficult, can only be feasible for certain bamboo species, and requires skilled operators, so that the commercially production is limited.

Another one can be called a derivative from the bamboo laminated lumber or bamboo scrimber technology, which plans or rotary cuts the squares of bamboo laminated lumber or bamboo scrimber. It can be enhanced by non-woven fabric during processing, or expand the size by lengthening process in the post-processing (Fig. 1.21).

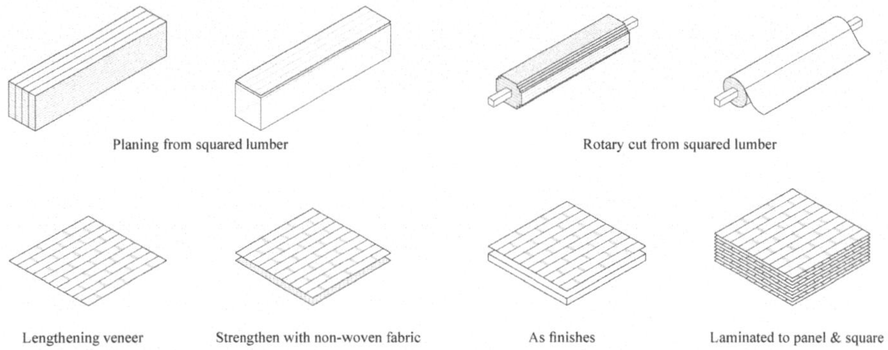

Fig. 1.21 Technical process diagram of bamboo veneer. *Source* Redrawn according to the related technical patents

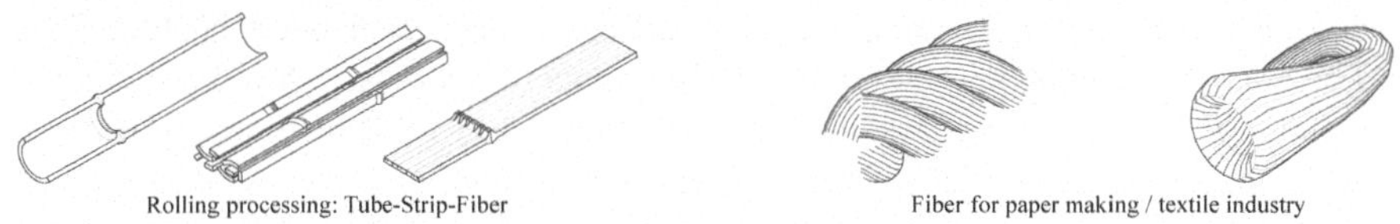

Fig. 1.22 Technical process diagram of bamboo fiber. *Source* Redrawn according to the related technical patents

1.1.2.9 Bamboo Fiber

The development and promotion of natural bamboo fiber attracted attention since the 1990s mainly in textile and composites industry in East Asian countries [39]. Specifically, it is composed of cellulose, hemicellulose, lignin pectin, and fat wax [40, 41], and extracted from raw bamboo mainly by mechanical rolling and flash explosion technologies [42, 43]. The fibers that are directly separated from the bamboo culms are natural bamboo fiber. Through further reprocessing, bamboo pulp fibers and bamboo charcoal fibers suitable for papermaking and textile applications can be extracted (Fig. 1.22).

1.1.2.10 Bamboo Charcoal

The development and application of bamboo charcoal caused concern since the 1990s mainly in East Asian and African countries. Specifically, bamboo charcoal is produced through the decomposition and carbonization of cellulose, hemicellulose, and a low amount of lignin [44]. Currently, direct firing by kiln, vertical destructive distillation, and pyrolysis by a retorting kettle constitute the three main industrial manufacturing methods [45, 46]. Generally, bamboo charcoal is widely used in air purification [47], soil improvement, water purification [48, 49], infills for moisture-control products [50], and composite industries [51, 52].

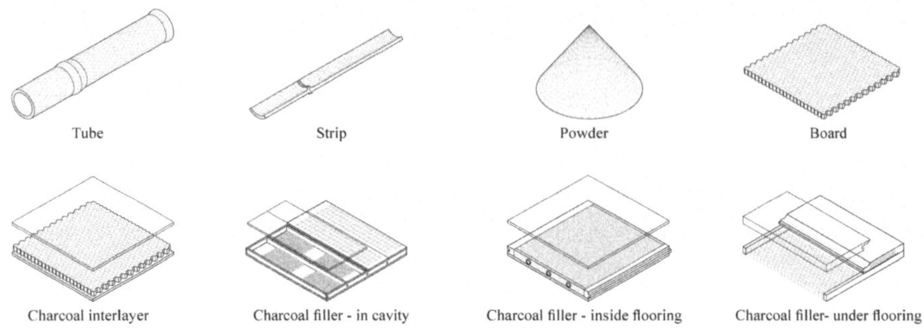

Fig. 1.23 Technical process diagram of bamboo charcoal. *Source* Redrawn according to the related technical patents

Bamboo charcoal can be processed into different forms as needed, such as flake and powder, and can be processed into a plate shape if necessary. Among them, the moisture-control bamboo charcoal is usually used as an infill in powdery form to regulate the moisture content of the targeted items. Bamboo charcoal flooring is a relatively mature application technique that attracts attention (Fig. 1.23).

The current state of the art shown in the existing patents is rich in both progress and results at the material level, basically from materials science and bamboo-based panel technology, which emphasizes the mechanical properties; fewer achievements at the building component level, mainly including the production and installation of profiles such as the bamboo flooring; and few techniques that emphasize the physical performance of the building components and their overall construction systems.

1.1.3 Application Status of Bamboo in Building Industry

Owing to differences in local forest resources, economic level, and construction techniques, there are regional differences in the research and application of bamboo in construction industries. Traditionally, large-diameter raw bamboos are broadly applied in the construction of buildings in tropical regions such as Southeast Asia and South America, as they are earthquake-resistant, cheap, and easy to obtain [53].

In the context of the forestry resources, materials technology, and building materials demand today, raw bamboo and its modified bamboo have received new attention. This process is influenced by the development of timber technology and the evolution of the timber construction system. Timber construction system has experienced the evolution successively from the log construction, traditional timbered structures, timber frame construction, to the skeleton construction and timber panel construction, of which the latter two were influenced by wood-based panel technology in the early twentieth century. This technology was successfully introduced to the bamboo industry in the 1980s, and a variety of bamboo-based panels were developed, which

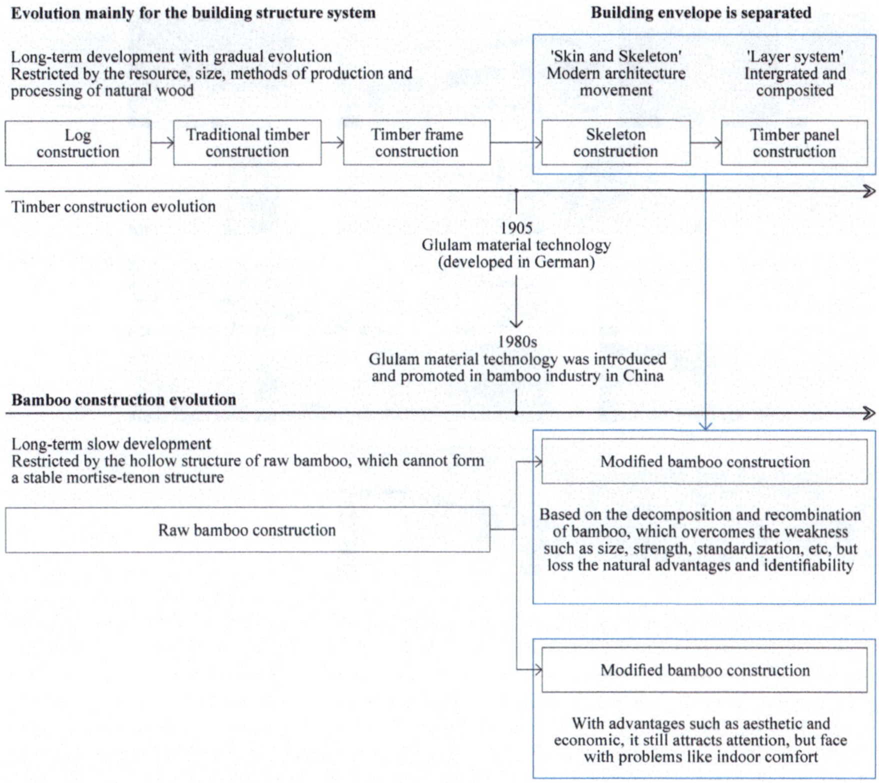

Fig. 1.24 Evolution of bamboo construction and the relation with the development of timber construction

have also attracted the attention of architects and scholars to carry out a series of practical engineering application (Fig. 1.24).

1.1.3.1 Practice Distribution

The practice of bamboo construction has attracted a lot of interest around the world, including some academic stars such as Frei Otto, Renzo Piano, Shigeru Ban, and Richard Rogers, exploring bamboo for construction from different perspectives. Since most of the construction practices is in small scale, they are dominated by architects, structural engineers, artists, or even fashion designers and bamboo-based panel experts. The construction cases show a clear individual difference, which are related to the professional background of the people who are in charge of the projects (Fig. 1.25).

From the perspective of the materials form used, these practical applications are analyzed as follows:

Fig. 1.25 Several representative cases of applying bamboo in construction. Raw bamboo: 1. Reconstruction of the Pereira Center Church Pereira; 2. Wind and Water Cafe in Binh Duong, Vietnam; 3. Bamboo Houses in Bali, Indonesia; 4. ZERI Pavilion on the EXPO 2000; 5. Children Activity and Learning Center, Koh Kood, Thailand; 6. The school made of earth and bamboo, Rudrapur, Dinajpur District, Bangladesh; 7. Helou Villa, Fangta Garden, Shanghai; 8. Ningbo Tengtou Case Hall of the EXPO 2010; Modified bamboo: 9. The ceilings project of the Madrid International Airport; 10. Beijing Bamboo House; 11. Wuxi Theater Interior design; 12. The 'bamboo anti-seismic living room' in Nanjing Forestry University; 13. Fast-prefabricated bamboo structural anti-seismic Resettlement Housing and Classroom in Sichuan earthquake; 14. Bamboo Exemplary Building in Hunan University; 15. INBAR Africa Exemplary Project; 16. China UNICEF KPMG Community Center in Sichuan earthquake; 17. The Prototype bamboo village house at Expo INTEGER, Kunming; Integrated use: 18. German-Chinese House at EXPO 2010 in Shanghai; 19. German-Chinese Bamboo Pavilions in Nanjing, Chongqing, Guangzhou, Shenyang, Wuhan. *Source* Photographs are taken from Internet

1.1.3.2 Application of Raw Bamboo

Due to its distinctive formal aesthetics, the raw bamboo attracts a lot of theoretical research and construction practices, which have accumulated a series of construction experiences that focus on the joints of round bamboo components, resulting in a unique form of architecture. The raw bamboo practices are integrated with the local construction, regionalization, economic society, and digital technology, which have a positive effect. However, the raw bamboo construction always has the limitations of poor durability, difficulty to select the standard building components, strong dependence on manual operation, and the poor quality of indoor environment. Therefore, it is used as a kind of inexpensive and temporary housing mainly in those countries and regions that are rich in bamboo resources but economically and technologically backward and fail to produce modified bamboo. The small-scale and experimental construction practice carried out by a small number of architects and artists, exploring the construction method of the original bamboo, is difficult to popularize.

As a building material, the raw bamboo has defects in size, standardization, and chemical properties. The raw bamboo is limited by the size of the plant itself, it will bend during the growth process, and the culm diameter is irregularly different, making it difficult to become a standardized building material. It has a high moisture content so that it is prone to cracking during it service life, and a large sugar content that makes it prone to mold and insect damage. Bamboo modification technology can make up for these defects to a certain extent and extend its service life in constructions [54, 55].

1.1.3.3 Application of Modified Bamboo

Compared with raw bamboo, the application range of modified bamboo has been greatly expanded. Some scholars with backgrounds in civil engineering and architectural design have been working for the application of modified bamboo to the building industry. Most of these people come from countries with mature bamboo/wood-based panel industries, such as China, Germany, and Japan, and adopt modified bamboo products to expand their usage. Zhang and Lu developed bamboo anti-seismic housing in 2009, in which plybamboo was used as wall framework and insulation material was filled inside to achieve favorable thermal performance [56]. Xiao carried out a series of housing construction with 'GluBam,' a kind of plybamboo developed by his team [57], of which 2440 × 1220 mm wall and ceiling panel units were composed of plybamboo board as partition, plybamboo square as inner support, and thermal insulation material as infill [58]. Referring to the structural form of light timber construction, Hao adopted the 2 × 4 construction method (2 × 4 in., 5 × 10 cm) in bamboo housing, in which bamboo laminated lumber and bamboo scrimber were used as wall framework, and thermal and acoustic insulation material as infill [59]. These studies focus on the physical and chemical properties improvement of the material, industrial prefabricated production and construction methods, and the improvement of indoor comfort. The results show the application advantages and technical development prospects of modified bamboo.

The existing enriched mechanical research provides strong support for the application of bamboo as load-bearing building elements. However, for the research and application of building envelope, bamboo is not regarded as an independent material class in those important database and reference in China, e.g., the '*Thermal Design Code for Civil Building*' and '*Building Materials.*' Due to the lack of hygrothermal property parameters, it is usually replaced by timber parameter, which brings uncertainty to the calculation results. Using bamboo laminated timber, bamboo scrimber, and bamboo composite panels as the raw materials, various bamboo panel profiles for flooring, ceiling, and wall are applied in building envelope, which form a parallel and competitive product series with timber in some regions. However, the current additional technical cost has caused some material varieties to lose their price advantage, and therefore it is difficult to be promoted to the other areas of the market. On the contrary, plybamboo and bamboo particleboard products have the advantage of price, but the added value of the products is low, so that they are mainly used in the building formwork and packaging industry. Many varieties of modified bamboo, affected by some treatment during the manufacturing processes, loss their appearance features and recognizability, resulting in the images that are easy to be confused with wood.

1.2 Research Scope Definition

1.2.1 Current Research Focus

The demand for sustainable development in building industry promotes the search and development of ecological building materials. As a traditional building material, bamboo has gained new attention in the context of the forestry resources, materials technology, and building materials demand today. The existing researches on bamboo are mostly carried out from the perspectives of botany, forestry, and materials science. By contrast, the theory and practice researches for architectural applications of bamboo are few, but gradually attract attention.

Worldwide, the studies on bamboo show uneven levels, which are influenced by the demand and support for carrying out the research in different regions. The study groups are distributed in a few countries in Asia-Pacific, the Americas, and Europe. The research contents among different groups overlap each other and meanwhile show different characteristics. The bamboo studies of almost all countries cover these aspects: bamboo botany, bamboo forestry, bamboo physical and chemical properties, where more literature in these directions can be found.

Countries and regions with bamboo utilization traditions, such as China, Japan, and Colombia, have a perspective of regional culture in their research on bamboo, where attention is paid to the expression of bamboo for national and regional culture. The construction practice of Chinese and Japanese architects will express an implicit oriental feature through the material language, while the Colombian architects use the unique round bamboo to express the American civilization of the wilderness. The USA is rich in wood forest resources, so that shows less sensitive to use bamboo in its own building industry. However, it expands its horizons to the developing countries such as Southeast Asia and Africa, exploring the sustainable potential that bamboo can achieve in these regions. Europe has no native bamboo, except a small amount of introduced species, and has also been interested in bamboo in recent years. By comparison, Germany and the Netherlands have conducted more research on bamboo. Relying on the technological advantages of their universities, enterprises, and research institutions, they carry out high-tech additional products research and development.

The existing architecture-related bamboo studies can be roughly classified as the sustainability of applying bamboo in building industry, the bamboo structure system, bamboo construction nodes, properties of raw bamboo and modified bamboo, and the modification process of bamboo, as well as the relevant industry technical standards, etc., which are grouped to the following three levels by the author (Fig. 1.26).

Firstly, macro level: the sustainability of bamboo in building industry. It includes specifically the sustainability of bamboo itself, the economic and social benefits of bamboo industry to bamboo-producing areas, and how bamboo building, as a kind of low-cost housing type, can help to improve the living situation of the economically and socially undeveloped areas. The existing researches on this level mainly reply to the issue why choose bamboo as a building material.

Secondly, meso-level: the structure system and joints of bamboo buildings. The former includes the raw bamboo structural system, steel bamboo structural system, concrete bamboo structural system, long-span bamboo structural system (mainly bamboo truss and bamboo grid), and the lectotype and seismic performance of the bamboo structures. The latter contains the construction nodes for the round bamboo components, the structural components of modified bamboo, the truss and grid in

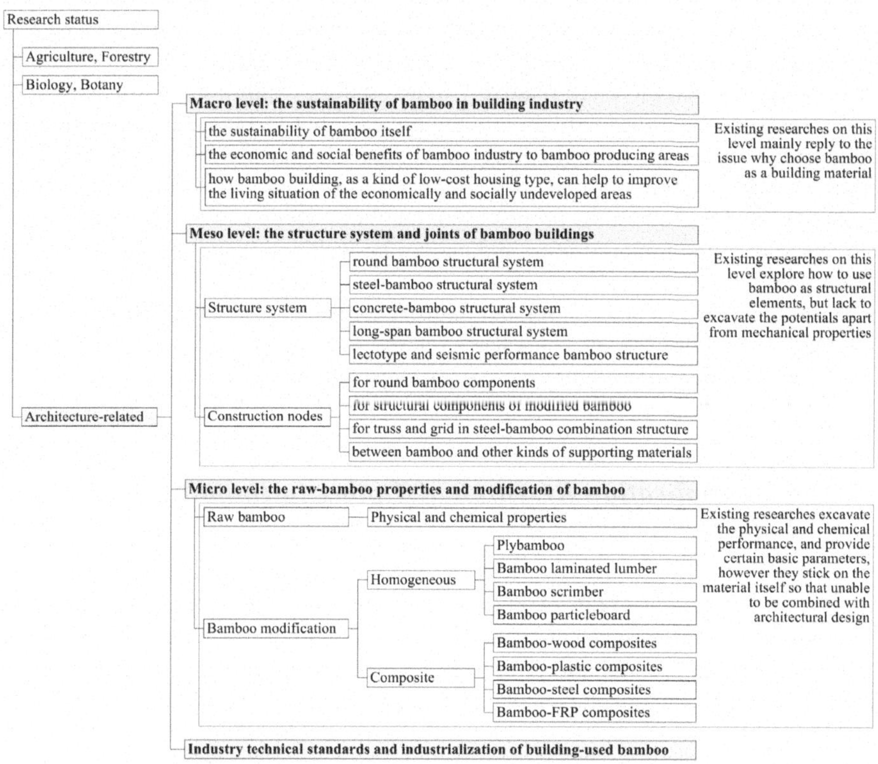

Fig. 1.26 Classification of the existing literature

steel bamboo combination structure, and between bamboo and other kinds of supporting materials. The existing researches on this level explore how to use bamboo as structural elements, but lack to excavate the potentials apart from mechanical properties.

Thirdly, microlevel: the material properties of raw bamboo and modification of bamboo. The former includes all kinds of the physical and chemical properties of raw bamboo. The latter consists of homogeneous modified bamboo and bamboo composite. The homogeneous modified bamboo includes a variety of bamboo-based panels or squares, which is decomposed into small-sized constituent units through modern material technology, and then reassembled into large-sized and standardized materials by gluing. In this way, the utilization efficiency of the raw bamboo, as well as physical and chemical properties of the material are improved. Based on the homogeneous bamboo, bamboo composite is a panel or square that is obtained through auxiliary bonding and special processing with and one or more complementary materials other than bamboo. The existing researches excavate the physical and chemical properties, and provide certain basic parameters; however, they stick on the material itself and are unable to be combined with architectural design.

To sum up, the above three levels of researches are not supposed to be separate, but limited by the research perspective, support, period, etc. Most studies focus only on a single level. For the first level, there has been a certain amount of research that largely answers the question of why bamboo is worthy of application in building industry. For the second level, there are relatively more studies, exploring the application of bamboo as a material for load-bearing components, but lacking the potential excavation of bamboo apart from its mechanical strength. For the third level, there are certain researches from the perspective of materials science, excavating and improving the physical and chemical properties of bamboo, which can provide some basic data, but stay at the material level and cannot be well combined with architectural design.

In addition, the research on related industry technical standards mainly belongs to the second and third levels. This part of work is extremely deficient, but recently it has attracted attention of many research groups around the world.

1.2.2 The Application Research of Timber in Building Industry

Bamboo and timber are both plant materials and have similar properties, so that examining the development law of timber construction can provide inspiration for the application research of bamboo in building industry. Worldwide, the construction system of timber is far more developed than that of bamboo. The application of timber has a long history and is well integrated with the current building industry in Europe, North America, and Japan. Traditionally, timber construction activities have mature systems in Asia, Europe, and the North America. In modern times, affected by the emerging new production and construction techniques, as well as new materials represented by steel and concrete, different development tracks are shown in these traditional timber construction zones.

1.2.2.1 Timber Construction System

The construction method is comprehensively determined by climatic conditions, available materials, tools, culture, and technical levels. In the northern and central parts of Europe, wood forest is abundant and timber construction is relatively prosperous. Since the nineteenth century, especially in the twentieth century, the timber construction in Europe has been largely affected by new technologies and new materials. Transportation methods, building energy efficiency, and comfort requirements promote the continuous evolution of timber construction [60]. The development of timber construction system has evolved from the log construction to the traditional timbered structure, from the timber frame construction to the skeleton construction, and to a new timber panel construction system. The development process of the timber construction system shows clear logic with the development and evolution of material technology, the separation, and decomposition of building elements such as the load-bearing components, building envelope and decorative elements, and creates new construction methods (Table 1.10).

Timber construction regions such as Europe, America, Australia, and Japan are rich in wood forest resources. With the industrial material modification and manufacturing technologies, the properties, production, transportation, and construction of timber are significantly improved. Driven by the higher requirements of material saving, energy conservation and indoor comfort, timber material technology and construction methods are constantly evolved, which results in a complete material and construction system [61]. At the material level, logs that traditional timber construction rely on have defects such as low raw material utilization rate, difficulty of standardization, limitation by size, and quality of the tree. During the transformation process from traditional to modern construction system, material technologies are developed synchronously to modify the timber for meeting higher requirements. Since the invention of plywood by Hetzer (German) in 1905, the wood-based panel technology helps to improve the utilization of the raw wood and overcome its defects in size, uniformity, strength, weather resistance, etc., through material gluing and compounding, which expand the applicable areas of timber in building industry. Currently, three mainstream products, namely the particleboard (mainly the OSB), fiberboard (mainly the MDF), and plywood, are developed and well promoted. At the construction level, the frame structural system separates the building envelope from the load-bearing components and develops an intensive, composite layered construction method that can facilitate the factory prefabrication and on-site dry work, and reduce the assembly time. With the aid of composite technology and modern structural theory, the properties of timber have been further and scientifically developed. New material and construction forms emerge, which break through the traditional image of timber construction, and show the technological aesthetics under new conditions.

Table 1.10 Development of timber construction (according to Ludwig Steiger '*Timber construction*')

(1) Log construction	(2) Traditional timbered structures
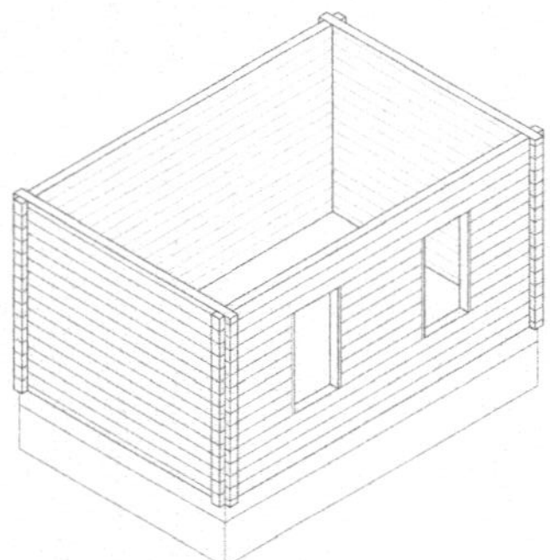	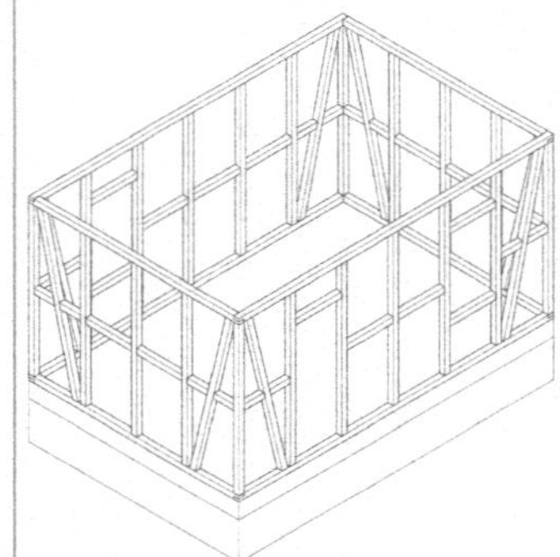
Due to the cross-connection at the end of the beam, it is often described as 'strickbau'; essentially it is a solid construction system that requires a large amount of timber, and the softwood with a straight trunk is most suitable; tenon and scarf joints are used for the connection between two members; the wall is difficult to meet the thermal requirements today, so it is necessary to add a separate insulation layer	Due to the clear load transfer, it is sometimes referred to as 'stil der Konstruktion'; the difference between the load-bearing structural column and the non-load-bearing interval infill wall is obvious; tenon joint is adopted; space between structural elements is filled with stone, or clay, etc., and now basically with insulation material; the standard cross-sectional size of the squared timber is 10/10, 12/12, or 14/14 cm, which is mostly hardwood, such as oak

(continued)

Table 1.10 (continued)

(3) Timber frame construction	(4) Skeleton construction
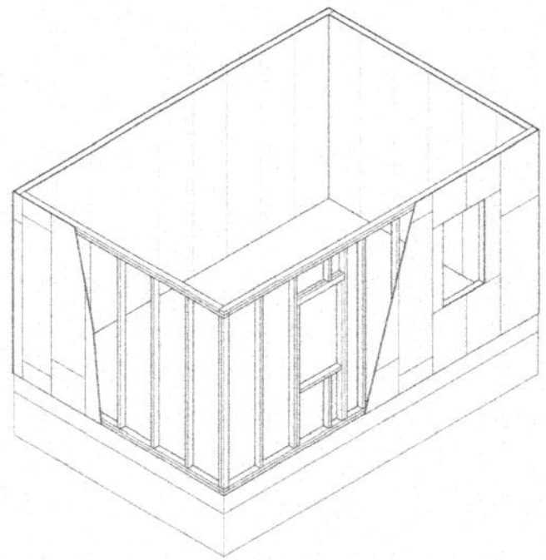	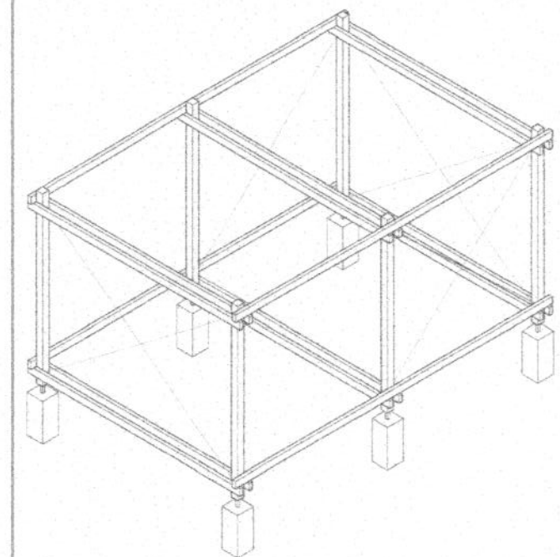
It originates from North America, where villages and towns along the new railway line require simple, economical, and fast-built houses; it is sometimes referred to as 'Rippenbau'; in the first half of the nineteenth century, industrial technology began to affect the timber construction, and European traditional timber construction was changed by steam engine sawmill and mechanical cutting nails; the timber is unified into a flat plate form, and the nailing joints are simple and easy, so that replace the manual joints that needs to be elaborated; in Europe, the frame structure was not widely accepted until 1980s. Different from the American '5 × 10 cm' (2 × 4 in.) timber cross section, Europe adopts '6 × 12 cm'; the common width of the small grilles is 62.5 cm	It adopts the column and main beam as a first load-bearing structure, which supports a secondary load-bearing structure composed of the secondary beam and raft; the non-load-bearing wall is further liberated, which realizes the modernist 'skin and skeleton' in twentieth century; the space between load-bearing columns is significantly larger than that of traditional timbered and timber frame construction systems, which is partly due to the plywood technology that enables the material strength to meet the requirements; multi-storey buildings use continuous columns instead of overlaying layers, and the horizontal beams attached to the columns become the ring beams; metal fittings are used to connect beams and columns without weakening the cross section of the wood

(continued)

Table 1.10 (continued)

(5) Timber panel construction	
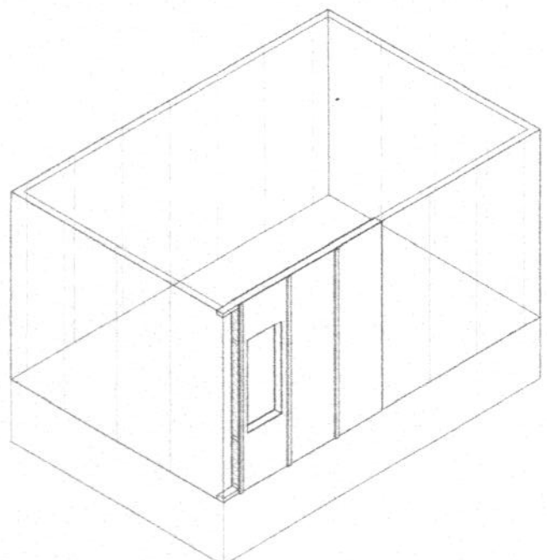	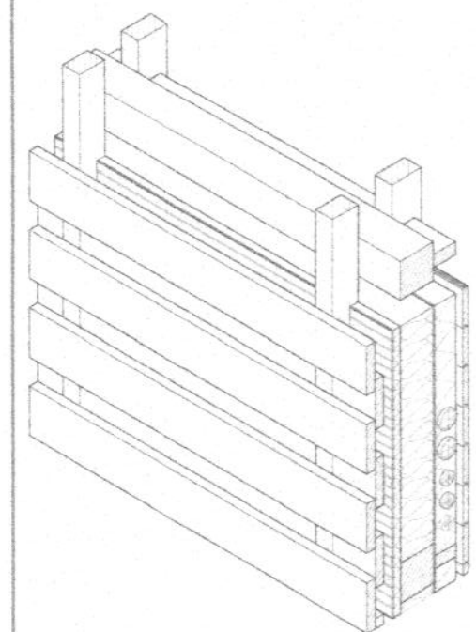
The timber panel construction can maximally transfer the workload to the factory; the prefabricated panel units are usually of a full storey height, completed with insulation treatment, and installed with necessary building elements, so that they need only to be erected, connected, and fixed on site; the load-bearing plate wall is made of solid laminated plywood or gluing unit to become a slab construction; the tendency of timber construction to use as little material as possible is being changed	

1.2.2.2 Case Study: 'Holz Der Zukunft' Joint Research Project

Located in the heart of Europe's traditional timber construction districts, Germany is still one of the European timber construction centers. Its wood-based panel production ranks third in the world, second only to China and the USA. Its research on wood science is integrated into the composite materials development, building structure, building energy efficiency, building safety, etc. The material potential is explored from multiple perspectives and fully oriented to architectural design and practical applications.

The project 'Holz Der Zukunft' (Timber construction of the future) was funded by the Bavarian Research Department of Germany, and finished through cooperation among the Technical University of Munich, Fachhochschule Rosenheim, and the Institute for Fenstertechnik, Rosenheim. Leading by 12 experts and 52 relevant scholars, around 20 subprojects were carried out on the application of timber in building industry from various perspectives, committing to the potential utilization of timber in building and environment, and promote the development of related disciplines and industries. At present, 18 subprojects have completed research and published research reports, presenting a multi-disciplinary exploration of timber-related issues in the building industry. The 'Holz Der Zukunft' joint research project has the following implications for this study (Table 1.11).

(1) Research object. Judging from the distribution of the research objects, the investigation on 'building with timber' that covers the application studies on materials, construction and components of timber, is more than the discussion on the 'timber construction system.' The former is integrated with material manufacturing and processing technologies to develop timber materials and building components products that are high-tech, high-quality, and easy to market.
(2) Research content. According to the distribution of research priorities, traditional themes, such as the construction and fire protection of timber, are still hot topics. At the same time, the application of high-quality timber materials and component products in building envelopes, as well as the potential exploration of timber in building energy conservation and improving indoor environment, attracts considerable attention.
(3) Research method. As can be seen from the distribution of the main methods used in the projects, the studies generally start from a rigorous typological analysis of those complex and large objects, on the basis of which a series of comparative experiments and calculations are performed to achieve the patterns and optimization methods. The judgments rely on the quantification of data, which break through the limitations in the sensory dimension, and unearth the technical potential of the materials.

Table 1.11 Eighteen subjects' distribution and analysis of 'Holz Der Zukunft'

Subproject	Research object						Research content		
	'Building with timber'			'Timber construction system'			Comprehensive	Construction	Fire protection
	Material level	Construction level	Component level	Load-bearing structure	Building envelope	Device system			
1					◑		●		●
2	●		●						●
3					●	●			
4	●					●			●
5		●			◑			●	
8							●		
9	●		●					●	
10	◑		●		●			●	
11			●			◑	◑		●
12			●			◑		●	
13		●	●			●		●	
14	●								
15	●								
16	●								
17	●			◑	●				
18		●							
19	◑	●	◑					●	
21 + 22							●		
Number	9	4	7	1	5	5	4	6	4

(continued)

Table 1.11 (continued)

Subproject	Research content						Main methods			
	Energy conservation	Indoor environment	Marketing	Standard and guidance	Digital technology	Material properties	Experiment	Computer simulation	Market survey	Typological analysis
1	●	◑								●
2				◑						●
3	●	●			◑		●			●
4							●			◑
5							●			◑
8			●						●	
9		◑		◑		◑	●			●
10	●	●					●			●
11				◑			◑			●
12		◑								●
13							◑			●
14			●			●			●	●
15						●	●	●		
16						●	●			◑
17	◑	●				●	●			◑
18	●	●					●	●		◑
19				◑			●	◑		◑
21 + 22					●			●		

(continued)

Table 1.11 (continued)

Subproject	Research content						Main methods			
	Energy conservation	Indoor environment	Marketing	Standard and guidance	Digital technology	Material properties	Experiment	Computer simulation	Market survey	Typological analysis
Number	5	7	2	4	2	5	12	4	2	15

● Large involved
◑ Small involved
Source Sort out according to the final reports of each subproject
1. Ganzheitliche Planungsstrategien: Konzeption und Umsetzung
2. Brandsicherheit im mehrgeschossigen Holzbau
3. Entwicklung von grundsätzlichen Strategien zur Energie- und Raumklimaoptimierung von Holzbauten für Büro- und Verwaltungsbau
4. Verknüpfung des anlagentechnischen Brandschutzes mit haustechnischen Installationen
5. Leichte Vorhangfassaden aus Holz Konstruktionsgrundlagen für eine definierte Feuchteabfuhr
8. Marktforschung & Markterschließung
9. Konstruktionsgrundlagen für den Einsatz von Leichtbauelementen im Innenausbau
10. Energetische Sanierung von Bestandsbauten in Holz- und Massivbauart unter Einsatz von Holz und Holzwerkstoffen
11. Mechanismen der Brandweiterleitung bei Gebäuden in Holzbauweise
12. Modulare, vorgefertigte Installationen in mehrgeschossigen Holzbauwerken
13. Entwicklung von Grundlagen für die Integration von Elektronik im Fenster-, Fassaden- und Türenbau
14. Hochwertige Bauprodukte aus Massivholz und Holzwerkstoffen aus starkem Stammholz
15. Flächen aus Brettstapeln, Brettsperrholz und Verbundkonstruktionen
16. Holzbeton
17. Holzleichtbeton im Hochbau
18. Anwendung des vertikalen Schiebemechanismus
19. Konstruktionsgrundlagen für Fenster, Türen und Fassadenelemente aus Verbundwerkstoffen und Holz
21 + 22. Integriertes Relationales Informationssystem für den Holzbau—IRIS

1.2.3 The Application Research of Bamboo in Building Load-Bearing Structure

On the one hand, compared with logs, raw bamboo has more prominent material defects in addition to the raw material utilization rate, standardization, and size limitation: The rapid growth of the plant makes the raw bamboo contains higher moisture and sugar than logs, and therefore it is more vulnerable to mold growth and insect damage; the microscopic fiber structure of bamboo lacks transverse tissue, which is more anisotropic than wood, and makes it is easier to crack; the shape of the members and joints of round bamboo makes its construction activities largely dependent on manual operations; the hollow cross section of the round bamboo makes it difficult to form a stable tenon joint connection, and the bundling joints commonly used in the conventional practice have weak reliability. On the other hand, the assistance of modern materials technology provides support for solving these problems, so as to improve the physical properties of bamboo, its standardization, and the utilization rate of the raw materials, and increase the size of the products, making it possible to be promoted to the application in high-quality and non-temporary construction projects.

1.2.3.1 Load-Bearing Bamboo and the Structural Application

Due to the high fiber strength, bamboo is considered to have a certain mechanical potential, so that it has been widely discussed as a load-bearing component in construction projects, of which the relevant literature is quite sufficient. Among these, there are the research of the raw bamboo system that has a long history, the study on the modern bamboo–wood-based panels and their composites, and the bamboo fiber reinforcement materials.

(1) Raw bamboo

These works focus on the mechanical performance of the round bamboo members and the joints between them. Colombian native architect López introduced the construction methods of the round bamboo joints in his book '*Bamboo: The Gift of the Gods*,' by listing various ways of connecting the round bamboo rods and the corresponding performance evaluation [1]. Colombian architect Velez uses the *Guadua*, which is a rich bamboo species in South America, as a low-cost building material for poor people in those underdeveloped and resource-poor areas. He has extensive research and practice experience in the design of roof, structure, and joint connections, and invents a round bamboo connection method with concrete infill, which has been promoted in many projects [62]. The German Institute for Lightweight Structures (IL) of Frei Otto has published a series of research articles, including the '*IL 31: Bamboo—Bamboo as a Building Material*,' which carried out an investigative research on the traditional bamboo constructions [63]. Janssen from TU Eindhoven in the Netherlands conducted a systematic study of the raw bamboo construction, and published four monographs from 1981 to 2000, introducing the growth and anatomy, physical, chemical, and mechanical properties of bamboo, experience in practical engineering, traditional construction method of raw bamboo wall, door, window, roof, floor, ceiling, roof truss, etc. [64–67]. German artist Heinsdorff develops connection joints with steel hardware for the round bamboo robs [68]. The team of Wang in China studies the mechanical properties of round bamboo and its prefabricated house system [69–71].

(2) Bamboo-based panel and its composites

China's bamboo-based panel production accounts for the vast majority of the world, and the application of bamboo-based panels in building industry has also attracted attention. The development of bamboo-based panels is dominated by the forestry profession, and the mechanical strength is the main indicator during the manufacturing process. Logically, the research groups of forestry and civil engineering have conducted research on the application of bamboo-based panels in building load-bearing structures.

'Anti-seismic bamboo house.' It is jointly developed by Chinese Academician Zhang who is an expert in wood processing and wood-based panel technology, and Academician Lu who is an architect. The project has been completed in the campus of Nanjing Forestry University in 2009, where various modified bamboos are applied to different components of the building according to the characteristics of each material: bamboo scrimber for columns, bamboo laminated lumber for beams and slabs, and plybamboo for exterior walls. The overall seismic capacity reaches 8 degrees, which can cope with earthquakes of about magnitude 7 [56].

'Modern bamboo structure.' Professor Xiao of the Institute of Modern Bamboo, Timber, and Composite Structure (IBTCS) in Hunan University leads a team to study the structural application of 'GluBam' in construction and bridge engineering. 'GluBam' is a kind of plybamboo, and Xiao owns its independent intellectual property rights. See Sect. 1.2.3.2 for details.

'Steel–bamboo composite member.' Professor Li of Ningbo University completed the National Natural Science Foundation of China (NSFC) project 'Study on the bonding and longitudinal shear performance of bamboo composite members' in 2010–2013, and an earlier project 'Study on bamboo composite members and their structural systems' in 2008–2011, using steel–bamboo composite member that combined plybamboo with cold-formed thin-walled steel through adhesive as the research object. The projects carried out investigation on the bonding characteristics of the steel–bamboo interface [72] and the mechanical performance of steel–bamboo composite as beams, columns, and wallboard members [73, 74], as well as the seismic test on the whole structural system [75], which resulted in patents of steel–bamboo composite beams, walls, and joints [76–78].

The Building Materials Institute in Building Environment Department of TU Eindhoven in the Netherlands has completed a series of studies on wooden materials, concrete, and fiber-reinforced composites. Among them, Janssen led a team to carry out research on bamboo-based panels, including the joints of the bamboo laminated lumber members and the bamboo–wood frames [79, 80], the mechanical performance of glued bamboo wall panels and their plank house [81]. The comparative test studies of the Bath University, School of Architecture of Cambridge University, and the Technical University of Braunschweig showed that the modified bamboo materials such as bamboo laminated lumber and bamboo scrimber approximated or exceeded the corresponding products of timber in mechanical strength [82–85].

(3) Bamboo fiber reinforcement

The application of bamboo fiber as an enhancement for concrete and plastic has attracted the attention of several research groups:

Bamboo–concrete composite. Wei from Nanjing Forestry University completed the NSFC project in 2012–2015: 'The mechanical behavior and design theory of a new bamboo–concrete composite structure with consideration to the interface slip,' to study the mechanical performance of reinforced bamboo structure and bamboo–concrete composite structure. Alireza of ETH Zurich, Switzerland, studied the composite bamboo and the enhancement of bamboo for concrete in his doctoral thesis [86].

Bamboo–plastic composite. Qiu from Fujian Agriculture and Forestry University completed the NSFC project in 2010–2013: 'Preparation of modified bamboo fiber/unsaturated polyester composite material and its interface compatibility mechanism,' using Moso bamboo and its processing residues as the raw materials for bamboo fiber, of which the surface was modified to be combinable with the unsaturated polyester to form a composite. In the earlier 2005–2008, Qiu finished another project: 'The interface characteristics and strengthening mechanism of bamboo fiber/thermoplastic composites,' revealing the strengthening mechanism of bamboo/plastic composites, in terms of the interface properties, the morphology and distribution of bamboo fiber, which formed the basis for the preparation of bamboo–plastic composite. Lu from Southeast University completed the NSFC project in 2010–2013: 'Mechanical performance and design theory research on a new type of BFRP reinforced bamboo structure,' using basalt fiber composite (BFRP) to enhance the bamboo structure, and carried out experimental test on the composite components [87–89]. Related projects included the 'Research on the rheological behavior of plant fiber/thermoplastic composites,' finished in 2007–2010 by Yang, and the 'Interface optimization design for the bamboo fiber reinforced polyamide resin composite,' completed in 2006–2009 by Chen from Fujian Agriculture and Forestry University.

1.2.3.2 Case Study: 'GluBam' Research Project

Professor Xiao from the Institute of Modern Bamboo, Timber, and Composite Structure (IBTCS) in Hunan University leads a team to study the structural application of 'GluBam,' which is a kind of plybamboo that Xiao owns the independent intellectual property rights. The team completed a key NSFC project in 2009–2013: 'Design theory and applied basic research of a new bamboo structure (50938002),' which carried out systematic studies on the application of plybamboo in load-bearing building structure on the 'material–component–structure system' levels (Table 1.12).

Table 1.12 Main content of the research project 'GluBam'

Material level [90, 91]	• Obtain mechanical properties parameters including tensile, compressive, shear, flexural strength, etc. • Give the allowable design stress recommended value • Experimental analysis of long-term creep law and aging
Component level [92, 93]	• Obtain the mechanical performance of the basic stressed members, including the plybamboo columns, plybamboo beams, plybamboo trusses, and bamboo–wood structural shear walls • Analysis of the relationship between plybamboo slenderness ratio and the load • Compare the effects of component connection, dimensions, bolting and FRP applications on the performance of plybamboo beams, the fatigue performance under repeated loading and the deformation performance under long-term loads • Inspective and full-span destructive tests on plybamboo roof trusses to establish semi-rigid node models • Test and analyze the plate nail connection and anti-lateral force performance of lightweight bamboo–wood shear wall panel nails, and obtain the bearing capacity formula of the connection and the wall
Structure system level [94, 95]	• Study the overall structure of prefabricated board houses, lightweight frame structures, etc. • Study on fire prevention, seismic performance, and indoor air quality of plybamboo houses • Construction of plybamboo demonstration buildings

Source Sort out according to the final report of the key NSFC project 'Design theory and applied basic research of a new bamboo structure (50938002)'

In addition to the theoretical achievement, the 'GluBam' research team also finished a number of construction practices, such as the fabricated plybamboo resettlement house after the Wenchuan earthquake, the plybamboo teahouse in Beijing Zizhuyuan Park, the plybamboo house in Cailun Bamboo Garden, the plybamboo demonstration building of Hunan University, the INBAR's demonstration housing project in Kenya and Uganda, Africa, as well as several bridges of pedestrian and vehicle, etc. In addition, they developed some patents including the construction of beams, wall components, and fabricated panels, which were of demonstration reference for the industry [96–99].

The 'GluBam' project has the following implications for this study:

(1) The systematicness of the research level. The project systematically studied the application of plybamboo from the progressive levels of 'material–component–structure system.' The mechanical parameters at the material level, the type of mechanical components at the component level, and the comprehensive performance evaluation at the structural system level were targeted completely, and therefore the systematic conclusions of the results are of sufficient reference value.

(2) Multi-disciplinary and modern experimental science. The project was based on the theories of modern mechanics, materials science, structural design, fire protection, building environment, etc., and obtained credible parameter values through large sample standardized experimental operation.

1.2.4 Scope Definition of the Research

The industrial utilization of bamboo largely refers to the material technology from the timber industry, and the resulting product system is also similar with timber to a certain extend. Existing studies on the application of timber in construction industry have covered a wide range of subjects, including the load-bearing structure, building envelope, fire protection, and equipment system at the building level, as well as the mechanical, physical, and chemical properties at the material level. In contrast, there are still wide areas waiting for improvement as to the application of bamboo in the construction industry. In this study, the research scope is defined as follows:

(1) Building level. There are already relatively sufficient studies on the application of bamboo in load-bearing structures. However, for the building envelope, the material consumption proportion is significant in the entire project, and it has important impact on the building energy consumption and indoor comfort, but still has not been sufficiently studied. Therefore, the building envelope is set as the research object in this study at the building level. As a reference, the research on timber building envelope has achieved systematic results, and this study is to improve the corresponding knowledge system for bamboo building envelope and provide a foundation for the promotion of bamboo in building envelope (Fig. 1.27).
(2) Material level. The research object is based on the product system of industrial utilization of bamboo, including raw bamboo, typical bamboo-based panels, bamboo fiber, and bamboo charcoal. According to the roles they play in the construction of building envelope, the former two can be classified as partition boards, and the latter two as construction infills. In the manufacturing process, the 'full bamboo utilization' series products are controlled by basic parameters, which guide the products to specific performance. In this study, the material parameters of hygrothermal properties related to the building envelopes are obtained through experimental test, and their correlation with the basic property parameters is analyzed to provide reference for the industrial production and application of bamboo in building envelope.
(3) Research orientation. Research on related topics can be carried out from various perspectives such as the thermal insulation, water and moisture proof, sound insulation, fire resistance, and durability. As a hygroscopic and organic building material of plant fiber, the hygrothermal properties of the bamboo and the hygrothermal performance of the bamboo building envelope are related to the cracking and mold growth damage caused by the heat and moisture stress and

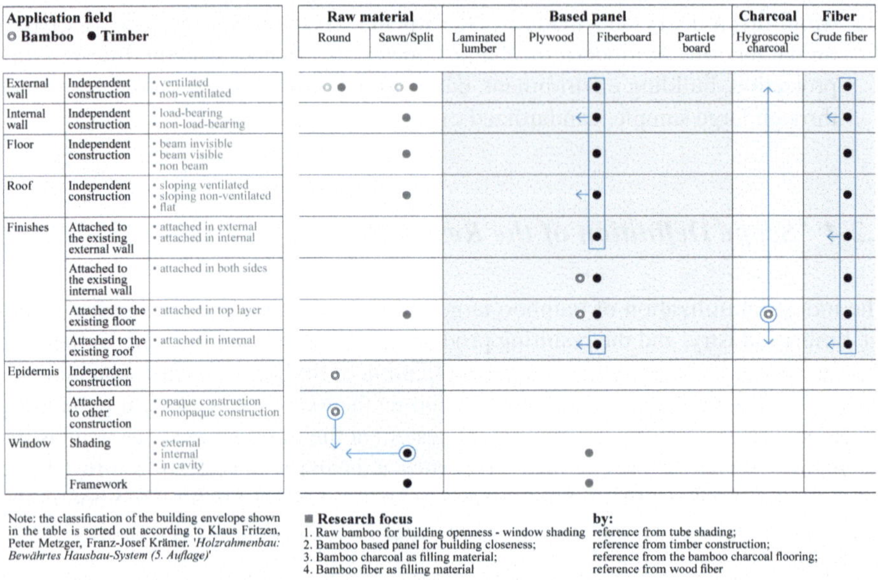

Fig. 1.27 Application status of bamboo and timber in building envelope and the potential extension space

the moisture content of the material itself, and have a key impact on the durability of the component, the indoor sanitation, comfort, and energy consumption. Therefore, the hygrothermal properties and hygrothermal performance are set as the research orientation correspondingly on the material and building envelope levels.

(4) Construction method. The timber layered construction system is a construction method that occurs along with the technology of the wood-based panel technology, and the separation of the building envelope and the load-bearing structure in construction system. Different from the solid structure, the layer construction distributes different functions to the appropriate material layer, and to achieve better comprehensive performance through the optimization of the layer order and connection mode. Due to the similarity of material forms, the timber layered construction system has good adaptability to bamboo, so that it is taken as the main construction method in this study. The material parameter database and structural design guidance are provided for its application (Fig. 1.28).

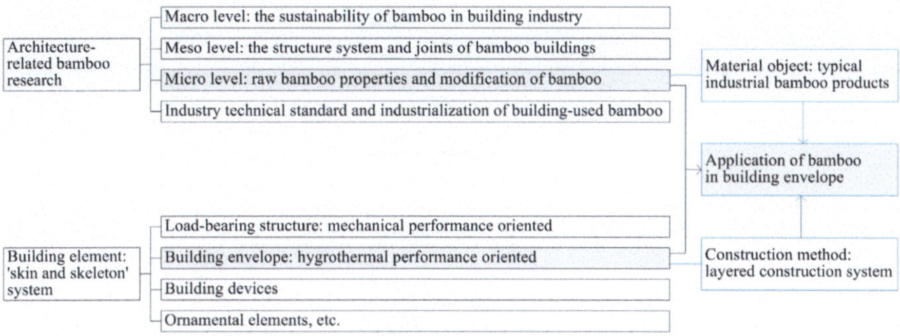

Fig. 1.28 Research scope definition and the relation with the existing knowledge

1.3 Research Status

1.3.1 Property Study on Bamboo

1.3.1.1 Bamboo Partition Board

Bamboo wall is the main part for industrial utilization, which has been studied from the perspectives of botany, materials and bamboo-based panels for the microstructure, surface morphology, chemical composition, density, mechanical properties, cracking, gluing ability, etc. Discussion has also been given to the natural life span, damage caused by non-biological and biological factors, as well as preventive treatment to the raw bamboo damage, such as prevention of cracking by controlling the moisture content and drying process, and physical and chemical methods of corrosion prevention [100–102].

The material parameters have an important impact on the calculation results of the building envelope, which means inaccurate material parameters can cause errors to the building envelope design, the indoor environment, and the calculation of building energy consumption. Some scholars have studied the basic physical properties of bamboo and compared it with timber products. The bulk density of timber products is generally between 400 and 800 kg/m^3 [103], while the raw bamboo and bamboo-based panels reach 650–1290 kg/m^3. Liese [12] and Grosser [104] from the Institute of Wood Research, University of Hamburg, Germany, López from Colombia [1], and Chinese bamboo-based panel expert and wood scholars such as Zhang, Jiang, Tang and Zhou [11, 105] observed the microstructure of bamboo and found that the fibers are almost totally longitudinally arranged in parallel, with extremely few radial fibers at the nodes, which was far different from wood. Considering that the basic properties significantly affect the hygrothermal properties [82], the difference of bulk density and microstructure between bamboo and wood provides a reference for the prediction of hygrothermal properties of bamboo and suggests the necessity of its research as an independent material system (Fig. 1.29).

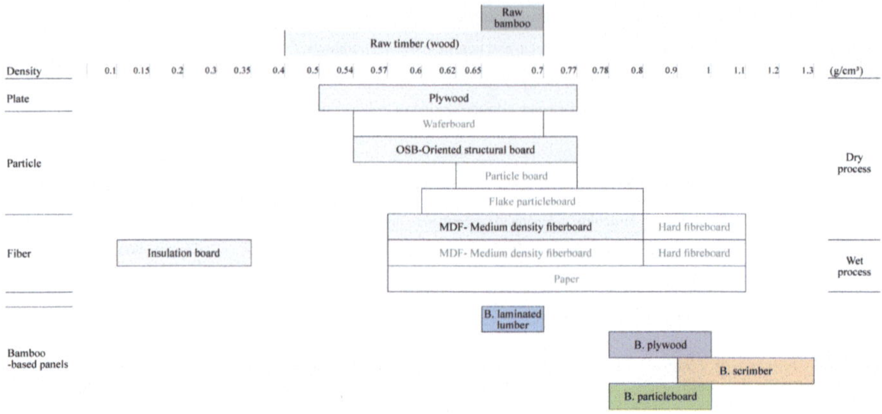

Fig. 1.29 Bulk density comparison among the main bamboo and timber products

The existing substantial mechanical studies on bamboo can support the application in load-bearing structure; however, owing to lack of hygrothermal properties, bamboo is not regarded as an independent material, such as the thermal design code in China [106]. There are previous studies on basic and hygrothermal properties of raw bamboo, mostly the Moso bamboo (Latin name: *P. heterocycla*), including bulk density, porosity, sorption isotherm, water vapor permeability, specific heat capacity, and thermal conductivity [107].

Wu, Huang et al. analyze the density distribution of bamboo nodes and internodes by computer tomography scanning and show that the bulk density of Moso bamboo is between 600 and 800 kg/m^3 [108, 109]. Huang et al. observe the pore structure of bamboo culms by computerized tomography and backscattered electron scanning and show that the average porosity of Moso bamboo internode is from 44.9 to 63.4% [110]. The bulk density and porosity tests have shown that the basic properties of bamboo have significant inhomogeneity on the scale of bamboo culm.

Ohmae and Nakano investigate the sorption isotherm of Moso bamboo in longitudinal direction by a saturated solution method [111]. Jiang et al. carry out sorption isotherm research on bamboo powder, bamboo blocks, parenchyma cells, and chemically macerated fibers of Moso bamboo and show that the parenchyma cells demonstrate the highest hygroscopic capability followed by bamboo fibers, bamboo powder, and bamboo blocks [112]. Wang compares the difference of moisture sorption among bamboo blocks, bamboo fiber, lignin, and hemicelluloses and finds that the equilibrium moisture content (EMC) of the hemicelluloses is much higher than the lignin under the same relative humidity, and the EMC of bamboo fiber is higher than the bamboo block [113]. Huang et al. measure the water vapor diffusion resistance factor of Moso bamboo by dry cup and find that the values in the radial and tangential directions range from 30 to 57 [114]. Huang also measures the liquid water properties and finds that the water adsorption coefficients are 0.014, 0.008, and 0.0019 kg/(m^2 s$^{0.5}$) for exterior, middle, and interior parts of bamboo culm wall, and the capillary saturation moisture contents are, respectively, 572, 479, and 385 kg/m^3 [107].

Previous studies by Wu and Huang show that the specific heat capacity values of Moso bamboo vary from 1.08 to 2.29 kJ/(kg K) [108], and from 1.7 to 2.3 kJ/(kg K) [115] at 40 °C using differential scanning calorimetry (DSC) method. Wu et al. measure the thermal diffusivity of Moso bamboo in longitudinal direction by a laser flash method [108], and Shah et al. measure the thermal conductivity of Moso bamboo veneer board and laminated board [116]. Huang et al. find that Moso bamboo is non-homogenous in terms of both morphological structure and thermal properties in the radial, longitudinal, and tangential directions [115, 117, 118].

1.3.1.2 Natural Bamboo Fiber

Extant studies focus on the microstructure, manufacturing process, hygric, and thermal comfort as clothing material and the mechanical properties as reinforcement in composites. The surface morphology of BF is investigated and indicates that the surface and pore structure are conducive to its rapid moisture adsorption and desorption ability [119]. The isothermal curve of BF is compared with the isothermals of flax and jute, and indicates that BF exhibits a higher adsorption and desorption rate [120]. The standard rate of moisture regain (equilibrium moisture content in $T = 20$ °C, RH = 65%) of BF reaches 12%, and this is close to that of ramie fiber and is more hygroscopic than cotton [121]. The fiber saturation point of BF is related to its species, and it approximately ranges from 13% (*P. pubescens*) to 20% (*Dendrocalamus strictus*), lower than that of wood fiber [1].

The length of bamboo fiber is determined as approximately ranging from 1.04 mm (*P. nigra*) to 2.64 mm (*Bambusa vulgaris* cv. Wamin), and this is between that of hardwood and softwood [12]. Given its favorable mechanical properties, BF is examined in several studies as a reinforced element in composites [122, 123]. In order to lower the manufacturing cost, a few scholars and enterprises develop a primary crude BF such that it is suitable for massive applications in the building industry. Among them a method to produce a type of 'hemp-shaped' crude BF that exhibits an average fineness in the range of 300–500 metric count is invented [124, 125], and an insulation material composed of BF is developed [126]. The microstructure of bamboo is almost fully parallel in the vertical direction with only a few radial cells observed around bamboo joints, and this is significantly different from that of wood microstructures. It costs less energy to separate and extract BF due to these characteristics (namely strong anisotropy and lack of radial cells) [127], and this is beneficial for its mass production.

Construction infill is a necessary constituent element in the multi-layered construction system of building envelope that plays the role in improving the acoustic and thermal performance of the entire component. Vegetable fibers are one of the most commonly used construction infills that possess the advantage of renewability and biodegradability [128, 129]. The thermal properties of straw, palm fiber, coconut fiber, and banana fiber as insulation infills are investigated by some scholars and indicated that the dry thermal conductivity among these materials are in the range of 0.038–0.065 W/(m K) in which the wood fiber is in the range 0.038–0.050 W/(m K) [130–135]. When compared with inorganic materials, vegetable fibers interact with the physical environment and especially the hygric environment. It is suggested that the thermal conductivity of palm fiber, barley straw, straw, flax, hemp fiber, and wood fiber is significantly affected by their own moisture content that depends on the ambient relative humidity [136, 137]. The aforementioned characteristics enable vegetable fiber to play its 'breathing effect' [138], that contributes to the hygrothermal regulation of the building component and the enclosed space, and conversely can negatively affect the thermal performance due to the change in the moisture content.

1.3.1.3 Bamboo Charcoal

The pore structure of BC without activation is mainly composed of large pores, and the micropores are formed and developed after 500 °C when the inner pore sizes shrink from 0.25–50 to 0.55–5.50 μm with increases in the temperature. The specific surface area increases significantly within 600–700 °C [139] and reaches 385 m^2/g when the carbonization temperature increases to 700 °C, and this is 2–5 times that of ordinary wood charcoal, and thereby exhibits the highest moisture-control capacity [140, 141]. The hygroscopic properties of BC are investigated and suggested that the particle size displays a very low effect, the specific surface area displays a disproportionate effect on the moisture adsorption capacity of BC, and the increase in the ambient temperature weakens the hygroscopic performance [142]. The equilibrium moisture content, moisture adsorption and desorption efficiency, and the humidity stability are considered as indicators to characterize the moisture-control performance of BC and analyze the impact of the processing parameters [143].

'Moisture-control material' is proposed in 1949, and it refers to those materials that automatically regulate the ambient air relative humidity through their adsorption and desorption processes that respond to changes in the hygric and thermal environment without artificial energy and mechanical equipment. Water vapor transmission coefficient and maximum equilibrium moisture content are used to characterize the moisture-control performance, and the material exhibits a better performance when both the indicators increase [144]. Based on the analysis of isothermal adsorption and desorption curve $u(X, T)$, $k = (\partial u/\partial x)T$ and $v = (\partial u/\partial T)x$ are selected to characterize the moisture content change rate responding to the air humidity when the air temperature T and the absolute air humidity X are in a constant state, respectively [145]. When compared with other types of building moisture-control materials (such as silica gel, inorganic salts, and organic polymer materials), porous materials (including bamboo charcoal, wood charcoal, zeolite, and kaolin) generally possess higher permeability, higher adsorption and desorption rate, relatively lower desorption hysteresis effect, simpler manufacturing process, lower cost, more favorable corrosion resistance, and a longer operation life such that they are widely used in civil industry [146]. In Asian countries, moisture-control BC is applied in civil construction to control the moisture content of the building components and the humidity of indoor air to prevent construction deformation and inhibit bacterial growth.

Qin et al. established a mathematical model suitable for heat and moisture transport in building envelope for different climatic zones, performed the simulation of moisture-control material used in Hong Kong and Paris, and suggested that the maximum air conditioning refrigeration load were reduced by 30 and 10%, respectively, and this indicated the improvement effect of moisture-control material on indoor air humidity and the energy efficiency. Ferreira simulated and measured the relative humidity (RH) amplitude of a storage room of a historical museum building to quantify the effect of moisture-control material on the indoor RH stability [147]. The test on the moisture-control material (spruce cotton board covered with mineral binder) showed that the EMC increased from 12 to 40 kg/m^3 in the process of increasing RH = 40% to RH = 80%. The water vapor permeation resistance factor resulted from the wet cups was 1.1, and that from the dry cups was 3.2. When the surface of ceiling and wall in the storage room, of which the base layer was gypsum board and lime mortar, was covered with this material, the annual indoor RH amplitude reduced from 59 to 33%. It showed that the hygroscopic inertia material helped to maintain the stability of the indoor RH, and this played a role in reducing the demand of the device system [148]. Similar study on moisture buffering included the simulation and test of the influence on the indoor thermal and hygric environment and HVAC demand in a historical church carried out by Erhardt et al. [149].

1.3.2 Performance Study on Bamboo Building Envelope

There is little literature on the study of bamboo in building components and space. In the 1960s and 1970s, Dunkelberg conducted a research on the typology of traditional bamboo buildings in the Asian region and collected the practices of the traditional bamboo-framed houses, walls, ceilings, roofs, doors, and windows, but the research was mainly within the scope of 'construction method,' rather than the 'performance' [63]. Yu et al. set up a solar house in 2012, with walls composed of bamboo partition boards and insulation infills, in which the discussion was focused on its primary energy consumption and the carbon emissions advantages [150].

Professor Li led a team to complete a NSFC project in 2008–2011: 'Coupled heat and moisture transfer model and experimental study on a new type of bamboo construction (50878078),' which carried out systematic studies on the effects of hygrothermal expansion properties and external environment on the hygrothermal stress of bamboo laminated lumber, plybamboo, and bamboo–concrete composite members. The project was aimed at the problem of the coupled heat and moisture deformation and hygrothermal stress of bamboo components, but it involved the porous media characteristics of bamboo [151], internal heat and moisture transfer principle and the coupling model [152, 153], simulation and measurement on the mold growth of bamboo plywood wall and the indoor air quality [154, 155] (Table 1.13).

Table 1.13 Main content of research project 'Coupled heat and moisture transfer model and experimental study on a new type of bamboo construction'

The main achievements of the project 'Coupled heat and moisture transfer model and experimental study on a new type of bamboo construction'	• Measured the heat transfer coefficient of bamboo wall under steady conditions and the law of heat flow under non-steady conditions • Theoretical derivation and experimental measurement to obtain the parameters of hygrothermal expansion of bamboo wall, hygrothermal stress, and its impact on the hygrothermal expansion coefficient • Established a theoretical model and numerical calculation model for the coupling of heat and moisture in bamboo wall, and proposed a transfer function analysis method for the dynamic prediction of the temperature and moisture content of bamboo composite wall • Explored the factors affecting the mold growth in bamboo walls • Measured the indoor air quality enclosed by bamboo walls

Source Sort out according to the final report of the NSFC project 'Coupled heat and moisture transfer model and experimental study on a new type of bamboo construction (50878078)'

In addition, there are a few discussions that concentrate simply on the thermal performance of bamboo components. Li et al. from Ningbo University constructed steel–bamboo composite walls and discussed the heat transfer coefficient, delay time, and attenuation coefficient. The measured results show that the heat transfer coefficient was improved by 26.1–48.4% compared with the two traditional walls, and corresponding simulation showed the energy saving was 27.1–40.9% in hot summer and warm winter areas [156].

1.3.3 Heat and Moisture Process Model of Building Envelope

1.3.3.1 Theory Study on HAM Model

The rational control of hygrothermal processes in building envelopes based on an accurate understanding of these processes will help extend the lifespan of building components [157–159], reduce energy consumption during building operations [160], and improve the comfort and cleanliness of indoor hygrothermal environments [161, 162]. In the field of building thermal engineering, the heat and moisture process model are usually used for calculation. Prior to the 1990s, Glaser proposed a steady-state water vapor permeation model, using vapor pressure as the driving potential, and determining the moisture transport rate of the pure vapor based on

Fick's law [163]. This method is of high operability and can be used for preliminary hygric performance evaluation on building components, predicting the condensation site and degree, and therefore it is broadly used, such as in the EN ISO 13788, DIN 4108, and GB 50176-93/2016 standards [106]. This model simply uses the vapor permeability coefficient to calculate moisture processes that occur in a one-dimensional steady state. However, actual moisture processes are often unsteady and simultaneously occur over multiple phases. Therefore, the Glaser simple steady-state evaluative method needs to be improved for the accurate simulation of heat and moisture processes.

In 1975, based on Fourier's law, Fick's law, and Darcy's law, Luikov established the governing equation according to the energy conservation law, the mass conservation law, and the momentum conservation law, which mechanically described the migration process of moisture in porous media [164, 165]. Over the past 20 years, Molend, Pedersen, Künzel, Häupl, Janssen, Mendes, Steeman, and Tariku have proposed coupled heat and moisture transfer models (HAM models) for analyzing the storage and transport of heat, moisture, and air in building envelopes [166–173]. These models have been validated from a number of perspectives. For example, the IEA Annex 41 project completed in 2008 compared 17 HAM models in detail using numerical simulation tools and carried out seven joint comparisons. It is confirmed that the results calculated using HAM models showed good agreement with the results of actual measurements [174, 175]. The description equations are the core of the corresponding HAM model theory, and the determining of driving potential for the heat and moisture process, as well as its requirements on the material parameters constitute the main content of the equations. The related themes of driving potentials and material parameters have been paid attention by scholars to derive the corresponding models.[4]

Janssen evaluated the efficiency and accuracy of different driving potentials for the simulation of moisture transport [176]. Van Belleghem et al. carried out verification of the coupled heat and multi-phase moisture models for porous materials in computational fluid dynamics (CFD) [177]. Van De Walle et al. developed a model for predicting the thermal conductivity of porous blocks [178]. Defraeye et al. discussed the influence of the uncertainty of heat and moisture transport properties on the drying characteristics of porous materials through numerical simulation [179] and investigated the heat and mass transfer on the outer surface of the building envelope [180]. The coupling between building dynamic simulation and behavioral patterns of indoor occupants has also received attention [181]. Nouidui and Geusen et al. developed a building hygrothermal model for the research on indoor thermal comfort [182, 183].

[4]The HAM model is a coupled model of heat, moisture, and air, and the coupling of air occurs only when there is a pressure difference between indoors and outdoors, so it is not considered in some models. The models that do not describe the air coupling are actually 'coupled heat and moisture model,' referred to as the HM model. In this project, the HM model is considered as a case of the HAM model, so they are collectively referred to as the HAM mode.

Deromea et al. investigated the wind-driven rain factor on building facade, summarized the key points of wind-driven rain prediction and the corresponding modeling methods, and estimated the wind-driven rain in different scales of the building environment [184]. Van den Brande et al. studied the implementation and application of a first-order runoff model for coupling HAM model with the rainwater runoff in porous building facades [185]. Peren and Kubilay et al. combined the wind-driven rain research with CFD models and carried out simulation and measurement for verification [186, 187]. The team of Blocken, Abuku, and Janssen finished a series of studies on wind-driven rain [188, 189], of which the themes related to the mechanism of the adsorption and desorption process of the rain on building facade and its impact on mold growth, indoor environment and energy consumption [162, 190, 191], simplified analysis of wind-driven rain as a boundary condition for HAM simulation [192], as well as the comparison and verification of the simulated and measured results [193, 194].

The study on HAM model has also become more specialized at the material level. Islahuddin and Janssen estimated the hygric properties including the isothermal moisture content curve and the moisture transport properties by analyzing the multi-scale pore structure of porous materials [195, 196]. Similarly, Todorovic and Janssen investigated the effect of gypsum that stuck in the pores on the moisture transport properties of bricks [197]. Feng et al. studied the hygric properties of porous materials and their test methods, discussed the impact of experimental environment, experimental procedures, and repeatability on test results [198–200]. Janssen and Vereecken et al. took the solid masonry walls as examples to analyze the influencing factors of the moisture transport in building materials [201, 202]. Different from the steady-state heat transfer, the new expressions of HAM model for the heat transport performance of building envelopes were investigated. Deconinck compared the thermal resistance description methods of building components by on-site measurement [203, 204]. The impact of heat storage performance of the materials on the indoor thermal environment and energy consumption has also been concerned [205].

1.3.3.2 Computer Program Development Based on HAM Model

HAM models usually adopt highly coupled nonlinear partial differential equations to describe the hygrothermal processes of building envelopes [206]. These equations are solved continuously through the finite difference method in the operation process, and the internal temperature and humidity distribution are updated by means of integration, which result in that the calculation time can reach 10^2–10^4 times of the common energy consumption model. Computer tools are the prerequisites for the practical application of HAM model, and some computer programs have been integrated with the HAM theory for specific analysis, e.g., the BES-HAM for building energy simulation analysis, WUFI series software for indoor hygrothermal environment, heat and moisture process in building components and mold growth analyses, etc. [168]. According to Delgado et al. in 2012, nearly 60 kinds of computer tools for heat and moisture process simulation used in the field of building physics were developed, but most of these were limited within the laboratory and could not be promoted to the market [207]. There are some other common building heat and moisture process simulation software, including the Delphin (Germany), HAM-Tools (Sweden), MOIST (USA), etc. Some comprehensive energy simulation software such as the EnergyPlus (USA) and Transys (USA and Europe) have also increased the HAM function (Table 1.14).

The credibility of HAM models shown in both laboratory and practical engineering has won the acceptance of an increasing number of practitioners in this field, and these HAM model-based approaches have also been accepted by the DIN4108-3 standard. With the aid of computers, the HAM model-based heat and moisture process simulation have the opportunity to replace the simple estimation, and become the mainstream calculation method of building envelope in the future.

1.3.3.3 Application of HAM Model Theory and Its Computer Program

HAM model involves different aspects of materials, building components, and the overall building systems. The tests, calculation and verification for the hygrothermal properties and hygrothermal performance based on a coupled heat and moisture process model require far more time and capital cost than those works based on a pure thermal process model, so that the HAM model has not become a research issue carried out widely. The main research groups on the heat and moisture process of building envelopes are mainly from the departments of building physics and building materials in universities, as well as some other building research institutes, but most of them are mainly aimed at one aspect and cannot be systematically developed at different levels.

The Hygrothermics Department of Fraunhofer IBP (Institute of Building Physics), the TU Dresden of Germany, the Department of Building Physics in Katholieke Universiteit Leuven of Leuven (KU Louven), the Oak Ridge National Laboratory (Oak), The Ridge National Laboratory and the University of Bath in England, etc., are representatives of the research groups with long-term and systematic works in the field of heat and moisture process in building envelopes [207].

In addition to the development of heat and moisture process models in the laboratory, the research areas have been extended to the levels of materials, building components, and building systems. At the material level, the works include the thermal properties and hygric properties' test of the building materials, the climate-related study on the material properties, as well as the damage analysis under the influence of the heat and moisture load; at the building component level, related works include the study on the heat, air, moisture, and salt transfer, the effects on wall, roof, and floor systems, building moisture assessment and the hygrothermal analysis, as well as bio-hygrothermics (preventing mold growth and the façade organisms), etc.; at the overall building system level, research topics include building energy consumption and its relationship with the related systems, indoor comfort, air quality and sustainable construction, historic building protection and old building renovation, as well as some case studies.

Table 1.14 List of the main computer programs based on HAM model

No.	Program name	Calculation dimension	Coupling factor	Producer	Country
1	1D-HAM	1D	Heat/Air/Moisture	Chalmers Technical University & Gothenburg, University of Lund	Sweden
2	BSim	1D	Heat/Moisture	Danish Building Research Institute	Denmark
3	Delphin	2D	Heat/Air/Moisture/Pollutant/Salt	Technical University of Dresden	Germany
4	EMPTIED	1D	Heat/Air/Moisture	Canada Mortgage & Housing Corporation	Canada
5	HygIRC	1D/2D	Heat/Air/Moisture	Concordia University	Canada
6	HAMLab	1D/2D/3D	Heat/Air/Moisture	Eindhoven University of Technology	Netherlands
7	HAM-Tools	1D	Heat/Air/Moisture	Technical University of Denmark & Chalmers Technical University	Sweden
8	IDA-ICE	1D	Heat/Air/Moisture	EQUA Simulation AB	Sweden
9	MATCH	1D	Heat/Air/Moisture	TUD-Thermal Insulation Laboratory	Denmark
10	MOIST	1D	Heat/Moisture	National Institute for Standards and Testing	USA
11	UMIDUS	1D	Heat/Moisture	Pontifical Catholic University of Parana Curitiba	Brazil
12	WUFI	1D/2D	Heat/Moisture	Fraunhofer Institute for Building Physics (IBP)	Germany

Note The programs 2, 10, 11, and 12 have not taken the air coupling into consideration, so that there are actually the 'coupled heat and moisture model,' which is regarded as a case of HAM model in this study
Source Sort out according to Delgado, Barreira, et al. Hygrothermal numerical simulation tools applied to building physics

The HAM model basic theory and its computer tools support the research of important topics such as hygroscopic materials and their construction systems, moisture buffering of building envelopes, and mold growth assessment.

The BRE—Center for Innovative Construction Materials in University of Bath—finished a series of tests on the plant materials [208, 209] and their building system [210–212]. It has researched and developed straw and hemp composite materials and their prefabricated panel building systems, built demonstration houses, and conducted tests on the heat transport and moisture buffering performance [213, 214]. Similar works on hygroscopic building materials include Langmans et al.'s experimental study on the thermal and hygric performance of straw walls [215], Maskell et al.'s study on the properties of plant fiber-based gypsum and its influence on the building indoor environment [216], Dubois and Latif et al.'s evaluation on the hygric properties, and the test on the moisture buffer value (MBV) of the clay-based blocks [217] and their moisture buffering effect of the walls [218], and Silva et al.'s study on the effects of biomass materials such as plant-based MDF boards on the quality of indoor air and hygrothermal environment [219]. Zillig completed the doctoral research on moisture transport in wood [220]. Langmans et al. carried out numerical simulation and measurement of a wooden frame wall with an exterior vapor barrier in a layered construction [221, 222]. Vanpachtenbeke et al. conducted a similar study on wooden walls with bricks as exterior cladding layer [223].

The moisture buffering effect of porous material such as plants and clay used in the building envelope attracts attention [224]. James et al. compiled numerical and experimental data sets as benchmarks for the moisture buffering models [225]. Vereecken et al. conducted an on-site measurement of the moisture buffering effect of an enclosed space [226]. Roels et al. compared and made judgment on the credibility of the material parameters used in moisture buffering studies [227]. Mcgregor et al. investigated the moisture buffering effect of the clay building materials and the unsintered clay blocks [228, 229].

The simulation results of building components and enclosed space output from the HAM model can be combined with specific theories for certain analysis, such as combining with mold biology research for analyzing mold growth conditions. Vereecken et al. studied the mold growth prediction models [230], discussed the role of various models in the assessment of mold damage [231], and evaluated the growth of molds on thermal bridges and their effects on building components, indoor environment, and energy consumption with the models [232]. Vanpachtenbeke et al. investigated the growth of molds in timber building envelope as the boundary condition and its effects on the damage of materials in the construction [233, 234].

1.3.3.4 Research Status in China

The basic theory of HAM model has also attracted the attention of many scholars in China. Chen of Hunan University completed a NSFC project in 2003–2006: 'Research on transient heat transfer analysis method of building envelope based on modern system identification theory,' which proposed a transient heat transfer analysis method for building envelope with frequency domain regression according to China's meteorological frequency range, and carried out corresponding adaptive analysis [235–237]. In 2010–2013, he finished another project 'Coupled heat and moisture dynamic transfer model and experimental research on the building walls in hot-humid climate region,' which was based on the theory of heat and mass transfer in porous media, targeted the heat and moisture migration and moisture accumulation in building envelope, and systematically studied the heat and moisture coupling process of multilayer walls in the hot-humid climate zone in southern China. The dynamic mathematical model of heat, moisture, and air migration in multilayer walls was established, and verification was carried out by in a test bench constructed in Changsha, which showed the significant impact of solar radiation on the heat and moisture migration through the wall [238, 239].

Meng of South China University of Technology completed a NSFC project: 'Research on passive cooling by water evaporation technology for the buildings in subtropical regions,' which determined the thermal conductivity with the change of moisture content for six kinds of porous building materials, and established an hourly rainfall model for the analysis of building energy consumption in Guangzhou area, and realized the expansion of meteorological model for calculation of building energy consumption in hot-humid climate region [240]. In this project, the first test bench for testing the conventional dynamic climate in China, the 'hot-humid climate wind tunnel' was constructed and the rationality of it was verified through simulation and experimental methods [241, 242]. In 1999–2001, he finished an earlier project: 'Research on passive evaporative cooling of ungreen space in subtropical cities,' which theoretically obtained the results of evaporative cooling coefficient, established the coupling model of porous materials under cyclic heat and moisture boundary conditions, and wrote a corresponding computer simulation software [243, 244].

Qin of Nanjing University completed a NSFC project in 2015: 'Study on the moisture buffering mechanism of porous moisture-control materials,' which carried out research and development of a composite phase change moisture-control material with both heat regulation and moisture-control functions. The material was used in building walls to regulate the indoor thermal and hygric environment and reduce the building energy consumption [245, 246]. In 2011–2014, he finished an earlier project: 'Study on the dynamic coupled heat and moisture transfer and the airflow model of the whole building,' perfecting the boundary condition measurement method from the perspective of the heat and moisture coupling transfer mechanism inside the porous building materials, and investigating the heat and moisture coupling, multi-zone airflow laws and coupling methods. In this project, program modules for rapid calculations of HAM, as well as programs for energy calculations of the whole building were developed. Through simulation, it was shown that in the humid areas of the hot summer and cold winter climate region, the moisture-control materials in building envelope can produce obvious moisture buffering effect on the indoor environment and reduce energy consumption [247, 248].

Liu from Xi'an University of Architecture and Technology completed a NSFC project in 2011–2015: 'Study on the quantitative impact of building envelope surface moisture transfer on the heating and cooling load of the building,' establishing a coupled heat and moisture transfer mathematics model for the surface and interior of the building envelope, as well as the indoor air. Through actual measurement, a correction method to quantify the effect of the moisture content on the thermal conductivity was proposed, and the heat transfer and mass transfer coefficient as a function of temperature and relative humidity were obtained. In addition, the project analyzed the impact of the moisture transfer of the building envelope on the heat transfer process, indoor thermal environment and HVAC demand in typical climate zone, and proposed a load correction method considering the moisture transfer [249, 250].

Yan from Xi'an University of Architecture and Technology completed a NSFC project in 2013: 'Study on the theories and techniques of thermal and hygric environmental control in Dunhuang Mogao Grottoes under multi-field coupling conditions,' which obtained systematic theoretical results, and provided technical solutions for the regulation of thermal and hygric environment in caves through continuous measurement of the temperature and relative humidity on site, hygrothermal properties' test of the interface materials and simulation of their impact on the microclimate environment in the cave area [251, 252]. In 2005–2007, he finished an earlier project: 'Research on indoor hygric environment regulation and control technology for energy-saving residential buildings,' which conducted research on the indoor hygric environment and its regulation technology through indoor environment measurement of energy-efficient residential buildings in different climate zones, hygric properties' test on the building materials and computer simulation [253, 254].

Shang from Xi'an University of Architecture and Technology completed a NSFC project in 2011–2015: 'Research on the composite mechanism and energy-saving technology based on temperature and moisture-control materials in the building,' using phase change materials with temperature control function, and hydrophilic materials with moisture-control function to study the composite mechanism, and the developed regulators are combined with gypsum to make changes to the heat and moisture storage properties of the latter, so that the modified gypsum as an interior finishing material obtained functions to adjust both the indoor thermal and hygric environment. Based on this, a heat and moisture transfer mathematical model containing the phase change material was proposed and the corresponding test verification by experimental room was carried out [255, 256].

Kong of Central South University completed a NSFC project in 2012–2015: 'Study on the principle of coupled heat and mass transfer of energy-saving building envelopes and its impact on energy consumption,' in which the principle of coupled heat and mass transfer of the building envelopes of new energy-saving building in typical climate zones in China was studied. Through theoretical calculation and actual measurement, a new coupled heat and mass transfer model of building envelopes was established, and the impact of initial moisture content with different material settings on the thermal insulation performance and energy consumption was compared [257].

Zhang of Shanghai Jiaotong University completed a NSFC project in 2014–2016: 'The influence of moisture-control building materials on the building envelope and indoor heat and moisture distribution under the moisture driving potential,' taking Shanghai as an example in the hot summer and cold winter area, and measuring the moisture-control effect of hygroscopic material on the indoor hygric environment through actual measurement. The project investigated the influencing factors on the moisture-control performance of the hygroscopic material, and on this basis, put forward the evaluation index of the adsorption and desorption effect. For the numerical method, a coupling model of the building envelope and the physical quantity in the air was established and verified by an artificial meteorological laboratory [258, 259].

Han of Shanghai University of Technology completed a NSFC project in 2012–2015: 'Study on the moisture transfer mechanism and the moisture adsorption and desorption dynamic characteristics of the interface in the moisture-control building envelope,' which carried out the basic theory and experimental study on the moisture transfer coefficient for modeling the heat and moisture transfer of the moisture-control building envelope. The heat and moisture coupling equations that could characterized the moisture transfer between the layers of the moisture-control building envelope were established and solved by numerical methods, and the test box was set to measure the actual effect of building envelope [260].

Guo of Nanchang University completed a NSFC project in 2012–2015: 'Dynamic model and experimental study of coupled heat, air and moisture of multilayer walls in hot-humid climate regions,' which established a coupled heat, air, and moisture transfer model for multilayer walls based on the heat and mass transfer theory of porous media as well as the energy and mass conservation laws, with consideration to the factors such as air infiltration, evaporation, and condensation of water inside the wall, and solar radiation. According to the meteorological conditions in the hot-humid climate regions of southern China, the sensitivity analysis of each parameter was carried out, and the main parameters affecting the accuracy of the model are obtained [261–263].

Li of Guangzhou University completed a NSFC project in 2011–2015: 'Simulation study of building energy consumption under complete meteorological parameters in Hot-Humid climate regions,' taking Guangzhou as an example to carry out systematic research on the heat and moisture transfer under complete meteorological parameter conditions including the rainfall. The actual measurement showed that using porous brick as the exterior cladding on the building facade, the evaporative cooling effect can be realized. The wall simulation was carried out in the Delphin computer program, which showed that the surface temperature and heat-flux density of the wall were reduced to different degrees by natural rainfall [264].

Feng from China Southwest Architectural Design and Research Institute Co., Ltd. completed a NSFC project in 2013–2016: 'Study on moisture diffusion coefficient of porous building materials,' which focused on the moisture diffusion coefficient of porous building materials, described and solved the model of the heat and moisture process under different pore connectivity conditions, developed, and improved the test methods for hygric properties [198, 199, 265, 266]. In 2006–2009, he finished an earlier project: 'Study on the thermal physical properties and thermal design calculation of new energy-saving building envelope,' which adopted measurement and simulation methods to study the hygric properties parameters such as the equilibrium moisture content, vapor permeability, and moisture diffusion coefficient of porous media in the new building materials, and carried out calculation and verification of the change law in the physical properties [267–269].

Wang from the Chinese Academy of Forestry completed a NSFC project in 2012–2015: 'Climate regulation mechanism and optimization design of wood frame exterior wall,' which conducted measurement on the interface air temperature, relative humidity, moisture content, and heat transfer coefficient of the timber construction located in Taihu Lake, indicating that the indoor relative humidity can be maintained within 60–80% in an environment where the outdoor relative humidity was higher than 80% for a long time [270].

There are some related research projects, for example, Ran from Huaqiao University investigated the moisture desorption properties of hygroscopic porous material and its improvement effect on the indoor hygrothermal environment [271]; Wang from Xi'an University of Architecture and Technology studied the coupled heat and moisture process of building envelope and its impact on the indoor hygric environment [272]; Chen from Nanhua University carried out model and experiment studies on the coupled heat and moisture transfer of the exterior walls in hot-humid climate regions [273]; Liu from Shandong University of Architecture conducted multi-scale research on the coupled heat, air, and moisture transfer mechanism of the exterior walls and the enclosed indoor environment [274]; Zhang from Tongji University performed study on the hygric properties of building materials and the coupled heat and moisture transfer within the walls [275–277].

1.3.4 Deficiencies of the Current Research

The existing research provides certain hygrothermal properties parameters of bamboo, which can be used to provide a certain reference for its applicability in building envelopes; the corresponding test specification provides a method basis for obtaining the material parameters; the theoretical model of heat and moisture process in building envelope and the development of the corresponding computer simulation tools can support the performance study of building components and enclosed space. However, there are still some shortcomings in the existing works on the application of bamboo in building envelope:

(1) Incomplete material objects. Under the background of 'industrial utilization of bamboo forest resources,' the bamboo product system has been greatly expanded, and the definition of 'bamboo' has broken through the original range of 'raw bamboo.' However, the test objects in previous studies are mainly raw bamboo, which cannot represent the more commonly used bamboo-based panels in the practical project. The raw bamboo samples are usually only 7–9 mm in thickness, which cannot represent the thicker bamboo-based panels. In addition,

during the manufacturing process of bamboo-based panels, the constituent units, the assembly methods, the protective treatments such as the anti-corrosion, fire prevention, and anti-cracking, as well as the gluing process will result in changes to the properties of the final products. Therefore, it is necessary to carry out independent material test for each of the bamboo-based panels. The feasibility of natural bamboo fiber and bamboo charcoal as building envelope materials has not been fully explored, and their potential accordingly as a thermal insulation and moisture-control infill is yet to be discovered.

(2) Imperfect material parameters. At present, bamboo is not considered as an independent type of building material, and there are only timber parameters in the database in related literatures, thermal design standard, computer simulation programs, etc. The existing test projects for bamboo could not provide the property parameters required to support a complete heat and moisture process model of building envelope, and most of them are directed to a certain basic physical property, thermal or hygric properties under steady-state conditions, and cannot describe the properties in the state where interaction occurs between heat and moisture. A large part of these tests are based on the methods from the micromaterials science, using insufficient samples or the ones that are much smaller than the actual size, which can have a non-negligible effect on the test results. Imperfect and inaccurate material parameters can cause errors in the calculation of the heat and moisture process to the building envelope.

(3) Non-systematic research. Whether it is the study of the partition boards or construction infills, works have focused on the microscopic properties of materials rather than the macroscopic performance of buildings; there is lack of correlation with actual conditions such as the climatic conditions and construction types; the long-period dynamic performance of the building components and enclosed space has not been sufficiently investigated.

The deficiency of the material parameters and the quantitative knowledge of properties lead to further impossibility of the construction calculations; the absence of performance research on building components and enclosed space, and the influence law of the material and construction parameters on the performance, have resulted in the lack of basis and guidance for the construction design of bamboo building envelopes. The insufficiency of these efforts has led to the lack of demonstration in the building industry to promote the 'substitute timber with bamboo,' and further, the foundation for the development of a modular and standardized bamboo construction system.

1.4 Research Content

In order to solve the above problems, this project carries out an application fundamental research on bamboo in building envelope.

1.4.1 Technical Route

(1) Considering that for bamboo as a hygroscopic building material, the moisture process has an important influence on the building envelope, the HAM model of building envelope is used as the theoretical model for the overall work. Based on this, the project combines experiment and simulation methods and carries out the works following a route of 'property study–performance evaluation–performance optimization.'
(2) Experimental method is adopted in the property study of material. Standard and practical material parameters are obtained according to a number of international building material test standards. Based on the investigation of manufacturing process, the existing industrialized bamboo products are classified, and six kinds of partition boards and two kinds of construction infills are selected as the material objects; the material parameters that can support the complete HAM model are measured by setting up nine standard experiments and one wind tunnel test, including the basic physical properties, the storage and transport properties of heat, gaseous moisture and liquid water; through the comparison between the modified bamboo and the raw bamboo, the correlation analysis between the parameters, and the correlation analysis of the heat and moisture transport properties with the moisture content, the understanding of the material properties is deepened.
(3) Simulation method is adopted in the performance study of building envelope. The performance' simulation of building components and enclosed space is carried out in the heat and moisture process simulation computer program based on HAM model. The boundary conditions are set in a layered construction which is suitable for the form of bamboo materials; the external climate conditions are set with reference to the Köppen climate classification method. Firstly, the meteorological parameters of the typical climate zones in North America are selected for the performance evaluation, and then the North American hot-humid climate zone is selected for the performance optimization; the simulation results of the annual hourly hygrothermal performance of the building component, indoor hygrothermal environment, and HVAC demand are output as indicators for the performance studies, and the factors' impact of the material and construction parameters of the layered construction are studied.

The system achievements of material study, performance evaluation, and performance optimization provide material parameter database and construction design guidance for the application of bamboo in building envelope (Fig. 1.30).

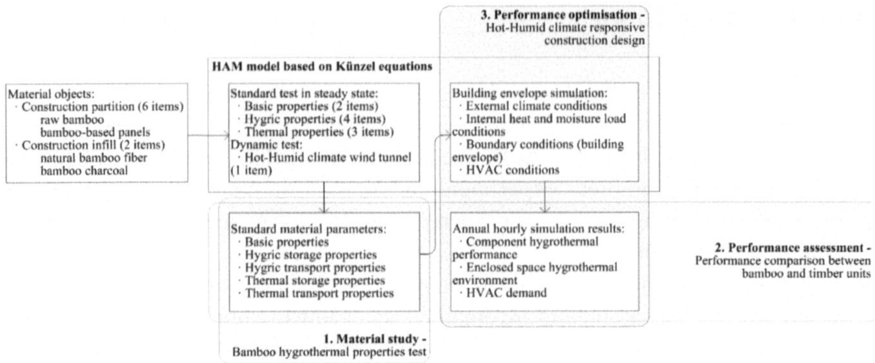

Fig. 1.30 Overall work structure diagram

1.4.2 Chapter Arrangement

The book consists of six parts. In addition to the introduction and conclusion chapters, the second to the fifth chapters are the demonstrations, which constitute the main body of the book (Fig. 1.31).

Chapter 2 is a discussion of the research model, in which the HAM model based on Künzel equation is analyzed; the HAM simulation of heat and moisture process

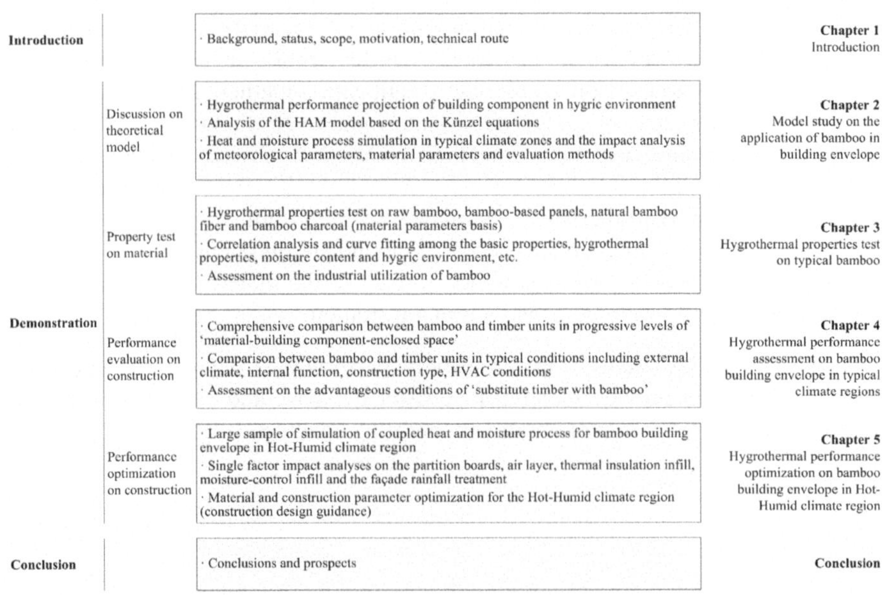

Fig. 1.31 Chapter arrangement

in the typical climatic zone and its comparison with the results of the steady-state calculation method provide suggestions to the meteorological parameters, material parameters, and evaluation methods for relevant research; a clear theoretical model for the study of bamboo applied to building envelope is determined.

Chapter 3 obtains the material parameters of typical bamboo products according to the corresponding building material test standards, which provides the basis for further computer simulation; the correlation between the obtained material parameters and the environmental parameters, and among different thermal and hygric properties, provide related laws at the material level; the comparison among different bamboo products provides a reference for the utilization of bamboo.

Chapter 4 simulates the performance of building components and enclosed space, which involves the typical conditions including external climate, internal function, construction type, HVAC conditions, etc.; through comparison with the corresponding timber unit at the progressive levels of 'material—building component—enclosed space,' the performance evaluation of the bamboo unit is formed, and the reference for the feasibility and advantageous conditions of 'substitute timber with bamboo' is provided.

Chapter 5 takes the hot-humid climate region as the external condition to carry out optimization study on the layered construction of bamboo building envelope, in terms of material and construction parameters; based on the large sample of simulation of the coupled heat and moisture process, single-factor impact analyses on the partition boards, air layer, thermal insulation infill, moisture-control infill, and the façade rainfall treatment are conducted, which provides guidance for the construction design in hot-humid climate region.

References

1. López OH (2003) Bamboo: the gift of the gods. O. Hidalgo-López, Bogotá
2. Das M, Bhattacharya S, Singh P et al (2008) Bamboo taxonomy and diversity in the era of molecular markers. Adv Bot Res 47:225–268
3. Austin R, Ueda K (1972) Bamboo. Weather Hill Publishing, New York
4. Rao AN, Ramanatha Rao V, Williams JT (1998) Priority species of bamboo and rattan. International Network for Bamboo and Rattan (INBAR), Beijing
5. Janssen JJA (1988) The importance of bamboo as a building material. In: Proceedings of the Third International Bamboo Workshop. Cochin, India, 14–18, November 1988, pp 235–241
6. De Vos V (2010) Bamboo for exterior joinery: a research in material and market perspectives. B.Sc. thesis, University of Applied Sciences Van Hall Larenstein (Wageningen UR)
7. Liese W (1985) Bamboos–biology, silvics, properties, utilization. Deutsche Gesellschaft für Technische Zusammenarbeit (GTZ) Schriftenreihe Nr. 180, TZ Verlagsges, Roßdorf
8. Yiping L, Henley G (2010) Biodiversity in the bamboo forests: a policy prospective for long term sustainability. International Network for Bamboo and Rattan (INBAR). Working paper 59
9. Liese W, Mende C (1969) Histometrische Untersuchungen uber den Aufbau der Spro-Bachse zweir Bambusarten. Holzforsch. Holzver-wert 21(5):113–117
10. Grosser D, Liese W (1974) Verteilung der Leitbündel und Zellarten in Sproßachsen verschiedener Bambusarten. Holz als Roh- und Werkstoff 32:473–482

11. Zhang QS, Jiang SX, Tang YY (2001) INBAR technical report no. 26—Industrial utilization on bamboo. International Network for Bamboo and Rattan (INBAR), Beijing
12. Liese W (1998) INBAR technical report no. 18—The anatomy of bamboo culms. International Network for Bamboo and Rattan (INBAR), Beijing
13. Li S, Fu S, Zeng QY et al (1995) Biomimicry of bamboo bast fiber with engineering composite materials. Mater Sci Eng 3(2):125–130
14. Mitch D, Harries KA, Sharma B (2010) Characterization of splitting behavior of bamboo culms. J Mater Civ Eng 22(11):1195–1199
15. Prawirohatmodjo S (1988) Comparative strengths or green and air dry bamboo. In: Proceedings of the Third International Bamboo Workshop. Cochin, India, 14–18, November 1988, pp 218–222
16. Sharma SN , Mehra ML (1970) Variation of specific gravity and tangential shrinkage in the wall thickness of Bamboo (Dendrocalamus strictus) and its possible influence on the trend of the shrinkage-moisture content characteristic. Indian For Bull 259(7)
17. Kumar S, Dobriyal PB, Rao IVR et al (1988) Preservative treatment of bamboo for structural uses. In: Proceedings of the Third International Bamboo Workshop. Cochin, India, 14–18, November 1988, pp 199–206
18. 萧江华 (2010) 中国竹林经营学. 科学出版社 / Xiao JH (2010) Chinese bamboo forest management. Science Press
19. Ganapathy PM, Zhu HM, Zoolagud SS et al (1999) INBAR technical report no. 12—Bamboo panel boards: a state-of-the-art review. International Network for Bamboo and Rattan (INBAR), Beijing, pp 3–45
20. LY/T 1660-2006 (2006) 竹材人造板术语. 国家林业局, 北京 / LY/T 1660-2006 (2006) Standard terminology for bamboo-based panel. The State Forestry Administration of the People's Republic of China, Beijing
21. 向仕龙, 蒋远舟 (2008) 非木材植物人造板(第2版). 中国林业出版社, 北京 / Xiang SL, Jiang YZ (2008) Non-wood plant based-panel, 2nd edn. China Forestry Publishing House, Beijing
22. Sun YM, Huang ZJ, Pan W (2015) Manufacturing classification of bamboo and comparative study on several physical properties, especially considering the application in building envelope. In: Proceedings of the 31st International PLEA Conference, vol 651. Building Green Futures, Bologna, pp 1–9
23. 王衍, 钟丽丹, 王正良 (2012) 基于专利文献的我国竹产业发展分析. 浙江林业科技 32(4):67–73 / Wang Y, Zhong LD, Wang ZL (2012) Analysis on patent of bamboo industry in China. J Zhejiang For Sci Technol 32(4):67–73
24. 肖岩, 单波 (2013) 现代竹结构. 中国建筑工业出版社, 北京 / Xiao Y, Shan B (2013) GluBam structures. China Architecture & Building Press, Beijing
25. Van der Lugt P, Vogtländer J, Brezet H (2008) INBAR technical report no. 30—Bamboo, a sustainable solution for Western Europe—design cases, LCAs and land-use. International Network for Bamboo and Rattan (INBAR), Beijing
26. 李晖, 朱一辛, 杨志斌 等 (2013) 我国竹材微观构造及竹纤维应用研究综述. 林业工程学报 27(3):1–5 / Li H, Zhu YX, Yang ZB et al (2013) Review on the studies of bamboo microstructure and bamboo fiber application in China. J For Eng 27(3):1–5
27. 吴秉岭, 余养伦, 齐锦秋 等 (2014) 竹束精细疏解与炭化处理对重组竹性能的影响. 南京林业大学学报 (自然科学版) 2014(6):115–120 / Wu BL., Yu YL, Qi JQ et al (2014) Effects of bamboo bundles treated with fine fluffing and carbonized treatment on the properties of bamboo scrimber. J Nanjing For Univ (Nat Sci Ed) 2014(6):115–120
28. 国家林业局知识产权研究中心 (2014) 木/竹重组材技术专利分析报告. 北京: 中国林业出版社 / Research Center of Intellectual Property Rights, the State Forestry Administration of the People's Republic of China (2014) Analysis of wood/bamboo scrimber patents. China Forestry Publishing House, Beijing
29. 余养伦, 刘波, 于文吉 (2014) 重组竹新技术和新产品开发研究进展. 国际木业 2014(7):8–13 / Yu YL, Liu B, Yu WJ (2014) Research progress on the new technology and new product development of bamboo international scrimber. Wood Ind 2014(7):8–13

30. 肖忠平, 张苏俊, 束必清 (2013) 碳化重组竹在竹结构建筑中的应用. 林产工业 40(6):44–45 / Xiao ZP, Zhang SJ, Su BQ (2013) Application of carbonized glued laminated bamboo structure construction. China For Prod Ind 40(6):44–45
31. 王春霞, 崔立东, 刘浩阳 等 (2013) 重组竹冷压工艺的研究. 林业机械与木工设备 2013(12):17–19 /Wang CX, Cui LD, Liu HY et al (2013) Study on cold press technology for reconsolidated bamboo. For Mach Woodwork Equip 2013(12):17–19
32. 唐连成 (2012) 丛生竹(木)重组材制造及性能研究. 南京林业大学博士学位论文, / Tang LC (2012) Study on manufacturing and properties of strand woven bamboo (wood) lumber made of sympodial species. Doctoral thesis, Nanjing Forestry University
33. 安鑫 (2012) 丛生竹及其复合材料的白蚁防治技术研究. 中国林业科学研究院硕士学位论文 / An X (2012) Study on control techniques of termite for sympodial bamboo and composite materials. Master thesis, Chinese Academy of Forestry
34. 李任 (2014) 阻燃重组竹的制备工艺与燃烧特性研究. 浙江农林大学硕士学位论文 / Li R (2014) Preparation and combustion characteristics of flame retardant reconstituted bamboo. Master thesis, Zhejiang Agriculture and Forestry University
35. 王戈 (2003) 毛竹/杉木层积复合材料及其性能. 中国林业科学研究院博士学位论文 / Wang G (2003) Moso bamboo/Chinese fir laminated composite and its properties. Doctoral thesis, Chinese Academy of Forestry
36. 虞华强, 袁东, 张训亚 等 (2011) 竹木复合地板变形有限元模拟. 建筑材料学报 14(06):793–797 / Yu HQ, Yuan D, Zhang XY et al (2011) Finite element analysis on the deformation of bamboo-wood composite flooring. J Build Mater 14(06):793–797
37. 羡瑜, 李海栋, 邓健超 等 (2015) 竹塑复合材料的冲击韧性. 东北林业大学学报 43(04):101–103+136 / Xian Y, Li HD, Deng JC et al (2015) Impact toughness characterization of bamboo plastic composites. J Northeast For Univ 43(04):101–103+136
38. 李文燕, 张双保, 任文涵 等 (2014) 不同改性方法对竹塑复合材料拉伸性能的影响. 南京林业大学学报 (自然科学版) 38(03):115–119 / Li WY, Zhang SB, Ren WH et al (2014) Effects of different modifying treatments on tensile properties of bamboo-plastic composites. J Nanjing For Univ (Nat Sci Ed) 38(03):115–119
39. Li H, Zhou GY, Zhang HY (2010) Research and utilization status of natural bamboo fiber. Adv Mater Res 159:236–241
40. 蔡再生 (2004) 纤维化学与物理. 中国纺织出版社, 北京 / Cai ZS (2004) Fiber chemistry and physics. China Textile & Apparel Press, Beijing
41. Wang YP, Wang G, Cheng HT et al (2010) Structures of natural bamboo fiber for textiles. Text Res J 80(1):334–343
42. Zakikhani P, Zahari R, Sultan MTH et al (2014) Extraction and preparation of bamboo fibre-reinforced composites. Mater Des 63:820–828
43. Khalil HPSA, Bhat IUH, Jawaid M et al (2012) Bamboo fibre reinforced biocomposites: a review. Mater Des 42:353–368
44. Yang L, Liu HB, Zhang DS et al (2011) Electron microscopy study on microstructure of bamboo charcoal. J Chin Electron Microsc Soc 30(2):137–142
45. 张文标 (2002) 竹炭·竹醋液的研究. 南京林业大学博士论文 / Zhang WB (2002) Study on the bamboo charcoal and bamboo vinegar. Doctoral thesis, Nanjing Forestry University
46. Yoshinari K (2003) Apparatus and method for producing bamboo or wood charcoal. JP200368
47. Isa SSM, Ramli MM, Hambali NAMA et al (2016) Adsorption properties and potential applications of bamboo charcoal: a review. MATEC Web Conf 2016(78):01097
48. Wu G, Wu G, Zhang Q (2016) Adsorptive removal of *p*-nitroaniline from aqueous solution by bamboo charcoal: kinetics, isotherms, thermodynamics, and mechanisms. Desalin Water Treat 57(55):26448–26460
49. Meng ZL, Zhang YH, An Q et al (2013) Removal of organic pollutants in solution by bamboo charcoal. Adv Mater Res 699:554–556
50. Fujiwara S, Shima K, Chiba K (2003) Fundamental characteristics and humidity control capacity of bamboo charcoal. Mokuzai Gakkaishi 49(5):333–341
51. Laohhasurayotin K, Pookboonmee S (2013) Multifunctional properties of Ag/TiO_2/bamboo charcoal composites: preparation and examination through several characterization methods. Appl Surf Sci 282:236–244

52. Yang L, Liu HB, Zhang DS et al (2011) Microstructure of graphitized bamboo charcoal and preparation and properties of its composite. Chin J Nonferrous Met 21(3):648–655
53. Liese W, Köhl M (2015) Bamboo: the plant and its use. Springer, New York
54. Wei DS (2014) Bamboo inhabiting fungi and their damage to the substrate. Dissertation, Department of Biology, Hamburg University
55. Wei DS, Schmidt O, Liese W (2013) Durability test of bamboo against fungi according to EN standards. Eur J Wood Wood Prod 71(5):551–556
56. 魏洋, 吕清芳, 张齐生 等 (2009) 现代竹结构抗震安居房的设计与施工. 施工技术 38(11):52–54 / Wei Y, Lu QF, Zhang QS et al (2009) Design and construction of modern bamboo anti-seismic living room. Constr Technol 38(11):52–54
57. 肖岩 (2015) 现代竹结构: 以竹代木 现代科技重新唤起我们的竹情结. 城市环境设计 2015(Z1):164 / Xiao Y (2015) Modern bamboo structure: substitute wood with bamboo. Urban Environ Des 2015(Z1):164
58. 肖岩, 佘立永, 单波 等 (2009) 装配式竹结构房屋的设计与研究. 工业建筑 2009(01):56–59 / Xiao Y, She LY, Shan B et al (2009) Research and design of prefabricated bamboo house. Ind Constr 2009(01):56–59
59. 郝琳 (2010) 竹化未来. 动感 (生态城市与绿色建筑) 2010(4):80–89 / Hao L (2010) Contemporary bamboo construction for a green future. Eco-city Green Build 2010(4):80–89
60. Steiger L (2011) Timber construction. 中国建筑工业出版社 / Steiger L (2011) Timber construction. China Architecture & Building Press
61. Holzrahmenbau Bewährtes Hausbau-System 5 (2014) Auflage. Bruderverlag, Köln
62. Velez S (2013) Grow your own house. Vitra Design Museum; Bilingual edn
63. Dunkelberg K (2005) IL 31: bamboo—bamboo as a building material. Krämer Verlag, Stuttgart
64. Janssen JJA (1991) Mechanical properties of bamboo. Springer
65. Janssen JJA (1995) Bamboo in building structures, 2nd edn. Practical Action
66. Janssen JJA (1995) Building with bamboo: a handbook. Practical Action
67. Janssen JJA (2000) INBAR technical report no. 26—Designing and building with bamboo. International Network for Bamboo and Rattan (INBAR), Beijing
68. Heinsdorff M (2011) Design with nature: the bamboo architecture. Hirmer Publishers; Blg Ltd edn
69. 张文福, 江泽慧, 王戈 等 (2013) 用环刚度法评价圆竹径向抗压力学性能. 北京林业大学学报 35(01):119–122 / Zhang WF, Jiang ZH, Wang G et al (2013) Radial compression mechanical properties of bamboo-culm by ring stiffness. J Beijing For Univ 35(01):119–122
70. 程海涛, 王戈, 王正 等 (2012) 太阳能圆竹预制房屋示范. 建设科技 2012(03):50–51 / Cheng HT, Wang G, Wang Z et al (2012) Prefabricated round bamboo solar house demonstration. Constr Sci Technol 2012(03):50–51
71. 高黎, 王正, 王戈 等 (2012) 原竹预制构件房屋设计与制造技术. 北京林业大学学报 34(05):153–156 / Gao L, Wang Z, Wang G et al (2012) Design and manufacture technology of prefabricated bamboo house. J Beijing For Univ 34(05):153–156
72. Zhang JL, Du YF, Yan ZM et al (2012) Study on bonding properties of bamboo-steel interface under static load. Appl Mech Mater 166–169:176–179
73. Yan ZM, Du YF, Huang JG et al (2013) Approach to steel-bamboo composite structure system. Appl Mech Mater 301(125):1263–1266
74. Li Y, Shen H, Shan W et al (2012) Flexural behavior of lightweight bamboo–steel composite slabs. Thin-Walled Struct 53(2):83–90
75. 蒋天元 (2012) 钢-竹组合梁柱边节点抗震性能试验研究. 宁波大学硕士学位论文 / Jiang TY (2012) Anti-seismic performance test on steel-bamboo composite beam-column exterior joints. Master thesis, Ningbo University
76. 蒋天元, 李玉顺, 杜永飞 等 (2013) 钢竹组合墙体与钢竹组合楼板的节点结构: CN 203066261 U. 中国国家知识产权局发明专利 / Jiang TY, Li YS, Du YF et al (2013) A construction note of steel-bamboo composite wall and steel-bamboo composite floor: CN 203066261 U. Invention Patent of the National Intellectual Property Administration, PRC

77. 蒋天元, 李玉顺, 杜永飞 等 (2013) 钢竹组合梁与钢竹组合柱的节点结构: CN 203066254 U. 中国国家知识产权局发明专利 / Jiang TY, Li YS, Du YF et al (2013) A construction note of steel-bamboo composite beam and steel-bamboo composite column: CN 203066254 U. Invention Patent of the National Intellectual Property Administration, PRC
78. 单炜, 李玉顺, 柳俊哲 (2012) 箱形钢竹或钢木组合楼板: CN 101775863 B. 中国国家知识产权局发明专利 / Shan W, Li YS, Liu JZ (2012) A box steel or steel wood composite floor: CN 101775863 B. Invention Patent of the National Intellectual Property Administration, PRC
79. Gijsbers R, Cox MGDM (2014) Structural strap connections for bamboo and wooden shelter frames. IFRC-SRU Cladding & Fixing Conference, 3–4 September 2014, Luxembourg
80. Jorissen AJM, Voermans J, Jansen MH (2007) Glued-laminated bamboo: node and joint failure in bamboo laminations in tension. J Bamboo Rattan 6(3–4):137–144
81. Gonzalez Beltran GE, Gutierrez JA, van Herwijnen F et al (2001) Selection criteria for a house design method using plybamboo sheets. J Bamboo Rattan 1(1):59–70
82. Sharma B, Gatoo A, Bock M et al (2015) Engineered bamboo: state of the art. Proc Inst Civ Eng: Constr Mater 168(2):57–67
83. Archila HA, Brandon D, Ansell M et al (2014) Evaluation of the mechanical properties of cross laminated bamboo panels by digital image correlation and finite element modelling. World Conference on Timber Engineering (WCTE)
84. Kumar A, Vlach T, Laiblova L et al (2016) Engineered bamboo scrimber: Influence of density on the mechanical and water absorption properties. Constr Build Mater 127:815–827
85. Sharma B, Gatóo A, Bock M et al (2015) Engineered bamboo for structural applications. Constr Build Mater 81:66–73
86. Alireza J (2017) Composite bamboo and its application as reinforcement in structural concrete. Dissertation, ETH Zurich
87. Zhang P, Wu G, Zhu H et al (2014) Mechanical performance of the wet-bond interface between FRP plates and cast-in-place concrete. J Compos Constr 18(6):04014016
88. 魏洋, 吴刚, 李国芬 等 (2014) 新型FRP-竹-混凝土组合梁的力学行为. 中南大学学报(自然科学版) 45(12):4384–4392 / Wei Y, Wu G, Li GF et al (2014) Mechanical behavior of novel FRP-bamboo-concrete composite beams. J Cent South Univ (Sci Technol) 45(12):4384–4392
89. 魏洋, 王晓伟, 李国芬 (2014) 配筋重组竹受弯试件力学性能试验. 复合材料学报 31(4):1030–1036 / Wei Y, Wang XW, Li GF (2014) Mechanical properties test of bamboo scrimber flexural specimens reinforced with bars. Acta Mater Compos Sin 31(4):1030–1036
90. Xiao Y, Shan B, Yang RZ et al (2014) Glue laminated bamboo (GluBam) for structural applications. In: Materials and joints in timber structures. Springer, Netherlands, pp 589–601
91. Xiao Y, Yang RZ, Shan B (2013) Production, environmental impact and mechanical properties of glubam. Constr Build Mater 44(3):765–773
92. Shan B, Chen J, Xiao Y (2012) Mechanical properties of glubam sheets after artificial accelerated aging. Key Eng Mater 517:43–50
93. Li Z, Xiao Y, Monti G et al (2016) Lightweight woodframe shear walls with ply-bamboo sheathings. Appl Mech Mater 847:529–536
94. 肖岩, 佘立永, 单波 等 (2009) 现代竹结构在汶川地震灾后重建中的应用. 自然灾害学报 18(3):14–18 / Xiao Y, She YL, Shan B et al (2009) Application of modern bamboo structure to reconstruction after Wenchuan earthquake. J Nat Disasters 18(3):14–18
95. Xiao Y, Ma J (2012) Fire simulation test and analysis of laminated bamboo frame building. Constr Build Mater 34(5):257–266
96. 肖岩, 单波, 张达威 (2009) 竹材梁构件: CN 101775863 B. 中国国家知识产权局发明专利 / Xiao Y, Shan B, Zhang DW (2009) A bamboo beam member: CN 101775863 B. Invention Patent of the National Intellectual Property Administration, PRC
97. 肖岩, 黄靓 (2008) 一种使用竹材加固墙体的方法: CN 101215930 A. 中国国家知识产权局发明专利 / Xiao Y, Huang J (2008) A method of reinforcing wall using bamboo: CN 101215930 A. Invention Patent of the National Intellectual Property Administration, PRC
98. 肖岩, 单波, 佘立永 (2008) 一种快速装配式竹木板房: CN 101338622 B. 中国国家知识产权局发明专利 / Xiao Y, Shan B, She YL (2008) A fast-assembled bamboo panel house: CN 101338622 B. Invention Patent of the National Intellectual Property Administration, PRC

99. 肖岩, 黄政宇, 陈国 (2008) 竹材住宅建筑: CN 1963056 A. 中国国家知识产权局发明专利 / Xiao Y, Huang ZY, Chen G (2008) A bamboo residential building: CN 1963056 A. Invention Patent of the National Intellectual Property Administration, PRC
100. Kim JS, Lee KH, Cho CH et al (2008) Micromorphological characteristics and lignin distribution in bamboo (*Phyllostachys pubescens*) degraded by the white rot fungus *Lentinus edodes*. Holzforsch 62(4):481–487
101. Schmidt G (2013) Ethiopian highland bamboo as raw material for industrial panel production: investigation on basic properties and raw material preparation. Masterarbeit, Hamburg: Univ Hamburg
102. Malanit P, Barbu MC, Liese W et al (2010) Macroscopic aspects and physical properties of *Dendrocalamus asper* Backer for composite panels. World Bamboo Rattan 7(3–4):151–163
103. Construction AIOT (2012) Timber construction manual, 6th edn. Wiley
104. Grosser D (1971) Beitrag zur Histologie und Klassifikation asiatischer Bambusarten. Kommissionsverlag Buchhandlung Max Wiedebusch Hamburg
105. Zou LH, Jin HLN, Lu WY (2009) Nanoscale structural and mechanical characterization of the cell wall of bamboo fibers. Mater Sci Eng, C 29:1375–1379
106. GB 50176-2016 (2016) 民用建筑热工设计规范. 中华人民共和国国家标准 / GB 50176-2016 (2016) Thermal design code for civil building. State Standard of the People's Republic of China
107. Huang P (2017) Hygrothermal performance of Moso bamboo based building material. Dissertation, University of Bath
108. Wu SC, Yu SY, Han J et al (2004) Testing and analysis of the thermodynamics parameters of *Phyllostachys edulis*. J Cent South For Univ 24(5):70–75
109. Huang P, Chang WS, Ansell MP et al (2015) Density distribution profile for internodes and nodes of *Phyllostachys edulis*, (Moso bamboo) by computer tomography scanning. Constr Build Mater 93:197–204
110. Huang P, Chang WS, Ansell MP et al (2017) Porosity estimation of *Phyllostachys edulis* (Moso bamboo) by computed tomography and backscattered electron imaging. Wood Sci Technol 51(1):11–27
111. Ohmae Y, Nakano T (2009) Water adsorption properties of bamboo in the longitudinal direction. Wood Sci Technol 43(5–6):415–422
112. Jiang ZH, Wang HK, Yu Y et al (2012) Hygroscopic behaviour of bamboo and its blocking units. J Nanjing For Univ 36(2):11–14
113. Wang HK (2010) Study of the mechanism of moisture affecting the bamboo fiber cell walls and macroscopic mechanical behavior. Cent South Univ For Technol
114. Huang P, Latif E, ChangW S et al (2017) Water vapour diffusion resistance factor of *Phyllostachys edulis* (Moso bamboo) [J]. Constr Build Mater 141(15):216–221
115. Huang P, Chang WS, Shea A et al (2013) Non-homogeneous Thermal Properties of bamboo. In: Aicher S, Reinhardt HW, Garrecht H (eds) Materials and joints in timber structures. Springer, Netherlands, pp 657–664
116. Shah DU, Bock MCD, Mulligan H et al (2016) Thermal conductivity of engineered bamboo composites. J Mater Sci 51(6):2991–3002
117. Huang P, Zeidler A, Chang WS et al (2016) Specific heat capacity measurement of *Phyllostachys edulis* (Moso bamboo) by differential scanning calorimetry. Constr Build Mater 125:821–831
118. Huang P, Chang WS, Ansell M et al (2014) Heat transfer simulation and experimental studies of bamboo in radial direction. In: World Conference on Timber Engineering (WCTE), 2014-08-10–2014-08-14
119. 田慧敏, 蔡玉兰 (2008) 竹原纤维微观形态及聚集态结构的研究. 棉纺织技术 36(9):544–546 / Tian HM, Cai YL (2008) Research of nature bamboo fibre microstructure and aggregation state structure. Cotton Text Technol 36(9):544–546
120. 万玉芹, 吴丽莉, 俞建勇 (2004) 竹纤维吸湿性能研究. 纺织学报 25(3):14–16 / Wan YQ, Wu LL, Yu JY (2004) Research on hygroscopic property of the bamboo fiber. J Text Res 25(3):14–16

121. 赵春梅 (2010) 竹原纤维性能研究及其产品开发. 东华大学硕士学位论文 / Zhao CM (2010) Research on performance of natural bamboo fiber and product development. Master thesis, Donghua University
122. Fei P, Xiong H, Cai J et al (2016) Enhanced the weatherability of bamboo fiber-based outdoor building decoration materials by rutile nano-TiO_2. Constr Build Mater 114:307–316
123. Ouyang WJ, Hu WP, Yang XP (2012) On the application of bamboo fiber in home furnishing design. Adv Mater Res 460:244–248
124. 姚文斌, 俞伟鹏, 张蔚 (2014) 一种麻形竹纤维的生产工艺: CN 102206873 B. 中国国家知识产权局发明专利 / Yao WB, Yu WP, Zhang W (2014) A production process of hemp bamboo fiber: CN 102206873 B. Invention Patent of the National Intellectual Property Administration, PRC
125. Yao WB, Zhang W (2011) Research on manufacturing technology and application of natural bamboo fibre. In: International Conference on Intelligent Computation Technology and Automation. IEEE, pp 143–148
126. 魏书元 (2014) 一种竹纤维建筑保温材料: CN 103553423 A. 中国国家知识产权局发明专利 / Wei SY (2014) A bamboo fiber building insulation material: CN 103553423 A. Invention Patent of the National Intellectual Property Administration, PRC
127. Defoirdt N, Biswas S, Vriese LD et al (2010) Assessment of the tensile properties of coir, bamboo and jute fibre. Compos A Appl Sci Manuf 41(5):588–595
128. Murphy RJ, Norton A (2008) Life cycle assessments of natural fibre insulation materials. Imperial College, London
129. Hurtado PL, Rouilly A, Vandenbossche V et al (2016) A review on the properties of cellulose fibre insulation. Build Environ 96(1):170–177
130. Wei KC, Lv CL, Chen MZ et al (2015) Development and performance evaluation of a new thermal insulation material from rice straw using high frequency hot-pressing. Energy Build 87(1):116–122
131. Ali ME, Alabdulkarem A (2017) On thermal characteristics and microstructure of a new insulation material extracted from date palm trees surface fibers. Constr Build Mater 138(1):276–284
132. Oushabi A, Sair S, Abboud Y et al (2015) Natural thermal-insulation materials composed of renewable resources: characterization of local date palm fibers (LDPF). J Mater Environ Sci 6(12):3395–3402
133. Alavez-Ramirez R, Chiñas-Castillo F, Morales-Dominguez VJ et al (2012) Thermal conductivity of coconut fibre filled ferrocement sandwich panels. Constr Build Mater 37:425–431
134. Manohar K, Adeyanju AA (2016) A comparison of banana fiber thermal insulation with conventional building thermal insulation. Br J Appl Sci Technol 17(3):1–9
135. Perino M, Volf M, Diviš J et al (2015) Thermal, moisture and biological behaviour of natural insulating materials. Energy Proc 78:1599–1604
136. Boukhattem L, Boumhaout M, Hamdi H et al (2017) Moisture content influence on the thermal conductivity of insulating building materials made from date palm fibers mesh. Constr Build Mater 148(1):811–823
137. Palumbo M, Lacasta AM, Holcroft N et al (2016) Determination of hygrothermal parameters of experimental and commercial bio-based insulation materials. Constr Build Mater 124(5):269–275
138. Holcroft N, Shea A (2013) Moisture buffering and latent heat effects in natural fibre insulation materials. Portugal SB13-Contribution of Sustainable Building to meet the EU, pp 221–228
139. Liese W, Silbermann S (2012) Bamboo charcoal: properties and utilization. Newsl Bamboo Soc EBS 31(6):15–22
140. 周建斌 (2005) 竹炭环境效应及作用机理研究. 南京林业大学研究生博士学位论文 / Zhou JB (2005) Study on the mechanism of action and in environmental protection. Doctoral thesis, Nanjing Forestry University
141. 左宋林, 高尚愚, 徐柏森, 等 (2004) .炭化过程中竹材内部形态结构的变化. 林产化学与工业 24(4):56–60 / Zuo SL, Gao SY, Xu BS et al (2004) Change of inner morphological texture of bamboo during carbonization. Chem Ind For Prod 24(4):56–60

142. 卢克阳, 张齐生, 蒋身学 (2006) 竹炭吸湿性能的初步研究. 木材工业 20(3):20–22 / Lu KY, Zhang QS, Jiang SX (2006) A primary study of bamboo charcoal moisture absorption. China Wood Ind 20(3):20–22
143. 蒋正武, 张静娟, 曾志勇 等 (2016) 炭化工艺参数对竹炭调湿性能的影响研究. 建筑材料学报 09:1–8 / Jiang ZW, Zhang JJ, Zeng ZY et al (2016) Effect of carbonization process parameters on humidity-controlling properties of bamboo charcoal. J Build Mater 09:1–8
144. Maki F, Norimoto M, Yamada T (1985) Humidity conditions in the house lined with wood based materials. Wood Res Tech Notes 21:87–95
145. Ikeda T (1992) Moisture absorption and desorption characteristics and the effects of porous materials (inorganic porous material). Gypsum Lime 240:355–363
146. 李国胜 (2005) 海泡石矿物材料的显微结构与自调湿性能研究. 河北工业大学硕士学位论文 / Li GS (2005) Microstructure and self-humidity-control performance of mineral material sepiolite. Master thesis, Hebei University of Technology
147. Qin MH, Belarbi R, Ait-Mokhtar A (2009) Simulation of coupled heat and moisture transfer in air-conditioned buildings. Autom Constr 18(5):642–631
148. Ferreira C, Freitas VPD, Ramos NMM (2014) Quantifying the influence of hygroscopic materials in the fluctuation of relative humidity in museums housed in old buildings. In: 2014: 10th Nordic Symposium on Building Physics, Lund University, pp 600–607
149. Erhardt D, Antretter F, Kilian R (2013) Measured and modeled influence of the moisture buffer effect in a historic stone church and its influence on possible HVAC measures. 11th REHVA world congress Clima 2013—Energy efficient, smart and healthy buildings, pp 4388–4397
150. Yu DW, Tan HW, Ruan YJ (2011) A future bamboo-structure residential building prototype in China: life cycle assessment of energy use and carbon emission. Energy Build 43(10):2638–2646
151. Li NP, Long JB, Lin SU et al (2012) Experimental and theoretical study on thermal and moisture characteristics of new-type bamboo structure wall. J Cent South Univ (english version) 19(3):600–608
152. Wang Q, Li NP, Zeng DJ et al (2010) Coupled heat and moisture transfer in bamboo composite wall. Sci Technol Rev 28(17):87–90
153. 王倩 (2011) 新型竹材结构墙体热湿耦合迁移特性研究. 湖南大学硕士学位论文 / Wang Q (2011) The research on coupled heat and moisture transfer characteristics of bamboo structure wall. Master thesis, Hunan University
154. 李念平, 李炳华, 胡锦华 (2010) 建筑墙体霉菌生长特性模拟分析. 科技导报 28(15):41–44 / Li NP, Li BH, Hu JH (2010) Simulation of mold growth on building walls. Sci Technol Rev 28(15):41–44
155. 肖书博, 李念平, 李靖 等 (2008) 现代竹结构房屋室内空气品质的实测与研究. 安全与环境学报 8(6):87–90 / Xiao SB, Li NP, Li J et al (2008) Measurement and analysis of the indoor air quality of the modern bamboo house in China. J Saf Environ 8(6):87–90
156. Li YS, Yao J, Li R et al (2017) Thermal and energy performance of a steel-bamboo composite wall structure. Energy Build 156:225–237
157. Geving S, Holme J (2010) The drying potential and risk for mold growth in compact wood frame roofs with built-in moisture. J Building Phys 33(3):249–269
158. Moonen P (2009) Continuous-discontinuous modelling of hygrothermal damage processes in porous media. Dissertation, LU Leuven
159. Mertens S (2009) Hysteresis, damage and moisture effects in quasi-brittle porous materials. Dissertation, LU Leuven
160. Mendes N, Winkelmann FC, Lamberts R et al (2003) Moisture effects on conduction loads. Energy Build 35(7):631–644
161. Zhang H, Yoshino H, Hasegawa K (2012) Assessing the moisture buffering performance of hygroscopic material by using experimental method. Build Environ 48(1):27–34
162. Abuku M, Janssen H, Roels S (2009) Impact of wind-driven rain on historic brick wall buildings in a moderately cold and humid climate: numerical analyses of mould growth risk, indoor climate and energy consumption. Energy Build 41(1):101–110

163. Glaser H (1959) Graphical method for investigation of diffusional process. Kalteteohnik 39(5):519–550
164. Luikov AV (1975) Systems of differential equations of heat and mass transfer in capillary-porous bodies (review). Int J Heat Mass Transf 18(6):1–14
165. Janssen H (2002) The influence of soil moisture transfer on building heat loss via the ground. Dissertation, LU Leuven
166. Molenda CHA, Crausse P, Lemarchand D (1992) The influence of capillary hysteresis effects on the humidity and heat coupled transfer in a non-saturated porous medium. Int J Heat Mass Transf 35(6):1385–1396
167. Pedersen CR (1992) Prediction of moisture transfer in building constructions. Build Environ 27(3):387–397
168. Künzel HM (1995) Simultaneous heat and moisture transport in building components. Fraunhofer IRB Verlag, Stuttgart
169. Häupl P, Grunewald J, Fechner H et al (1997) Coupled heat air and moisture transfer in building structures. Int J Heat Mass Transf 40(7):1633–1642
170. Janssen H, Blocken B, Carmeliet J (2007) Conservative modelling of the moisture and heat transfer in building components under atmospheric excitation. Int J Heat Mass Transf 50(5–6):1128–1140
171. Mendes N, Philippi PC (2005) A method for predicting heat and moisture transfer through multilayered walls based on temperature and moisture content gradients. Int J Heat Mass Transf 48(1):37–51
172. Steeman HJ, Van Belleghem M, Janssens A et al (2009) Coupled simulation of heat and moisture transport in air and porous materials for the assessment of moisture related damage. Build Environ 44(10):2176–2184
173. Tariku F, Kumaran K, Fazio P (2010) Transient model for coupled heat, air and moisture transfer through multilayered porous media. Int J Heat Mass Transf 53(15–16):3035–3044
174. Kumaran MK (1996) IEA Annex 24: heat, air and moisture transfer in insulated envelope parts. Final Report, vol 3, Task 3: Material Properties
175. Woloszyn M, Rode C (2008) IEA Annex 41: whole building heat, air, moisture response. Subtask 1: modeling principles and common exercises. International Energy Agency, Executive committee on Energy Conservation in Buildings and Community Systems
176. Janssen H (2014) Simulation efficiency and accuracy of different moisture transfer potentials. J Build Perform Simul 7:379–389
177. Van Belleghem M, Steeman M, Janssen H et al (2014) Validation of a coupled heat, vapour and liquid moisture transport model for porous materials implemented in CFD. Build Environ 81:340–353
178. Van De Walle W, Janssen H (2016) A thermal conductivity prediction model for porous building blocks. In: Proceedings of the CESBP Central European Symposium on Building Physics and Bausim 2016. Central European Symposium on Building Physics. Dresden, 14–16 September 2016, pp 67–74
179. Defraeye T, Blocken B, Carmeliet J (2013) Influence of uncertainty in heat-moisture transport properties on convective drying of porous materials by numerical modelling. Chem Eng Res Des 91(1):36–42
180. Defraeye T (2011) Convective heat and mass transfer at exterior building surfaces. Dissertation, LU Leuven
181. Parys W, Saelens D, Hens H (2011) Coupling of dynamic building simulation with stochastic modelling of occupant behaviour in offices—a review-based integrated methodology. J Build Perform Simul 4(4):339–358
182. Stephane Nouidui T, Nytsch-Geusen C, Holm A et al (2008) Object-oriented hygrothermal building physics library as a tool to predict and to ensure a thermal and hygric indoor comfort in building construction by using a Predicted-Mean-Vote (PMV) control ventilation system. 8th Nordic Symposium on Building Physics in the Nordic Countries 2008, Copenhagen, Denmark, pp 825–832

183. Nytschgeusen C, Nouidui T, Holm A et al (2005) A hygrothermal building model based on the object-oriented modeling language Modelica. Ninth International IBPSA Conference, Montreal, Canada
184. Deromea D, Kubilaya A, Defraeyea T et al (2017) Ten questions concerning modeling of wind-driven rain in the built environment. Build Environ 114:495–506
185. Van den Brande T, Blocken B, Roels S (2013) Rain water runoff from porous building facades: implementation and application of a first-order runoff model coupled to a HAM model. Build Environ 64:177–186
186. Peren J, van Hooff T, Leite B et al (2016) CFD simulation of wind-driven upward cross ventilation and its enhancement in long buildings: impact of single-span versus double-span leeward sawtooth roof and opening ratio. Build Environ 96:142–156
187. Kubilay A, Derome D, Blocken B et al (2015) Wind-driven rain on two parallel wide buildings: field measurements and CFD simulations. J Wind Eng Ind Aerodyn 146:11–28
188. Blocken B (2004) Wind-driven rain on buildings: measurements, numerical modelling and applications. Dissertation, LU Leuven
189. Abuku M (2009) Moisture stress of wind-driven rain on building enclosures. Dissertation, LU Leuven
190. Blocken B, Carmeliet J (2015) Impact, runoff and drying of wind-driven rain on a window glass surface: numerical modelling based on experimental validation. Build Environ 84:170–180
191. Abuku M, Janssen H, Poesen J et al (2009) Impact, absorption and evaporation of raindrops at building facades. Build Environ 44(1):113–124
192. Janssen H, Blocken B, Roels S et al (2007) Wind-driven rain as a boundary condition for HAM simulations: analysis of simplified modelling approaches. Build Environ 42(4):1555–1567
193. Briggen P, Blocken B, Schellen H (2009) Wind-driven rain on the facade of a monumental tower: Numerical simulation, full-scale validation and sensitivity analysis. Build Environ 44(8):1675–1690
194. Abuku M, Blocken B, Roels S (2009) Moisture response of building facades to wind-driven rain: field measurements compared with numerical simulations. J Wind Eng Ind Aerodyn 97(5–6):197–207
195. Islahuddin M, Janssen H (2017) Hygric property estimation of porous building materials with multiscale pore structures. Energy Proc 132:273–278
196. Islahuddin M, Janssen H (2016) Numerical estimation of isothermal moisture storage and transport properties. In: Grunewald J (ed) Proceedings of the Cesbp Central European Symposium on Building Physics and BauSIM 2016. Central European Symposium on Building Physics. Dresden, Germany, 14–16 September 2016, pp 301–308
197. Todorovic J, Janssen H (2016) Impact of gypsum pore clogging on moisture transfer properties of bricks. In: Grunewald J (ed) Central European Symposium on Building Physics. Dresden, Germany, 14–16 September 2016
198. Feng C, Janssen H (2016) Hygric properties of porous building materials (II): analysis of temperature influence. Build Environ 99:107–118
199. Feng C, Janssen H, Feng Y et al (2015) Hygric properties of porous building materials: analysis of measurement repeatability and reproducibility. Build Environ 85:160–172
200. Feng C, Janssen H (2017) The influence of temperature on the capillary absorption coefficient—a confrontation of two recent papers in building and environment. Build Environ 116:257–258
201. Janssen H, Scheffler G, Plagge R (2016) Experimental study of dynamic effects in moisture transfer in building materials. Int J Heat Mass Transf 98:141–149
202. Vereecken E, Roels S (2013) Hygric performance of a massive masonry wall: How do the mortar joints influence the moisture flux. Constr Build Mater 41:697–707
203. Montazeri H, Blocken B (2017) New generalized expressions for forced convective heat transfer coefficients at building facades and roofs. Build Environ 119:153–168
204. Deconinck A, Roels S (2016) Comparison of characterisation methods determining the thermal resistance of building components from onsite measurements. Energy Build 130:309–320

205. Reynders G, Diriken J, Saelens D (2017) Generic characterization method for energy flexibility: applied to structural thermal storage in residential buildings. Appl Energy 198:192–202
206. Van Belleghem M, Steeman HJ, Steeman M et al (2010) Sensitivity analysis of CFD coupled non-isothermal heat and moisture modelling. Build Environ 45(11):2485–2496
207. Delgado JMPQ, Barreira E, Ramos NMM et al (2013) Hygrothermal numerical simulation tools applied to building physics. Springer
208. Palumbo M, Lacasta AM, Holcroft N et al (2016) Determination of hygrothermal parameters of experimental and commercial bio-based insulation materials. Constr Build Mater 124:269–275
209. Palumbo M, McGregor F, Heath A et al (2016) The influence of two crop by-products on the hygrothermal properties of earth plasters. Build Environ 105:245–252
210. Barclay M, Holcroft N, Shea A (2014) Methods to determine whole building hygrothermal performance of hemp-lime buildings. Build Environ 80:204–212
211. Shea AD, Wall K, Walker P (2013) Evaluation of the thermal performance of an innovative prefabricated natural plant fibre building system. Build Serv Eng Res Technol 34(4):369–380
212. Shea A, Lawrence M, Walker P (2012) Hygrothermal performance of an experimental hemp-lime building. Constr Build Mater 36:270–275
213. Lawrence R, Walker P, Latif E et al (2016) In situ assessment of the fabric and energy performance of five conventional and non-conventional wall systems using comparative coheating tests. Build Environ 109:68–81
214. Walker P, Thomson A, Maskell D (2017) Nonwood biobased materials: straw. In: Jones D, Brischke C (eds) Performance of bio-based building materials, 1st edn. Woodhead Publishing, Duxford, pp 112–119
215. Langmans J, Versele A, Roels S (2014) On the hygrothermal performance of straw bale wall elements in Belgium. International Conference on Sustainability in Energy and Buildings. Sustainability in Energy and Buildings. Cardiff, 25–27 June 2014
216. Maskell D, Church M, Thomson A et al (2017) Development of a bio-based plasterboard. International Conference on Bio-Based Building Materials, France, 21 June 2017 to 23 June 2017
217. Dubois S, McGregor F, Evrard A et al (2014) An inverse modelling approach to estimate the hygric parameters of clay-based masonry during a Moisture Buffer Value test. Build Environ 81:192–203
218. Latif E, Lawrence R, Shea A et al (2015) Moisture buffer potential of experimental wall assemblies incorporating formulated hemp-lime. Build Environ 93(2):199–209
219. Silva CFD, Stefanowski B, Maskell D et al (2017) Improvement of indoor air quality by MDF panels containing walnut shells. Build Environ 123:427–436
220. Zillig W (2009) Moisture transport in wood using a multiscale approach. Dissertation, LU Leuven
221. Langmans J, Klein R, Roels S (2013) Numerical and experimental investigation of the hygrothermal response of timber frame walls with an exterior air barrier. J Building Phys 36(4):375–397
222. Langmans J, Klein R, Roels S (2011) Hygrothermal response of highly insulated timber frame walls with an exterior air barrier system: laboratory investigation. 9th Nordic Symposium on Building Physics, Tampere University of Technology
223. Vanpachtenbeke M, Langmans J, Van den Bulcke J et al (2017) Hygrothermal behaviour of timber frame walls finished with a brick veneer cladding. Nordic Symposium on Building Physics. Trondheim, Norway, 11–14 June 2017, pp 1–6
224. McGregor F, Heath A, Maskell D et al (2016) A review on the buffering capacity of earth building materials. Proc Inst Civ Eng: Constr Mater 169(5):241–251
225. James C, Simonson C, Talukdar P et al (2010) Numerical and experimental data set for benchmarking hygroscopic buffering models. Int J Heat Mass Transf 53(19–20):3638–3654
226. Vereecken E, Roels S, Janssen H (2011) In situ determination of the moisture buffer potential of room enclosures. J Building Phys 34(3):223–246

227. Roels S, Talukdar P, James C et al (2010) Reliability of material data measurements for hygroscopic buffering. Int J Heat Mass Transf 53(23–24):5355–5363
228. McGregor F, Heath A, Fodde E et al (2014) Conditions affecting the moisture buffering measurement performed on compressed earth blocks. Build Environ 75:11–18
229. Mcgregor F, Heath A, Shea A et al (2014) The moisture buffering capacity of unfired clay masonry. Build Environ 82:599–607
230. Vereecken E, Vanoirbeek K, Roels S (2015) Towards a more thoughtful use of mould prediction models: a critical view on experimental mould growth research. J Building Phys 39(2):102–123
231. Vereecken E, Saelens D, Roels S (2011) A comparison of different mould prediction models. In: Proceedings of the 12th International Conference of the International Building Performance Simulation Association. BS 2011. Sydney, Australia, 14–16 November 2011, pp 1934–1941
232. Vereecken E, Roels S (2014) Mould risk assessment fot thermal bridges: what is the impact of the mould prediction model. In: XIII International Conference on Durability of Building Materials and Components. International Conference on Durability of Building Materials and Components. Sao Paulo, 02–05 September 2014, pp 616–623
233. Vanpachtenbeke M, Van den Bulcke J, De Windt I et al (2017) An experimental study of mould growth and wood decay in timber frame walls. International Research Group on Wood Protection. Gent, 4–8 June 2017
234. Vanpachtenbeke M, Van den Bulcke J, De Windt I et al (2016) An experimental set-up to study mould growth and wood decay under dynamic boundary conditions. WCTE 2016 World Conference on Timber Emgineering. World Conference on Timber Engineering. Vienna, Austria, 22–25 August 2016, pp 1589–1597
235. 陈友明, 王盛卫, 张泠 (2004) 系统辨识在建筑热湿过程中的应用. 中国建筑工业出版社 / Chen YM, Wang SW, Zhang L (2004) Application of system identification in building heat and moisture process. China Architecture & Building Press
236. 陈友明, 王盛卫 (2004) 建筑围护结构非稳定传热分析新方法. 科学出版社 / Chen YM, Wang SW (2004) New method for unsteady heat transfer analysis of building envelope. Science Press
237. 陈友明, 周娟, 左政 (2003) 用频域回归方法计算墙体周期反应系数. 建筑热能通风空调 22(1):8–10 / Chen YM, Zhou J, Zuo Z (2003) Calculating wall periodic response factors using frequency-domain regression method. Build Energy Environ 22(1):8–10
238. Guo XG, Chen YM, Deng YQ (2012) Development and experimental validation of a one-dimensional dynamic hygrothermal modeling based on air humidity ratio. J Cent South Univ 19(3):703–708
239. 张乐, 陈友明 (2010) 太阳辐射对南方多层墙体热湿性能影响的研究. 建筑科学 26(10):332–336 / Zhang L, Chen YM (2010) Study on the influence of solar radiation on the thermal and hygric performance of multi-layer walls in South China. Build Sci 26(10):332–336
240. 孟庆林 (2003) 屋面含湿砂层温度波衰减研究. 暖通空调 33(2):110–111 / Meng QL (2003) Study on temperature wave decaying of wet sand layer of roof. Heating Ventilating Air Conditioning 33(2):110–111
241. 张玉, 孟庆林, 陈渊睿 (2008) 动态热湿气候风洞实验台的研制. 华南理工大学学报 (自然科学版) 36(3):99–103 / Zhang Y, Meng QL, Chen YR (2008) Development of dynamic laboratory of hot-humid climate wind tunnel. J South China Univ Technol (Nat Sci Ed) 36(3):99–103
242. Zhang L, Feng Y, Meng Q et al (2015) Experimental study on the building evaporative cooling by using the Climatic Wind Tunnel. Energy Build 104:360–368
243. 孟庆林 (2001) 建筑表面被动蒸发冷却. 华南理工大学出版社 / Meng QL (2001) Passive evaporative cooling of building surfaces. South China University of Technology Press
244. 孟庆林, 陈启高, 冉茂裕 等 (1999) 关于蒸发换热系数he的证明. 太阳能学报 20(2):216–219 / Meng QL, Chen QG, Ran MY et al (1999) Derivation of the vaporising heat transfer coefficient h_e. Acta Energiae Solaris Sin 20(2):216–219

245. 张明杰, 秦孟昊, 陈智 (2016) 复合相变调湿材料对室内热湿环境影响的研究. 建筑科学 32(12):72–79 / Zhang MJ, Qing MH, Chen Z (2016) Study on impact of composite phase change moisture-controlled materials on indoor thermal and humid environment. Build Sci 32(12):72–79
246. 刘奕彪, 秦孟昊 (2015) 多孔调湿材料湿缓冲特性的实验研究. 土木建筑与环境工程 2015(5):129–134 / Liu YB, Qing MH (2015) Experimental analysis of the moisture buffering properties of different porous hygroscopic materials. J Civ Archit Environ Eng 2015(5):129–134
247. Chen Z, Qin M, Yang J (2015) Synthesis and characteristics of hygroscopic phase change material: composite microencapsulated phase change material (MPCM) and diatomite. Energy Build 106:175–182
248. Qin M, Zhang H (2015) Analytical methods to calculate combined heat and moisture transfer in porous building materials under different boundary conditions. Sci Technol Built Environ 21(7):993–1001
249. Wang Y, Liu Y, Wang D et al (2016) The correction coefficients for the building air-conditioning load accounting for the whole building hygrothermal transfer process for major cities in China. Indoor Built Environ 26(5):642–661
250. 陈飞, 刘艳峰, 马超 等 (2017) 湿迁移对围护结构传热系数地域差异性影响研究. 建筑热能通风空调 36(1):6–11 / Chen F, Liu YF, Ma C et al (2017) Influence of moisture transfer on building enclosure's heat transfer coefficient among regional differences. Build Energy Environ 36(1):6–11
251. 李哲伟 (2014) 莫高窟洞窟前室对窟内热湿环境调控机理研究. 西安建筑科技大学硕士学位论文 / Li ZW (2014) Study on the atria regulatory mechanism of thermal and moisture environment in the Mogao grottoes. Master thesis, Xi'an University of Architecture and Technology
252. 樊夏玮, 闫增峰, 尚瑞华 等 (2016) 敦煌莫高窟区室外下垫面温湿度现场实测研究. 建筑与文化 2016(11):145–146 / Fan XW, Yan ZF, Shang RH et al (2016) Contrast and analysis of Dunhuang Mogao Grottoes Area Outdoor underlying surface temperature and humidity. Archit Cult 2016(11):145–146
253. 马斌齐, 闫增峰, 尚建丽 等 (2007) 被动式调湿建筑材料的开发与应用研究. 建筑科学 23(8):72–75 / Ma BQ, Yan ZF, Shang JL et al (2007) Research on development and application of passive humidity controlling materials. Build Sci 23(8):72–75
254. 闫增峰, 林海燕, 周辉 等 (2007) 建筑围护结构中热桥稳态传热计算研究. 暖通空调 37(7):11–14 / Yan ZF, Lin HY, Zhou H et al (2007) Research on steady-state heat transfer calculation of thermal bridge in building envelopes. Heating Ventilating Air Conditioning 37(7):11–14
255. 尚建丽, 宋冬毅, 麻向龙 等 (2016) 建筑石膏复合材料的调温调湿性能研究. 西安建筑科技大学学报 (自然科学版) 48(4).601–605 / Shang JL, Song DY, Ma XL et al (2016) Study on properties of temperature controlling and humidity controlling of gypsum compound materials. J Xi'an Univ Archit Technol (Nat Sci Ed) 48(4):601–605
256. 尚建丽, 张浩 (2016) 癸酸-棕榈酸/SiO_2相变储湿复合材料的制备与表征. 复合材料学报 33(2):341–349 / Shang JL, Zhang H (2016) Preparation and characterization of decanoic acid-palmitic acid/SiO_2 phase change and humidity storage composites. Acta Mater Compos Sin 33(2):341–349
257. 孔凡红, 赵强, 郭小强 等 (2016) 建筑材料多孔介质热物性参数测试实验研究. 太阳能学报, 2016(11):2883–2888 / Kong FH, Zhao Q, Guo XQ et al (2016) Study on test experiment of thermal physical parameters of building porous medium material. Acta Energiae Solaris Sin 2016(11):2883–2888
258. Zhang H, Yoshino H, Iwamae A et al (2015) Investigating simultaneous transport of heat and moisture in hygroscopic materials by a semi-conjugate CFD-coupled approach. Build Environ 90:125–135
259. 史路阳, 刘京, 张会波 (201) 稻草板与常见建筑材料调湿特性对比实验. 建筑科学 33(2):42–46 / Shi LY, Liu J, Zhang HB (2017) Experiments for comparing the hygroscopicity of straw panel with other hygroscopic building materials. Build Sci 33(2):42-46

260. 韩星, 陈秋火 (2014) MgO板的吸湿性能实验及在建筑环境中的调湿效果. 全国暖通空调制冷2014年学术文集 2014:180–183 / Han X, Chen QH (2014) Moisture absorption performance of MgO board and its humidity control effect in building environment. Heating Ventilating Air Conditioning 2014(S):180–183
261. Liu X, Chen Y, Ge H et al (2015) Determination of optimum insulation thickness for building walls with moisture transfer in hot summer and cold winter zone of China. Energy Build 109(4):361–368
262. 张景欣, 郭兴国, 陈友明 等 (2015) 墙体内热湿耦合过程的时域递归展开算法. 土木建筑与环境工程 2015(6):147–152 / Zhang JX, Guo XG, Chen YM et al (2015) A time domain recursive algorithm for solving the model of coupled heat and moisture transfer in building wall. J Chongqing Jianzhu Univ 2015(6):147–152
263. 郭兴国, 陈友明, 陈国杰 等 (2014) 热湿气候地区湿传递对吸放湿墙体热湿性能的影响. 建筑技术 45(8):747–750 / Guo XG, Chen YM, Chen GJ et al (2014) Effect of moisture transfer on hydrothermal performance of multilayer wall subjected to hot humid climate. Archit Technol 45(8):747–750
264. 邓小飞, 徐瑾, 李丽 (2017) 佛甲草绿化屋面的隔热和降温增湿效应研究——以广州市羊城创意产业园屋顶为例. 西北林学院学报 32(4):279–282 / Deng XF, Xu J, Li L (2017) Thermal insulation, temperature-lowering and humidity-promotion effects of sedum lineare green roof: a case study of Yangcheng creative industrial park. J Northwest For Univ 32(4):279–282
265. Feng C, Janssen H, Feng Y et al (2014) Experimental analysis of the repeatability of vacuum saturation and capillary absorption tests. Nordic Symposium on Building Physics. Lund, Sweden, pp 394–401
266. 冯驰, 冯雅, 孟庆林 (2013) 加气混凝土蒸汽渗透系数的变物性取值方法. 土木建筑与环境工程 35(5):132–136 / Feng C, Feng Y, Meng QL (2013) Approach to determine value of variable vapor permeability of autoclaved aerated concrete. J Chongqing Jianzhu Univ 35(5):132–136
267. 钟辉智, 雷波, 冯雅 (2010) 多孔材料湿扩散系数研究情况. 建筑环境科学与技术国际学术会议, 2010:842–847 / Zhong HZ, Lei B, Feng Y (2010) Study on the moisture diffusion coefficient. International Conference on Building Environmental Science and Technology, pp 842–847
268. 南艳丽, 冯雅, 谷晋川 等 (2007) 建筑热桥的数值模拟与实验研究. 低温建筑技术 2007(3):110–111 / Nan YL, Feng Y, Gu JC et al (2007) Numerical simulation and experiment study on thermal bridge structure. Low Temp Archit Technol 2007(3):110–111
269. 高庆龙, 杨柳, 刘大龙 等 (2007) 建筑热环境模拟分析用逐时相对湿度生成方法. 四川建筑科学研究 33(6):203–206 / Gao QL, Yang L, Liu DL et al (2007) The method of getting hourly relative humidity for building thermal condition simulation. Sichuan Build Sci 33(6):203–206
270. 王晓欢, 费本华, 王永兵 (2013) 木框架外墙保温性能试验研究. 木材加工机械 2013(6):11–15 / Wang XH, Fei BH, Wang YB (2013) Field testing of insulation performance of wood-frame external wall. Wood Process Mach 2013(6):11–15
271. 冉茂宇, 曹学功 (2001) 周期性热湿作用下多孔吸湿体吸解湿性能的实验研究. 西安建筑科技大学学报 (自然科学版) 33(4):313–315 / Ran ZM, Cao XG (2001) Tentative study on adsorption-desorption characteristics of porous adsorbent under periodic heat and moisture impulsion. J Xi'an Univ Archit Technol (Nat Sci Ed) 33(4):313–315
272. 王莹莹 (2013) 围护结构湿迁移对室内热环境及空调负荷影响关系研究. 西安建筑科技大学博士学位论文 / Wang YY (2013) Research on the effect of the palisade structure moisture transfer on the indoor thermal environment and air-conditioning load. Doctoral thesis, Xi'an University of Architecture and Technology
273. 陈国杰, 刘向伟, 陈友明 等 (2013) 多孔介质墙体热湿耦合迁移实验方法. 南华大学学报 (自然科学版) 2013(4):90–95 / Chen GJ, Liu XW, Chen YM et al (2013) Experimental method for coupled heat and moisture transfer in porous building envelope. J Univ South China (Sci Technol) 2013(4):90–95

274. 李玮, 刘芳, 陈宝明 等 (2017) 湿传递对建筑墙体传热性能的影响. 建筑节能 2017(1):57–61 / Liu W, Liu F, Chen BM et al (2017) Moisture transfer effects on heat transfer performance of building wall. Build Energy Effi 2017(1):57–61
275. 李魁山, 张旭, 韩星 等 (2009) 建筑材料等温吸放湿曲线性能实验研究. 建筑材料学报 12(1):81–84 / Li KS, Zhang X, Han X et al (2009) Experimental research of isothermal sorption curve of building materials. J Build Mater 12(1):81–84
276. 李魁山, 张旭, 韩星 等 (2009) 建筑材料水蒸气渗透系数实验研究. 建筑材料学报 12(3):288–291 / Li KS, Zhang X, Han X et al (2009) Experimental research of water vapor permeability through building materials. J Build Mater 12(3):288–291
277. 李魁山, 张旭, 高军 (2009) 周期性边界条件下多层墙体内热湿耦合迁移. 同济大学学报 (自然科学版) 37(6):814–818 / Li KS, Zhang X, Gao J (2009) Coupled heat and moisture transfer in multi-wall under periodic boundary conditions. J Tongji Univ (Nat Sci) 37(6):814–818

Chapter 2
Model Study on the Application of Bamboo in Building Envelope

In the study on the application of bamboo, as a hygroscopic material, in building envelope, particular attention should be paid to the moisture process. As described in Chap. 1, the accurate description of moisture process requires method change from the static evaluation with Glaser model to the dynamic simulation with HAM model. In order to clarify the difference between HAM and Glaser models and determine the impact of meteorological data and material parameter on the simulation results of the heat and moisture process in building envelope, the HAM model based on the Künzel equations is analyzed in this chapter. Model groups for comparison are constructed and simulated in WUFI Plus, a coupled heat and moisture process simulation computer program based on the Künzel equations, which is developed by Fraunhofer IBP. The impact of the meteorological data, material parameters, and evaluation methods are shown in the results, which gives reference to the determination of theoretical model for the study on the application of bamboo in building envelope (Fig. 2.1).

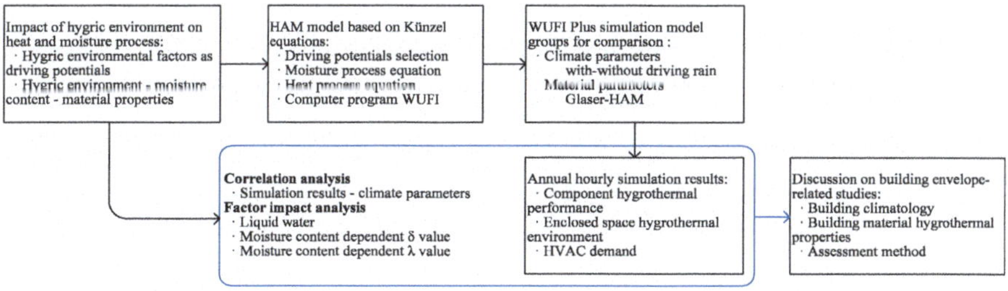

Fig. 2.1 Framework of Chap. 2

Z. Huang, *Application of Bamboo in Building Envelope*,
Green Energy and Technology, https://doi.org/10.1007/978-3-030-12032-0_2

2.1 Hygrothermal Performance Projection of Building Component in Hygric Environment

The impact of hygric environment of the heat and moisture process includes two cases. On the one hand, the hygric environmental factors, as driving potentials, directly affect the transport process of heat and moisture. In this case, various description equations are proposed due to the different selection of driving potentials, which in turn requires different material parameters to be the equation coefficients. On the other hand, the hygric environment affects the moisture content of the materials, changes their hygrothermal properties, and finally affects the heat and moisture process of the construction, and this phenomenon is particularly obvious to the hygroscopic building materials. The impact of hygric environment on the hygrothermal performance of the bamboo construction would be discussed below.

2.1.1 Curve Fitting Between Material Hygrothermal Properties and the Hygric Environment

(1) Construction partition

Curve fitting is carried out for the test results of the construction partition groups in Chap. 3, of which the relation of the mass rate equilibrium moisture content u, water vapor transfer coefficient δ and thermal conductivity λ correspondingly with ambient relative humidity φ are shown in the table below. Among them, the δ and λ are first fitted with u value and then are associated indirectly to the φ values (Tables 2.1, 2.2 and 2.3).

(2) Construction infill

Curve fitting is carried out for the test results of natural bamboo fiber (BF) in Chap. 3, of which the mass rate equilibrium moisture content u and thermal conductivity λ

Table 2.1 Bamboo boards: Curve fitting with function '$u = a \cdot e^{b \cdot \varphi}$' between equilibrium moisture content u and the ambient relative humidity φ, in which a and b are the fitting coefficients

Material	a	b	R^2	Sig.
FB—flattened bamboo panel	1.052861	0.028257	0.985951	9.6181E^{-08}
BSB—bamboo laminated lumber	1.409074	0.025695	0.983999	1.5176E^{-07}
BMB—plybamboo	1.310343	0.025194	0.983424	1.7177E^{-07}
BFB—bamboo scrimber	0.520628	0.031226	0.957729	0.000005
BPB—bamboo particleboard	1.510761	0.025196	0.979499	3.6196E^{-07}
BOSB—bamboo oriented strand board	1.406688	0.024654	0.979828	3.4197E^{-07}

Table 2.2 Bamboo boards: Curve fitting with function '$\delta = a \cdot e^{b \cdot u}$' between vapor permeability coefficient δ and equilibrium moisture content u, in which a and b are the fitting coefficients

Material	a	b	R^2	Sig.
FB—flattened bamboo panel	6.9969E^{-13}	25.901810	0.972738	0.000006
BSB—bamboo laminated lumber	9.4511E^{-13}	24.486998	0.941027	0.000066
BMB—plybamboo	5.1624E^{-13}	26.974973	0.997189	6.9458E^{-09}
BFB—bamboo scrimber	1.1541E^{-13}	56.169734	0.976291	0.000030
BPB—bamboo particleboard	4.3498E^{-12}	16.285983	0.908253	0.000250
BOSB—bamboo oriented strand board	2.1042E^{-13}	27.584925	0.973023	0.000006

Table 2.3 Bamboo boards: Curve fitting with function '$\lambda = a \cdot u + b$' between thermal conductivity λ and equilibrium moisture content u, in which a and b are the fitting coefficients

Material	a	b	R^2	Sig.
FB—flattened bamboo panel	0.258728	0.108800	0.969227	0.000359
BSB—bamboo laminated lumber	0.213729	0.147500	0.996019	0.000006
BMB—plybamboo	0.218498	0.173300	0.977901	0.000184
BFB—bamboo scrimber	0.328903	0.162500	0.986426	0.000069
BPB—bamboo particleboard	0.458253	0.080033	0.976924	0.000201
BOSB—bamboo oriented strand board	0.688971	0.119667	0.999661	4.3146E^{-08}

can be fitted with the ambient relative humidity φ with general exponential function $Y = a \times \mathrm{EXP}(b \times \varphi) + c$. For u–φ relation, there is:

$$u = 0.0088e^{3.3637\varphi} \left(R^2 = 0.949500, \text{sig} = 0.000009\right)$$

For the BF with filling density of $\rho = 110$ kg/m^3, the λ–φ fitting results are:

$$\lambda = 0.000163e^{0.059683\varphi} + 0.0432 \left(R^2 = 0.943838, \ \text{sig} = 0.000259\right)$$

2.1.2 Building Component Hygrothermal Performance in Equilibrium with the Hygric Environment

2.1.2.1 Settings of the Construction Groups

The construction type, material and thickness of the each layer are arranged corresponding to the boundary conditions (namely the exterior walls) in Chap. 5, where BFB is set as the exterior board, BFB and BSB as interior boards, BMB, BPB, and BOSB as interlayer boards, and BF ($\rho = 110$ kg/m^3) as infill layer. As a result of permutation and combination, 4 groups of 2-layer construction and 12 groups of

Table 2.4 Construction conditions arrangement for the projections of the hygrothermal performance

2-layer construction (2 partition boards with 1 cavity)	Construction layer (from outside to inside) [mm]		
	18	30	12
2-1	BFB	Air	BFB
2-2	BFB	Air	BSB
2-3	BFB	BF	BFB
2-4	BFB	BF	BSB

3-layer construction (3 partition boards with 2 cavities)	Construction layer (from outside to inside) [mm]				
	18	30	12	30	12
3-1	BFB	Air	BMB	Air	BFB
3-2	BFB	Air	BPB	Air	BFB
3-3	BFB	Air	BOSB	Air	BFB
3-4	BFB	Air	BMB	Air	BSB
3-5	BFB	Air	BPB	Air	BSB
3-6	BFB	Air	BOSB	Air	BSB
3-7	BFB	Air	BMB	BF	BFB
3-8	BFB	Air	BPB	BF	BFB
3-9	BFB	Air	BOSB	BF	BFB
3-10	BFB	Air	BMB	BF	BSB
3-11	BFB	Air	BPB	BF	BSB
3-12	BFB	Air	BOSB	BF	BSB

3-layer construction are generated, of which the arrangement in the construction cavity includes two conditions, i.e., the enclosed air layer (Air) and the natural bamboo fiber (BF) (Table 2.4).

2.1.2.2 Hygrothermal Performance Projection of the Construction Groups

(1) Moisture storage

The moisture storage performance of the construction is characterized by the moisture storage capacity per unit area, namely the ξ/A [kg/m^2], which is equal to the accumulation of ξ/A values from each construction layer:

$$\frac{\xi}{A} = \sum_{i=1}^{n} u_i \cdot \rho_i \cdot d_i$$

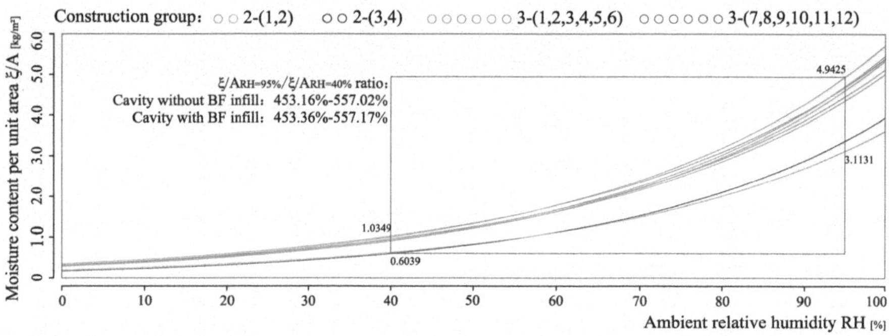

Fig. 2.2 Relation between moisture content of the construction and the ambient relative humidity

where ξ is the volume moisture content of the construction layers; $A = 1\ \text{m}^2$; for bamboo-based panels and natural bamboo fiber, there is both $u = a \cdot e^{b \cdot \varphi}$ functional relation between u and φ; for air layer, there is $u = 0$. The fitting results are substituted into the equations to obtain the ξ/A–φ relation (Table 2.5, Fig. 2.2).

(2) Moisture transport

The moisture transport performance is described with the vapor diffusion thickness of the construction s_d [m], which is the product of the water vapor diffusion resistance factor μ and the thickness of the construction layer d. The total s_d of the construction is the accumulation of s_d values from each construction layer:

$$s_d = \sum_{i=1}^{n} \mu_i \cdot d_i$$

where for bamboo-based panels, there is $\mu = \delta_a/\delta$. According to

$$\delta_a = \frac{0.083}{R_v \cdot T \cdot p}\left(\frac{T}{273}\right)^{1.81}$$

The calculated δ_a value when $T = 23$ °C, $p = 1013$ hPa is 1.9510E^{-10} kg/(m s Pa); the δ–u relation of the panels is $\delta = a \cdot e^{b \cdot u}$, where u is substituted by its fitting function; the μ values of the air and BF are set constant as 0.38 and 1.63, respectively. The above settings are substituted into the equations to obtain the s_d–φ relation (Table 2.6, Fig. 2.3).

(3) Heat storage

The heat storage performance of the construction is characterized by the thermal inertia index D [–], which is the product of the thermal resistance R and the thickness of the thermal effusivity S. Here, S is simplified as the 24-h thermal

Table 2.5 Calculation results of the ξ/A–φ relation between moisture content of the construction and the ambient relative humidity (corresponding to the $\varphi = 40$–95%)

φ (%)	ξ/A [kg/m^2]															
	2-1	2-2	2-3	2-4	3-1	3-2	3-3	3-4	3-5	3-6	3-7	3-8	3-9	3-10	3-11	3-12
40	0.604	0.629	0.605	0.630	0.938	0.913	1.009	0.963	0.938	1.034	0.939	0.915	1.010	0.964	0.939	1.035
45	0.706	0.727	0.707	0.728	1.085	1.057	1.164	1.106	1.078	1.185	1.086	1.058	1.165	1.107	1.079	1.186
50\	0.825	0.840	0.827	0.841	1.255	1.223	1.343	1.270	1.238	1.358	1.257	1.225	1.345	1.271	1.239	1.359
55	0.965	0.971	0.966	0.972	1.453	1.416	1.551	1.458	1.422	1.557	1.454	1.418	1.553	1.460	1.424	1.559
60	1.128	1.122	1.130	1.124	1.681	1.640	1.791	1.675	1.634	1.785	1.683	1.642	1.793	1.678	1.637	1.787
65	1.318	1.297	1.321	1.300	1.946	1.899	2.068	1.925	1.879	2.048	1.948	1.902	2.071	1.928	1.881	2.050
70	1.541	1.501	1.544	1.504	2.253	2.200	2.390	2.212	2.160	2.349	2.256	2.203	2.393	2.216	2.163	2.352
75	1.801	1.736	1.805	1.739	2.609	2.549	2.761	2.543	2.483	2.696	2.612	2.553	2.765	2.547	2.487	2.699
80	2.106	2.008	2.110	2.012	3.022	2.954	3.192	2.924	2.856	3.094	3.026	2.958	3.196	2.928	2.861	3.098
85	2.461	2.324	2.467	2.329	3.500	3.424	3.690	3.363	3.286	3.552	3.505	3.429	3.695	3.368	3.291	3.557
90	2.877	2.689	2.883	2.695	4.056	3.969	4.267	3.868	3.781	4.079	4.062	3.975	4.273	3.874	3.787	4.085
95	3.364	3.113	3.371	3.120	4.700	4.601	4.935	4.450	4.351	4.685	4.707	4.608	4.943	4.457	4.358	4.692

Table 2.6 Calculation results of the s_d–φ relation between vapor diffusion thickness of the construction and the ambient relative humidity (corresponding to the $\varphi = 40–95\%$)

φ (%)	s_d [m]															
	2-1	2-2	2-3	2-4	3-1	3-2	3-3	3-4	3-5	3-6	3-7	3-8	3-9	3-10	3-11	3-12
40	18.304	11.931	18.342	11.969	20.038	18.590	22.247	13.665	12.217	15.874	20.075	18.627	22.284	13.702	12.255	15.912
45	15.409	10.077	15.446	10.115	16.932	15.671	18.850	11.601	10.339	13.519	16.970	15.708	18.888	11.638	10.377	13.556
50	12.600	8.276	12.637	8.314	13.916	12.837	15.550	9.593	8.514	11.227	13.953	12.875	15.588	9.630	8.551	11.264
55	9.959	6.580	9.996	6.617	11.074	10.171	12.438	7.695	6.792	9.060	11.111	10.209	12.476	7.733	6.830	9.097
60	7.565	5.038	7.603	5.075	8.490	7.753	9.602	5.962	5.225	7.075	8.527	7.790	9.640	6.000	5.263	7.112
65	5.487	3.693	5.524	3.730	6.235	5.650	7.118	4.441	3.856	5.324	6.272	5.688	7.156	4.478	3.894	5.362
70	3.770	2.575	3.808	2.612	4.359	3.910	5.040	3.163	2.714	3.845	4.396	3.947	5.078	3.201	2.752	3.882
75	2.433	1.696	2.471	1.733	2.882	2.550	3.390	2.145	1.813	2.653	2.919	2.588	3.427	2.182	1.850	2.690
80	1.460	1.048	1.497	1.085	1.791	1.556	2.155	1.379	1.144	1.743	1.828	1.594	2.193	1.416	1.182	1.781
85	0.806	0.604	0.843	0.641	1.041	0.883	1.292	0.839	0.681	1.089	1.078	0.921	1.329	0.876	0.719	1.127
90	0.405	0.323	0.442	0.361	0.566	0.466	0.730	0.484	0.385	0.649	0.603	0.504	0.767	0.522	0.422	0.686
95	0.185	0.162	0.222	0.200	0.290	0.232	0.392	0.268	0.210	0.370	0.328	0.270	0.430	0.306	0.248	0.408

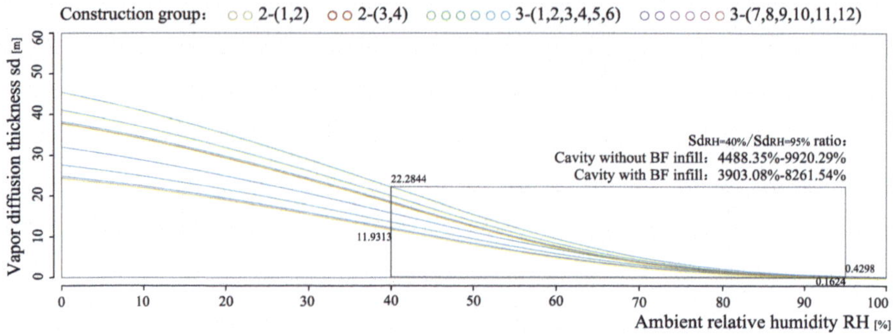

Fig. 2.3 Relation between vapor diffusion thickness of the construction and the ambient relative humidity

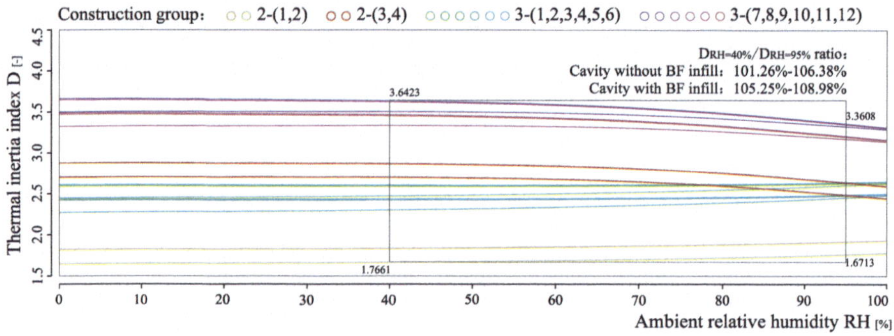

Fig. 2.4 Relation between thermal inertia index of the construction and the ambient relative humidity

effusivity $S_{24h} = 0.51(\rho \cdot c \cdot \lambda)^{0.5}$. The total D is equal to the accumulation of D values from each construction layer:

$$D = \sum_{i=1}^{n} R_i \cdot S_i$$

where there is $\rho = \rho_0 \times (1 + u)$, $c = c_0 \times [1 + (4.2 \times u)/c_0]$, of which the c_0 is the specific heat capacity of the bamboos in dry state, and approximately valued as 2.0 kJ/(kg K) here. Then, there is $c = c_0 \times [1 + (4.2 \times c)/2.0] = 2.0 + 4.2u$ [kJ/(kg K)]. These values and relations are substituted into the equations to obtain the D–φ relation (Table 2.7, Fig. 2.4).

(4) Heat transport

The heat transport performance is described with the thermal resistance R [m^2 K/W], which is the quotient of the thickness of the construction layer d and the thermal conductivity λ. When the heat transfer resistance between the construction surface

Table 2.7 Calculation results of the D–φ relation between thermal inertia index of the construction and the ambient relative humidity (corresponding to the φ = 40–95%)

φ (%)	D [–]															
	2-1	2-2	2-3	2-4	3-1	3-2	3-3	3-4	3-5	3-6	3-7	3-8	3-9	3-10	3-11	3-12
40	1.841	1.671	2.875	2.706	2.475	2.608	2.595	2.306	2.438	2.426	3.510	3.642	3.630	3.340	3.473	3.460
45	1.844	1.675	2.871	2.702	2.481	2.608	2.595	2.312	2.439	2.427	3.508	3.635	3.622	3.339	3.466	3.454
50	1.847	1.680	2.865	2.698	2.487	2.607	2.595	2.320	2.440	2.428	3.505	3.625	3.613	3.337	3.458	3.445
55	1.851	1.685	2.857	2.691	2.494	2.607	2.596	2.328	2.441	2.429	3.499	3.613	3.601	3.333	3.447	3.435
60	1.856	1.691	2.845	2.681	2.502	2.608	2.596	2.337	2.443	2.432	3.491	3.597	3.586	3.327	3.433	3.421
65	1.861	1.698	2.830	2.667	2.512	2.609	2.598	2.348	2.446	2.435	3.480	3.578	3.567	3.317	3.415	3.404
70	1.868	1.706	2.811	2.649	2.522	2.611	2.600	2.361	2.450	2.438	3.465	3.554	3.543	3.304	3.393	3.382
75	1.875	1.715	2.786	2.627	2.535	2.614	2.603	2.375	2.454	2.443	3.446	3.525	3.514	3.286	3.366	3.355
80	1.883	1.725	2.756	2.599	2.549	2.618	2.607	2.391	2.460	2.449	3.422	3.491	3.480	3.264	3.333	3.323
85	1.893	1.737	2.721	2.565	2.565	2.624	2.613	2.409	2.468	2.457	3.393	3.452	3.441	3.237	3.296	3.285
90	1.905	1.751	2.682	2.528	2.583	2.631	2.620	2.429	2.477	2.466	3.360	3.408	3.397	3.207	3.254	3.243
95	1.918	1.766	2.638	2.487	2.604	2.640	2.629	2.453	2.488	2.478	3.325	3.361	3.350	3.173	3.209	3.198

and the air is ignored, the total R of the construction is the accumulation of R values from each construction layer:

$$R = \sum_{i=1}^{n} \frac{d_i}{\lambda_i}$$

where for bamboo-based panels, the λ–u relation is $\lambda = a \cdot e^{b \cdot u}$, in which u is substituted by its fitting function; for BF, the λ–u relation is $\lambda = a \cdot e^{b \cdot \varphi} + c$; the λ of the air is set constant as 0.23 W/(m K). The above values and relations are substituted into the equations to obtain the R–φ relation (Table 2.8, Fig. 2.5).

The hygrothermal performance of construction in equilibrium state with the hygric environment is discussed within the range $\varphi = 40$–95%, which is divided into two groups, namely the construction with BF layer (BF groups) and construction with enclosed air layer (NF groups), according to the arrangement in the cavity.

The ratio between the maximum values and the minimum values is used to analyze the impact of hygric environment on the hygrothermal performance of the building components. For hygric performance, the ratio between $\xi/A_{\mathrm{RH}=95\%}$ and $\xi/A_{\mathrm{RH}=40\%}$ is in the range 453.36–557.17 and 453.16–557.02%, respectively, for the BF and NF groups. Correspondingly, the ratio between $s_{d\mathrm{RH}=40\%}$ and $s_{d\mathrm{RH}=95\%}$ is 3903.08–8261.54 and 4488.35–9920.29%. For thermal performance, the ratio between $D_{\mathrm{RH}=40\%}$ and $D_{\mathrm{RH}=95\%}$ is in the range 105.25–108.98% and 101.26–106.38%, respectively, for the BF and NF groups. And correspondingly, the ratio between $R_{\mathrm{RH}=40\%}$ and $R_{\mathrm{RH}=95\%}$ is 154.38–174.31 and 106.86–114.16%. From the above analysis, significant influence of hygric environment on the hygrothermal performance can be observed, which indicates that the constant value of material parameters would inevitably bring about non-negligible errors to the simulation results.

Similarly, an extreme value difference $D_{\mathrm{ev}} = (\text{max. value} - \text{min. value})/\text{mean. value} \times 100\%$ is defined to describe the amplitude of the hygrothermal performance. Results show that for moisture storage performance, the D_{ev} of ξ/A is in the range

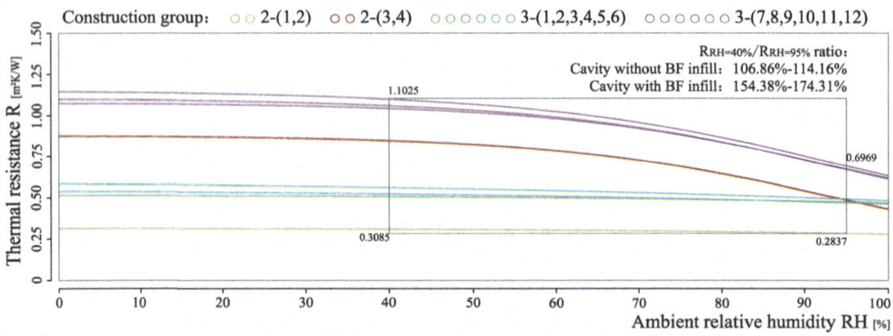

Fig. 2.5 Relation between thermal resistance of the construction and the ambient relative humidity

Table 2.8 Calculation results of the R–φ relation between thermal resistance of the construction and the ambient relative humidity (corresponding to the φ = 40–95%)

φ (%)	R [m^2 K/W]															
	2-1	2-2	2-3	2-4	3-1	3-2	3-3	3-4	3-5	3-6	3-7	3-8	3-9	3-10	3-11	3-12
40	0.309	0.314	0.845	0.851	0.505	0.560	0.521	0.511	0.566	0.527	1.042	1.097	1.058	1.048	1.103	1.064
45	0.307	0.313	0.835	0.841	0.504	0.556	0.518	0.509	0.562	0.524	1.031	1.084	1.046	1.037	1.089	1.052
50	0.306	0.312	0.822	0.827	0.502	0.552	0.515	0.508	0.557	0.521	1.018	1.067	1.031	1.023	1.073	1.036
55	0.305	0.310	0.805	0.811	0.500	0.547	0.512	0.506	0.552	0.517	1.001	1.047	1.012	1.006	1.053	1.017
60	0.303	0.308	0.784	0.790	0.498	0.541	0.508	0.503	0.546	0.513	0.979	1.022	0.989	0.984	1.028	0.994
65	0.301	0.306	0.758	0.763	0.496	0.536	0.503	0.501	0.541	0.508	0.952	0.992	0.960	0.957	0.997	0.965
70	0.299	0.304	0.726	0.731	0.493	0.529	0.499	0.498	0.534	0.503	0.920	0.956	0.925	0.925	0.961	0.930
75	0.297	0.302	0.688	0.693	0.490	0.523	0.493	0.494	0.527	0.498	0.881	0.914	0.885	0.886	0.918	0.889
80	0.294	0.299	0.644	0.648	0.486	0.515	0.488	0.491	0.520	0.493	0.836	0.865	0.837	0.840	0.870	0.842
85	0.291	0.296	0.594	0.599	0.482	0.508	0.482	0.487	0.512	0.487	0.785	0.811	0.785	0.790	0.815	0.789
90	0.288	0.292	0.540	0.545	0.478	0.500	0.475	0.482	0.504	0.480	0.731	0.753	0.728	0.735	0.757	0.733
95	0.284	0.288	0.485	0.489	0.473	0.491	0.469	0.477	0.496	0.473	0.674	0.692	0.670	0.679	0.697	0.674

205.35–232.92 and 205.29–232.89%, respectively, for the BF and NF groups. For moisture transport performance, the corresponding value of s_d is 205.35–232.92 and 205.29–232.89%. For heat storage performance, the D_{ev} of D is in the range 5.04–8.64 and 1.26–7.08%, respectively, for the BF and NF groups. And for heat transport performance, the corresponding value of R is 41.23–50.80 and 7.37–15.34%. Calculation results show that the amplitude of hygric performance is far greater than that of the thermal performance. Meanwhile, since the main contribution of BF layer in the construction is to the thermal performance, the amplitude of D and R values of BF groups is significantly larger than the NF groups, but this is not obvious in the ξ/A and s_d values.

2.2 HAM Model Based on the Künzel Equations

2.2.1 *Künzel Equations of the Coupled Heat and Moisture Process*

As described in Chap. 1, in the field of building physics, the heat and moisture process model is usually adopted in the calculation of building envelope. Prior to the 1990s, the steady-state moisture diffusion model proposed by Glaser was broadly used, which simply adopts vapor pressure as the driving potential, and the vapor permeability coefficient to calculate moisture processes that occur in a one-dimensional steady state. The Glaser model is highly operative and can perform preliminary hygric performance evaluation on building components. However, actual moisture processes are often unsteady and simultaneously occur over multiple phases. Therefore, the Glaser simple steady-state evaluative method is unable to meet the requirements of a more accurate calculation.

There are many differences between the HAM model and the Glaser model. The simulation in the former one is performed by means of coupling heat and moisture, rather than separating the heat process from the moisture process. In addition to temperature, relative humidity, and solar radiation, the HAM model requires rainfall (in the form of wind-driven rain to the building facade) as meteorological parameters; apart from bulk density, specific heat capacity, thermal conductivity, and water vapor diffusion resistance factor, the HAM model also requires isothermal adsorption and desorption curves, porosity, and liquid water transfer coefficients as material parameters. Considering the effects of multi-phase moisture processes and the moisture storage properties (moisture content) of materials on the heat transport properties, as well as the liquid water and gaseous moisture transport properties, the HAM model requires the moisture-content-dependent transport properties of heat and moisture.

Among the studies on HAM model by different scholars, due to the differences in the basic description equations, various material parameters are required as inputs, and even for the same material parameters, due to differences in driving potential selection and valuing methods, as well as different judgments on the environmental factors, material moisture content, etc., there are various testing requirements. The use of temperature as a driving potential for heat transport is consistently recognized, but there is no uniform view on the driving potential selection of the moisture transport. Commonly used moisture transport driving potentials include the temperature, moisture content, vapor pressure, and suction stress. German building physicist Künzel believes that it is necessary to use two independent factors as the driving potentials for the transport of gaseous moisture and liquid water, respectively, and considers that the temperature and moisture content are not direct moisture-driving potentials, and the transfer coefficient obtained by combining these two factors is difficult to determine and would lead to relatively complex functions, and meanwhile the capillary suction stress cannot be applied to the dry and non-capillary-active materials, and can not be directly measured in wet materials. Therefore, he selects vapor pressure and relative humidity as the driving potentials for gaseous moisture and liquid water transfer, which are widely recognized and easy to be measured [1].

Through theoretical calculations and actual measurements, Künzel simplifies the material parameters required for one-dimensional and two-dimensional coupled heat and moisture process calculations, in which the processes are described as two coupled differential equations, as shown below[1]:

Moisture process function, and its terms denote a material's moisture storage, liquid water transport, and gaseous moisture transport:

$$\rho_{\mathrm{l}}\frac{\partial u}{\partial \varphi}\cdot\frac{\partial \varphi}{\partial t}=\frac{\partial}{\partial x}\left(\rho_{\mathrm{l}}D_{\mathrm{l}}\frac{\partial u}{\partial \varphi}\frac{\partial \varphi}{\partial x}\right)+\frac{\partial}{\partial x}\left(\frac{\delta}{\mu}\frac{\partial p}{\partial x}\right)$$

Heat process function, and the three terms in this equation denote a material's heat storage, heat transport, and latent heat effect:

$$\frac{\partial H}{\partial \vartheta}\cdot\frac{\partial \vartheta}{\partial t}=\frac{\partial}{\partial x}\left(\lambda\frac{\partial \vartheta}{\partial x}\right)+h_{\mathrm{v}}\frac{\partial}{\partial x}\left(\frac{\delta}{\mu}\frac{\partial p}{\partial x}\right)$$

The left side of both equations is comprised of storage terms: moisture storage is described by a material's isothermal moisture adsorption and desorption curves and their derivatives. Heat storage includes the heat capacity of the dry material and the heat capacity of the moisture within the material. The right side of both equations consists of transport terms: liquid water transport includes surface diffusion and capillary conduction driven by a relative moisture gradient, and it is slightly affected by temperature. Gaseous moisture diffusion, on the other hand, is strongly influenced by temperature fields. Heat transport is described by the thermal conductivity and vapor enthalpy flow that is related to the moisture content. Heat transport induced

[1]As described in Chap. 1, HAM model is coupled of heat, moisture, and air, and the coupling of air only occurs when there is a pressure difference between indoors and outdoors, which is not considered in some models such as the Künzel model. These models without air coupling are actually the 'coupled heat and moisture model', referred to as HM model, which is regarded as a case of HAM model in this study.

by vapor enthalpy flows occurs via evaporation-mediated heat absorption in one location, followed by condensation-mediated heat release after diffusion to a different location; this form of heat transfer is often referred to as the latent heat effect. The operation of these equations requires a number of control parameters, including material data, meteorological data, internal conditions, data grids, time steps, and so on. The material data, in particular, include basic physical properties such as bulk density, porosity, and other interrelated parameters that characterize the storage and transport properties of heat and moisture [1].

The research area of the Fraunhofer IBP's Hygrothermics Department covers the development of heat and moisture process models in laboratories, thermal parameter test, moisture test, climate-related material properties, meteorological data collection, climate simulation, outdoor test, building hygrothermal analysis, moisture proof assessment, and biological hygrothermics (prevention of mold growth and formation on building façade), etc. Based on the Künzel equations, the Fraunhofer IBP Hygrothermics Department has developed the computer program WUFI (Wärme- Und Feuchtetransport Instationär, which means transient heat and moisture transfer) for the simulation of the heat and moisture processes of building envelopes (Fig. 2.6).

Due to the capacity of outputting instantaneous data such as the moisture content of building components, the temperature and moisture field of enclosed space throughout the year, WUFI can be integrated with other theories to achieve specialized research. For example, the overlapping with the temperature and humidity environmental requirements for mold growth can be used for the mold growth assessment [2–4]; the combination with HVAC system engineering and building heat and moisture process simulation [5, 6]; the integration with the building airflow model to optimize the natural ventilation strategy [7]; and the studies on construction parameters, such as the thermal bridges and air layers that affect the hygric and thermal performance of building envelopes with the annual dynamic simulation results [8, 9].

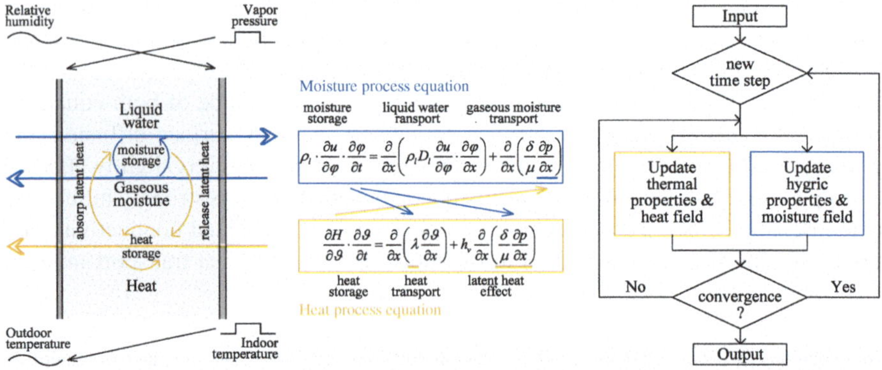

Fig. 2.6 Building envelope coupled heat and moisture process mechanism, Künzel equations and the WUFI simulation process

WUFI Plus can correlate the hygrothermal performance of the building envelope and its influence on the hygrothermal performance of the enclosed space, which is of particular significance for the study of hygroscopic materials such as bamboo and timber. Winkler et al. studied the effects of moisture buffering and related latent heat exchange on the energy consumption and thermal comfort of the room. In the study, factors impact analyses were carried out on related parameters including the surface material properties, area, moisture generation period, climatic conditions and air-exchange rate of the targeted unit room. Results showed that the moisture buffering effect of wood materials significantly affected the relative humidity in the room, which could weaken the relative humidity fluctuations, to form a more stable indoor hygric environment, thus reducing the energy consumption and humidification and dehumidification demand [10].

2.2.2 Climate and Material Parameters Specially Considered

For all HAM models, the simulation of coupled heat and moisture process requires interrelated heat and moisture properties' parameters, and external meteorological parameters as the driving potentials. Compared to the traditional uncoupled models, HAM requires the isothermal adsorption and desorption curve, the moisture-content-dependent vapor permeability coefficient and thermal conductivity. In addition, since the HAM model considers multi-phase moisture processes, the liquid water adsorption process caused by rainfall on the façade requires additional treatment. The HAM model based on the Künzel equation and its computer program WUFI have their own treatment method for the driving rain and liquid water transport.

(1) Driving rain

Fraunhofer IBP conducts research on the moisture sources such as the rainfall in the building climatology [11]. Typical meteorological data, such as the Thermal Reference Years and the Test Reference Years (TRY), lack of rainfall parameters, resulting in the inability to support the construction types that are susceptible to external rainfall. Based on this, Fraunhofer IBP develops a new type of meteorological data, the Hygrothermal Reference Years (HRY), for Germany and carries out verification on it [12]. At present, the cities with complete meteorological parameters accumulated by Fraunhofer IBP are mainly distributed in Europe, North America, and Japan.

The driving rain coefficient (DR) is used to describe the driving rain load on the outer surface of building components. This parameter describes the amount of rain that acts on different directions of the building and is therefore different from the precipitation. The value of DR in WUFI is estimated by precipitation, wind speed, and average wind direction:

$$\mathrm{DR} = \mathrm{precipitation} \cdot (R_1 + R_2 \cdot \mathrm{wind\ speed})$$

The 'precipitation' is the rainfall intensity [mm/h] on a horizontal plane. 'Wind speed' is the component of the average wind speed orthogonal to the surface of the building component in an open area at a height of 10 m. R_1 and R_2 depend on the specific location on the facade of the building. For vertical surfaces in open positions that are unaffected by surrounding buildings, R_1 is 0 and R_2 is approximately 0.2 s/m (which is smaller at the center of the facade and larger at the edges and corners of the building). WUFI provides a number of preset DR values for this according to the different building heights, including a group of low-rise buildings and three groups of high-rise buildings, and these coefficients are determined by three-dimensional computational fluid dynamics (CFD) simulation [13]. For those facades with an inclination of greater than 90°, WUFI cannot provide a detailed option, but default to $R_1 = 1$, and $R_2 = 0$, which means DR is equal to the normal precipitation. Nevertheless, this parameter can be modified by the users.

Another way to estimate the driving rain load on vertical facade is the ASHRAE Standard 160 'Design Criteria for Moisture Control in Buildings' [14]:

$$\text{DR} = \text{precipitation} \cdot \text{FE} \cdot \text{FD} \cdot 0.2 \cdot \text{wind speed}$$

The definitions of 'precipitation' and 'wind speed' here are the same as the former ones. FE is the rain exposure factor, which depends on the surrounding terrain and building height, and FD is the rain deposition factor, which describes liquid water adsorption capacity of the external surface of the building. The valuing of both refers to the recommendations in ASHRAE Standard 160P.[2]

(2) Liquid water transport

The liquid water transport involves surface diffusion and capillary conduction. Surface diffusion is the moisture transport of the water molecules on the pore walls of the hygroscopic materials and microcapillary. Capillary conduction occurs when the adhesion between moisture and the capillary wall is greater than the cohesion within the water itself, and the water is drawn into the porous material.

Through measurement and calculation, Künzel realizes that the independent calculation of surface diffusion and capillary conduction from the vapor diffusion is feasible. Similar to the feasibility of describing the Fick diffusion and Knudsen transfer for water vapor transfer uniformly with the water vapor diffusion resistance factors (see Chap. 3, Sect. 3.1), the surface expansion and capillary conduction phenomena occur simultaneously in liquid water transport and can be obtained overall by experimental tests.

Capillary water absorption coefficient and liquid water transfer coefficient are important material parameters for describing the liquid water transport. The water absorption coefficient A_{cap} [kg/(m^2 s$^{0.5}$)] is a standard parameter for characterizing the capillary suction characteristics of a building material which is in contact with water. The capillary transport coefficient D_w [m^2/s] is significantly affected by the moisture

[2]A detailed introduction of WUFI's explanation on driving rain can be seen in the documents from the Fraunhofer IBP Hygrothermics Department.

content, and also influenced by temperature due to the influence of temperature on the surface tension σ and water viscosity η, and 20 °C in the laboratory environment is generally used as a temperature reference. The D_w value is divided into two types, describing separately the suction and redistribution processes, of which the latter is usually 1–2 orders of magnitude smaller than the former.

The theoretical model adopted in this study is based on Künzel's coupled equations for heat and moisture processes, which is mainly due to the following factors:

(1) Widely recognized research results. Building physicist Künzel continues to study the hygrothermal properties of building materials, as well as the heat and moisture processes of building envelopes, and is one of the contributors to the several mainstream HAM models.
(2) Comparable parameter database of other building materials. The Hygrothermics Department of Fraunhofer IBP, where Künzel is from, has been testing the hygrothermal properties of building materials for decades, accumulating a hygrothermal properties database of common building materials including timber, which can be used for reference and comparison in this study [15].
(3) The combination with computer programs. Based on the Künzel equations, Fraunhofer IBP has developed the WUFI series software, which is one of the worldwide-recognized computer tools for building envelope heat and moisture process simulation.
(4) Continuous verification and correction through measurement. The Fraunhofer IBP's Hygrothermics Laboratory in Holzkirchen, Germany, has comprehensive indoor and outdoor testing field and facilities, and has been conducting long-term tracking of building materials and components, and has accumulated large amounts of data to verify and correct its theoretical models and computer programs.

However, there are also limitations to use this model and its simulation tool WUFI, including some general defects of the HAM models and the computer programs, mainly shown as follows:

(1) For the material. The storage and transport of the moisture as well as the heat primarily target the heat and moisture processes within the material. The characteristics of heat and moisture exchange of the surface with the outside air or the adjacent construction layers cannot be described and quantified, so empirical values are normally given in the simulation tools.
(2) For the boundary condition of the model. The heat and moisture processes are established strictly between an external and an internal condition, and their transport and storage processes are driven by the heat and moisture load difference. The conditions with unclear distinction of heat and moisture load difference due to the communication of indoor and outdoor environment through the openings can not be described.
(3) For the external climatic conditions. The Künzel equations and its WUFI simulation tool require driving rain as an external meteorological parameter, which is determined by the annual hourly rainfall and wind speed, wind direction

and other conditions, and needs to be measured independently or converted by some other sufficient detailed meteorological data. At present, only some cities in Europe and America, and several cities in Japan can provide these meteorological data.

(4) Other functional limitations include: the ventilated air layer that is helpful to the hygrothermal performance optimization cannot be simulated; WUFI Plus (the one of the WUFI series software that can be used for the whole building system simulation) can only conduct simulation with one-dimensional boundary conditions, therefore it is unable to simulate the thermal bridge in the construction and the related influence caused by the bridge. The study on the thermal bridge requires an additional use of WUFI 2D, which works only at the building component level.

2.3 Simulation and Analysis with HAM Model for Building Envelope in North American Typical Climate Zones

The coupled heat and moisture process simulation for building envelope is carried out in WUFI Plus, of which the simulation results from comparison model groups are analyzed to demonstrate the impact of the meteorological data, material parameters, and evaluation methods, which gives reference to the determination of theoretical model for the study on the application of bamboo in building envelope.

2.3.1 Simulation Model Design

The simulation model constructed in WUFI Plus requires the arrangement of the external conditions, internal conditions, boundary conditions, and HVAC conditions, as well as some control parameters such as the grid and time step (Fig. 2.7).

2.3.1.1 External Conditions

The currently available meteorological data in WUFI Plus mainly come from North America, Europe, and Japan. According to the Köppen climate classification, there are rich types of climate zones in North America, therefore, the meteorological data of the 20 representative cities from 10 typical climate zones of North America are selected as the exterior conditions. Each meteorological data consist of annual hourly meteorological parameters covering solar radiation (including total counter radiation and cloud index), air temperature, relative humidity, and driving rain (including precipitation and wind) (Figs. 2.8 and 2.9, Table 2.9).

2.3.1.2 Internal Conditions

Enclosed space units of 3.0 m × 3.0 m × 3.0 m (width × depth × height) are set, of which the room occupancy period is set as 8:00–17:00. Standard office indoor heat and moisture load is given including convective heat: 33.3 W, radiant heat: 25.2 W, moisture: 17.55 g/h, CO_2: 20.79 g/h, and human activity: 1.2 met. The PMV/PPD-related additional data are given including clothing: 0.7 clo and air velocity: 0.1 m/s.

2.3.1.3 Boundary Conditions

The exterior walls (each 9 m^2) are selected as the study objects, while the floor and ceiling are set as partitions between the same interior conditions to avoid heat and moisture exchange. Three commonly used construction types are considered,

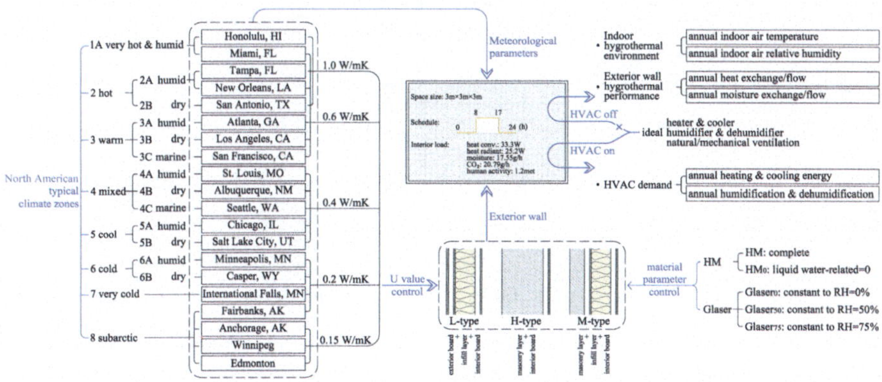

Fig. 2.7 WUFI Plus model design (external, internal, boundary, and HVAC conditions)

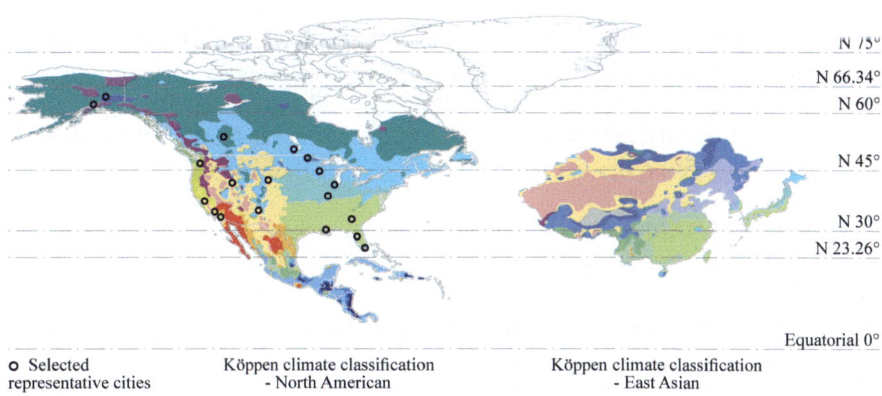

Fig. 2.8 North American and East Asian map of Köppen climate classification and the selected representative cities

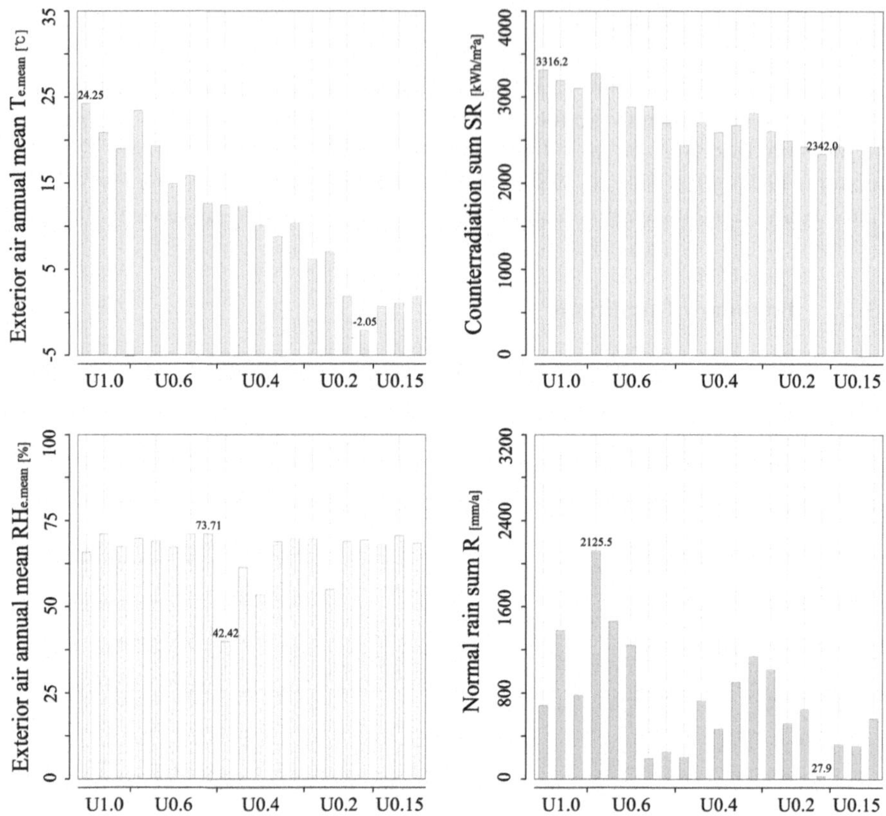

Fig. 2.9 Meteorological data of the selected 20 representative cities from the North America, including annual outdoor air temperature mean value, total solar radiation, relative humidity mean value, and the total rainfall

including L-type (exterior board + infill layer + interior board), H-type (masonry layer + interior board), and M-type (masonry layer + infill layer + interior board). The constructions are designed by controlling the heat transfer coefficient *U* value according to the thermal requirement of each climate zones in North America, which are divided into five groups corresponding to U = 0.15/0.2/0.4/0.6/1.0 W/(m^2 K) (Table 2.10).

Two groups of HAM model (HAM and HAM_0) and three groups of Glaser model ($Glaser_0$, $Glaser_{50,}$ and $Glaser_{75}$) are set up by adjusting the material parameters, of which the HAM groups are given the complete parameters. On the basis of the material parameters of HAM groups, the liquid water process-related parameters including the water absorption coefficient and liquid water transport coefficient of the HAM_0 groups are set as 0, meanwhile the rainfall load on the external walls is closed. The transport property parameters of heat and gaseous moisture constantly

Table 2.9 Selected 20 representative cities from North America

No.	Zone	Type	Representative city	Latitude [°N]
U1.0-1	1A	Very hot—humid	Honolulu, HI	21.30
U1.0-2	2A	Hot—humid	Tampa, FL	27.97
U1.0-3	2B	Hot—dry	San Antonio, TX	29.53
U0.6-1	1A	Very hot—humid	Miami, FL	25.80
U0.6-2	2A	Hot—humid	New Orleans, LA	29.95
U0.6-3	3A	Warm—humid	Atlanta, GA	33.65
U0.6-4	3B	Warm—dry	Los Angeles, CA	33.93
U0.6-5	3C	Warm—marine	San Francisco, CA	37.77
U0.4-1	4B	Mixed—dry	Albuquerque, NM	35.05
U0.4-2	4A	Mixed—humid	St. Louis, MO	38.75
U0.4-3	5B	Cool—dry	Salt Lake City, UT	40.77
U0.4-4	5A	Cool—humid	Chicago, IL	41.78
U0.4-5	4C	Mixed—marine	Seattle, WA	47.45
U0.2-1	6A	Cold—humid	Minneapolis, MN	44.88
U0.2-2	6B	cold—dry	Casper, WY	42.92
U0.2-3	7	Very cold	International Falls, MN	48.56
U0.2-4	8	Subarctic	Fairbanks, AK	64.82
U0.15-1	8	Subarctic	Anchorage, AK	61.22
U0.15-2	8	Subarctic	Winnipeg	49.90
U0.15-3	8	Subarctic	Edmonton	53.55

corresponding to RH = 0%, RH = 50%, and RH = 75% are given to $Glaser_0$, $Glaser_{50}$, and $Glaser_{75}$, respectively. The moisture storage and liquid-water-related material parameters are set to 0 for all the three groups of Glaser model, so that the transport processes of heat and moisture are separated.

2.3.1.4 HVAC Conditions

Based on the models that consist of the above conditions, the HVAC is firstly turned off to evaluate the indoor hygrothermal environment. Subsequently, the HVAC is turned on, and the indoor air temperature is maintained within 20–26 °C; the relative humidity is maintained within 40–60%; the max. CO_2 concentration is set as 1550 ppmv; natural ventilation is closed; the infiltration air change rate is set as 0, while the mechanical ventilation is set as 0.6 times per hour. The operation data including the annual heating and cooling, humidification and dehumidification demand are calculated. The operation period is set from 2015-01-01 to 2017-01-01, in which the hourly data for the second complete year are collected for further analysis.

Table 2.10 Boundary condition (construction arrangement of the exterior walls)

U value	Type	Construction layer arrangement (mm)										Construction diagram
		BFB	Aerated concrete	Lime silica brick	PE	Air layer	BOSB	Cellulose fiber	Air layer	PE	BSB	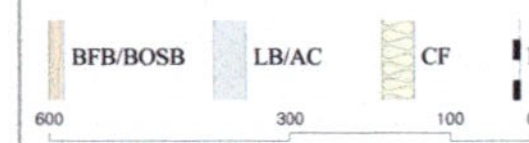
U1.0	L	18			0.2			20	40		12.5	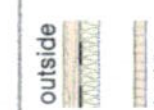
	H		60		0.2				40		12.5	
	M			120	0.2			20	40		12.5	
U0.6	L	18			0.2	40	12.5	40	40		12.5	
	H		150		0.2	40	12.5		40		12.5	
	M			120	0.2	40	12.5	40	40		12.5	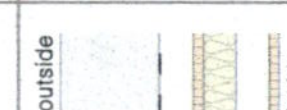
U0.4	L	18				40	12.5	80	40	0.2	12.5	
	H		240			40	12.5		40	0.2	12.5	
	M			120		40	12.5	80	40	0.2	12.5	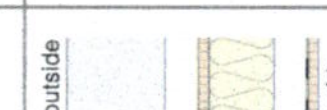

(continued)

Table 2.10 (continued)

U value	Type	Construction layer arrangement (mm)										Construction diagram
		BFB	Aerated concrete	Lime silica brick	PE	Air layer	BOSB	Cellulose fiber	Air layer	PE	BSB	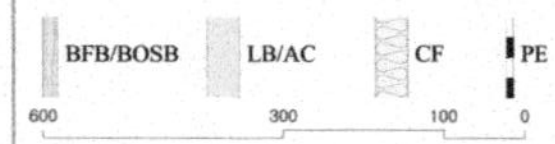
U0.2	L	18				40	12.5	160	40	0.2	12.5	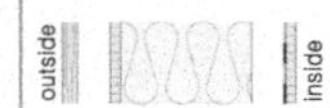
	H		450			40	12.5		40	0.2	12.5	
	M			120		40	12.5	160	40	0.2	12.5	
U0.15	L	18				40	12.5	240	40	0.2	12.5	
	H		600			40	12.5		40	0.2	12.5	
	M			120		40	12.5	240	40	0.2	12.5	

2.3.2 Results Analysis

2.3.2.1 Correlation Analysis Between Simulation Results and the Climate Parameters

The annual heating and cooling demand P, heating peak $P_{heating.max}$ and cooling peak $P_{cooling.max}$ values, the humidification and dehumidification H, humidification peak $H_{humid.max}$ and dehumidification peak $H_{dehumid.max}$ are chosen as indicators to characterize the HVAC demand of the enclosed space in the HVAC on conditions. When the HVAC off, the annual indoor air temperature mean value $T_{i.mean}$ and amplitude $T_{i.amp}$, annual relative humidity mean value $RH_{i.mean}$ and amplitude $RH_{i.amp}$ are indicators to characterize the indoor hygrothermal environment. The annual heat flow H_{flow} and moisture flow M_{flow} through the exterior walls, and the corresponding peak values $H_{flow.max}$ and $M_{flow.max}$ are selected as indicators to characterize the hygrothermal performance of the exterior walls. The outdoor meteorological parameters include the annual total solar radiation SR, air temperature T_e, relative humidity RH_e, and the total driving rain DR. Correlation analyses are carried out between the simulation results and the meteorological parameters.

Considering that the simulation results and the meteorological parameters do not match the bivariate normal distribution, and there is not linear correlation between variables, a nonparametric statistical method, the Spearman Correlation Coefficient analysis, is adopted in the following discussion, with SPSS statistical analysis software for operation. The related indicators are divided into two groups:

Group 1 is the correlation analyses for annual total/mean values between simulation results and the meteorologic parameters.
Group 2 is the correlation analyses for annual amplitude/peak values between simulation results and the meteorologic parameters.
(Note: for group 2, considering that the annual max. value of outdoor relative humidity is 100%, and the min. value of driving rain is 0, the corresponding min. value of outdoor relative humidity and the max. value of driving rain are used as the representatives of amplitude/peak values).

The correlation analyses results of the annual total/mean values show that the heat process-related indicators, including P, $T_{i.mean}$ and H_{flow} values, exhibit significant correlation with both $T_{e.mean}$ and SR at 0.01 level (2-tailed), while $T_{i.mean}$ and H_{flow} have significant correlation with DR at 0.05 level (2-tailed) and at 0.01 level (2-tailed), respectively, which proves that the heat process of building envelope depends not only on the thermal meteorological parameters but also the hygric parameters. For the moisture-process-related indicators, there are significant correlation at 0.01 level (2-tailed) and at 0.05 level (2-tailed), respectively, between $RH_{i.mean}$ and $RH_{e.mean}$, and between M_{flow} and DR. The H value simulation results do not show significant correlation with any outdoor meteorological parameters[3] (Table 2.11, Fig. 2.10).

[3]Note: there are normally two levels to judge the strength of significance correlation, namely the significant correlation at 0.01 level and at 0.05 level. The former means that the confidence level

Table 2.11 Correlation analyses between HAM model group simulation results and the meteorologic parameters—Group 1 annual total/mean values

Simulation results		$T_{e.mean}$			SR			$RH_{e.mean}$			DR		
		L	H	M	L	H	M	L	H	M	L	H	M
P	*r*	*0.693***	*0.683***	*0.710***	*0.598***	*0.579***	*0.623***	0.123	0.017	0.148	0.250	0.198	0.277
	Sig. (2-tailed)	*0.001*	*0.001*	*0.000*	*0.005*	*0.007*	*0.003*	0.604	0.942	0.533	0.289	0.402	0.238
H	*r*	0.174	0.158	0.174	0.105	0.087	0.105	0.041	0.044	0.041	0.176	0.170	0.176
	Sig. (2-tailed)	0.462	0.506	0.462	0.659	0.715	0.659	0.863	0.853	0.863	0.458	0.474	0.458
$T_{i.mean}$	*r*	*0.996***	*0.997***	*0.997***	*0.929***	*0.928***	*0.928***	0.011	0.023	0.023	*0.537**	*0.535**	*0.535**
	Sig. (2-tailed)	*0.000*	*0.000*	*0.000*	*0.000*	*0.000*	*0.000*	0.962	0.922	0.922	*0.015*	*0.015*	*0.015*
$RH_{i.mean}$	*r*	0.560*	0.420	0.466*	0.419	0.287	0.323	*0.653***	*0.662***	*0.680***	0.017	0.150	0.093
	Sig. (2-tailed)	0.010	0.065	0.038	0.066	0.219	0.164	*0.002*	*0.001*	*0.001*	0.942	0.529	0.698
H_{flow}	*r*	*0.734***	*0.722***	*0.541**	*0.812***	*0.803***	*0.654***	0.363	0.327	0.350	*0.630***	*0.591***	*0.586***
	Sig. (2-tailed)	*0.000*	*0.000*	*0.014*	*0.000*	*0.000*	*0.002*	0.115	0.159	0.131	*0.003*	*0.006*	*0.007*

(continued)

Table 2.11 (continued)

Simulation results		$T_{e.mean}$			SR			$RH_{e.mean}$			DR		
		L	H	M	L	H	M	L	H	M	L	H	M
M_{flow}	r	0.409	0.281	0.400	0.481*	0.392	0.477*	0.285	0.287	0.317	*0.498**	*0.502**	*0.514**
	Sig. (2-tailed)	0.073	0.230	0.081	0.032	0.087	0.034	0.223	0.221	0.173	*0.026*	*0.024*	*0.020*

Abbreviation

$T_{e.mean}$—annual outdoor air temperature mean value; SR—annual solar radiation total value; $RH_{e.mean}$—annual outdoor relative humidity mean value; DR—annual driving rain total value

P—annual heating and cooling demand total value; H—annual humidification and dehumidification demand total value

$T_{i.mean}$—annual indoor air temperature mean value; $RH_{i.mean}$—annual indoor relative humidity mean value

H_{flow}—annual heat flow through exterior wall total value; M_{flow}—annual moisture flow through exterior wall total value

r—correlation coefficient; Sig.—statistical significance

The italic emphasized area in the table is the conditions where the simulation results of all the three types (L, H, M) of construction have significant correlation with the outdoor meteorological parameters; Group with '**' means the correlation is at 0.01 level (2-tailed), while with '*' means at 0.05 level (2-tailed)

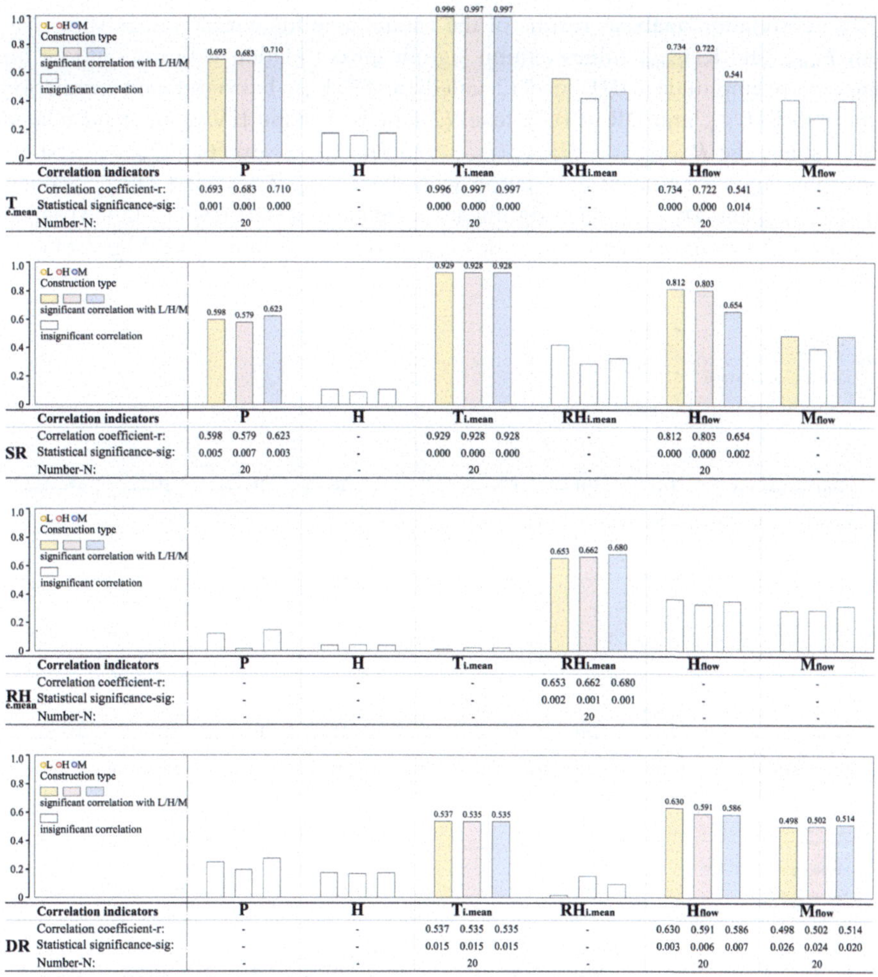

Fig. 2.10 HAM model group correlation analyses between meteorologic parameters (annual average exterior air temperature $T_{e.mean}$, annual total solar radiation SR, annual average exterior relative humidity $RH_{e.mean}$, and annual total driving rain DR) and the simulation results (Group 1 annual total/mean values)

---

is 99%, and the latter 95%. The smaller level of significance means stronger judgment confidence. As to the hypothesis testing, the abnormal value of the test statistic usually has two sides, of which the left side of the probability distribution curve corresponds to the excessively small values, and the right side to the excessively large values. In general, small probability events at both tail ends of the probability distribution curve should be considered, which is called the 'two-tailed test'. But if it can be determined in advance that a small probability event does not exist on one side, than the hypothesis testing can be carried out only on the other side, which is called the 'single-tailed test'. The two-tailed test is generally considered to be more rigorous and persuasive.

The correlation analyses results of the annual amplitude/peak values show that both $T_{i.amp}$ and $RH_{i.amp}$ values exhibit significant correlation with the outdoor air temperature amplitude at 0.01 level (2-tailed), and $RH_{i.amp}$ has significant correlation with both $RH_{e.min}$ and DR at 0.01 level (2-tailed). For the HVAC demand-related peak values, the $P_{cooling.max}$, $H_{dehumid.max,}$ and $H_{humid.max}$, except $P_{heating.max}$, show correspondingly significant correlation with the $T_{e.max}$, $T_{e.min,}$ and DR_{max} values. The $H_{flow.max}$ and $M_{flow.max}$ have significant correlation at 0.01 level (2-tailed) and at 0.05 level (2-tailed), respectively, with $T_{e.min}$ and DR_{max} (Table 2.12; Fig. 2.11).

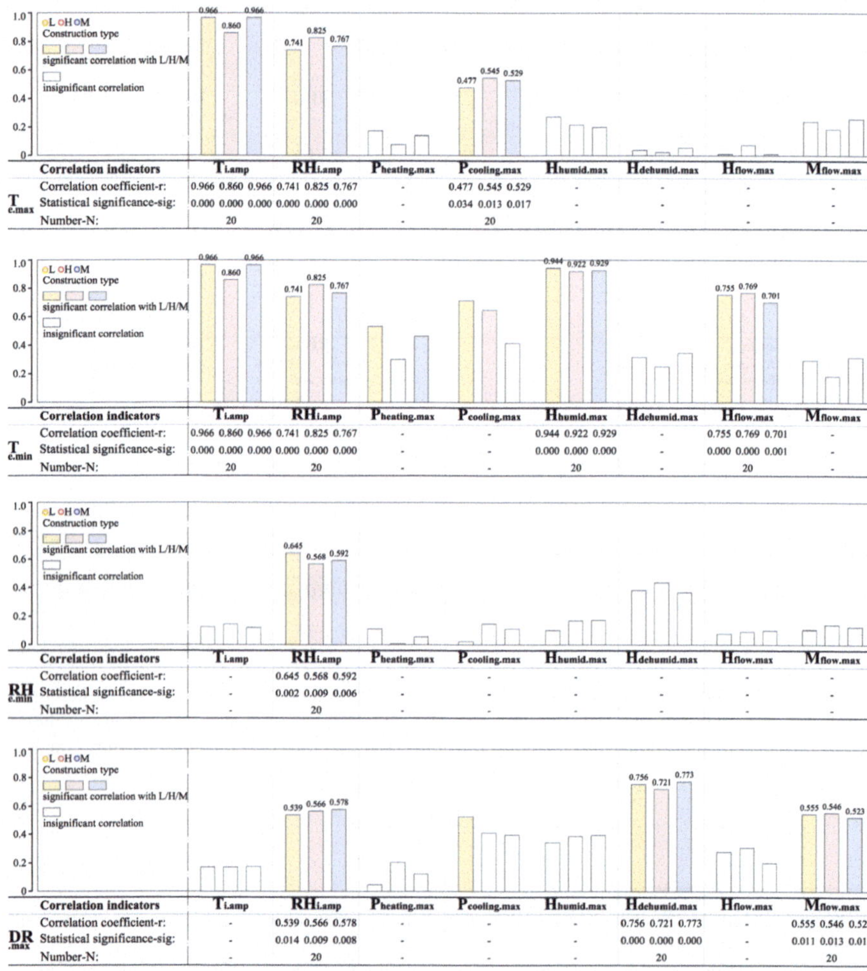

Fig. 2.11 HAM model group correlation analyses between meteorologic parameters (annual maximum exterior air temperature $T_{e.max}$, annual minimum exterior air temperature $T_{e.min}$, annual minimum exterior relative humidity $RH_{e.min}$, and annual maximum driving rain DR_{max}) and the simulation results (Group 2 annual amplitude/max. values)

Table 2.12 Correlation analyses between HAM model group simulation results and the meteorologic parameters—Group 2 annual amplitude/peak values

Simulation results		$T_{e.max}$			$T_{e.min}$			$RH_{e.min}$			DR_{max}		
		L	H	M	L	H	M	L	H	M	L	H	M
$T_{i.amp}$	r	*0.966***	*0.860**=*	*0.966***	*0.966***	*0.860***	*0.966***	0.125	0.145	0.119	0.171	0.170	0.175
	Sig. (2-tailed)	*0.000*	*0.000*	*0.000*	*0.000*	*0.000*	*0.000*	0.600	0.541	0.616	0.470	0.474	0.462
$RH_{i.amp}$	r	*0.741***	*0.825***	*0.767***	*0.741***	*0.825***	*0.767***	*0.645***	*0.568***	*0.592***	*0.539**	*0.566***	*0.578***
	Sig. (2-tailed)	*0.000*	*0.000*	*0.000*	*0.000*	*0.000*	*0.000*	*0.002*	*0.009*	*0.006*	*0.014*	*0.009*	*0.008*
$P_{heating.max}$	r	0.173	0.074	0.140	0.533*	0.30	0.464*	0.110	0.006	0.056	0.047	0.206	0.122
	Sig. (2-tailed)	0.467	0.757	0.555	0.016	0.199	0.039	0.645	0.980	0.813	0.845	0.384	0.609
$P_{cooling.max}$	r	*0.477**	*0.545**	*0.529**	0.713**	0.647**	0.414	0.023	0.146	0.111	0.527*	0.414	0.398
	Sig. (2-tailed)	*0.034*	*0.013*	*0.017*	0.000	0.002	0.070	0.925	0.538	0.640	0.017	0.069	0.082
$H_{humid.max}$	r	0.274	0.216	0.200	*0.944***	*0.922***	*0.929***	0.102	0.169	0.173	0.344	0.388	0.397
	Sig. (2-tailed)	0.243	0.359	0.397	*0.000*	*0.000*	*0.000*	0.670	0.475	0.467	0.137	0.091	0.083
$H_{dehumid.max}$	r	0.036	−0.020	0.053	0.317	0.252	0.347	0.381	0.436	0.368	*0.756***	*0.721***	*0.773***
	Sig. (2-tailed)	0.879	0.932	0.825	0.173	0.285	0.134	0.098	0.055	0.110	*0.000*	*0.000*	*0.000*

(continued)

Table 2.12 (continued)

Simulation results		$T_{e.max}$			$T_{e.min}$			$RH_{e.min}$			DR_{max}		
		L	H	M	L	H	M	L	H	M	L	H	M
$H_{flow.max}$	*r*	0.008	0.075	0.010	*0.755***	*0.769***	*0.701***	0.078	0.093	0.101	0.277	0.307	0.200
	Sig. (2-tailed)	0.975	0.752	0.967	*0.000*	*0.000*	*0.001*	0.743	0.698	0.673	0.238	0.188	0.398
$M_{flow.max}$	*r*	0.240	0.186	0.259	0.294	0.183	0.315	0.105	0.139	0.124	*0.555**	*0.546**	*0.523**
	Sig. (2-tailed)	0.308	0.434	0.269	0.208	0.441	0.176	0.661	0.560	0.602	*0.011*	*0.013*	*0.018*

Abbreviation
$T_{e.max}$—annual outdoor air temperature max. value; $T_{e.min}$—annual outdoor air temperature min. value; $RH_{e.min}$—annual outdoor relative humidity min. value; DR_{max}—annual driving rain max. value
$P_{heating.max}$—annual heating peak value; $P_{cooling.max}$—annual cooling peak value; $H_{humid.max}$—annual humidification peak value
$H_{dehumid.max}$—annual dehumidification peak value
$T_{i.amp}$—annual indoor air temperature amplitude; $RH_{i.amp}$—annual indoor relative humidity amplitude
$H_{flow.max}$—annual heat flow through exterior wall peak value; $M_{flow.max}$—annual moisture flow through exterior wall peak value
r—correlation coefficient; Sig.—statistical significance
The italic emphasized area in the table is the conditions where the simulation results of all the three types (L, H, M) of construction have significant correlation with the outdoor meteorological parameters; Group with '**' means the correlation is at 0.01 level (2-tailed), while with '*' means at 0.05 level (2-tailed)

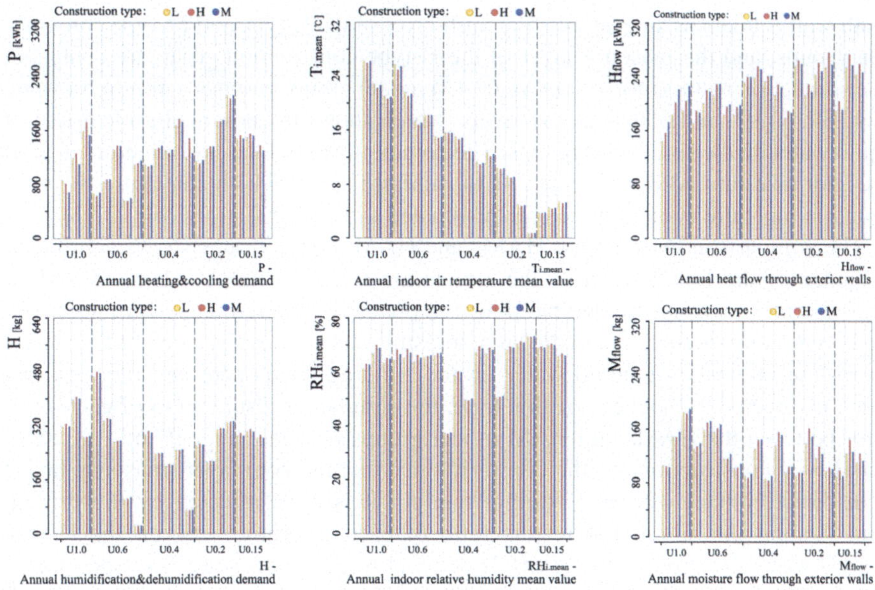

Fig. 2.12 HVAC demand, indoor hygrothermal environment, and the exterior walls hygrothermal performance' simulation results of the HAM model group

The simulation results from the model groups of the same thermal zone and different hygric subzone differ significantly from each other, which indicates that the influence of hygric meteorologic parameters, such as the outdoor relative humidity and driving rain, can not be ignored. Therefore, there are certain defects in the pure thermal-zoning methods according to the outdoor air temperature and solar radiation, and the single control of the heat transfer coefficient in each zone as the design basis for building envelope (Fig. 2.12).

The climate-zoning method of North America adopts the temperature and solar radiation as the primary zoning indicators, and relative humidity and rainfall as the secondary indicators, ensuring the thermal meteorological parameters as the dominant indexes and meanwhile considering the hygric parameters also as the design basis, which has reference significance for the building climate-zoning method in other countries.

2.3.2.2 Impact of Material Parameters on the Simulation Results

(1) Liquid water

The simulation results ratio between the HAM_0 model group (without driving rain) and the HAM model group (with driving rain) is chosen as indicator to characterize the impact of liquid water on the annual moisture content of the exterior walls. The south wall is taken as an example, the infill layer of the L-type and M-type construction groups, and the masonry layer of the H-type construction group are analyzed. The simulation results ratio of HAM_0/HAM model groups reaches a minimum value of 92.84, 31.93, and 5.86%, respectively, which shows the non-negligible effect of liquid water on the simulation results. The commonly used Glaser evaluation method does not take into account the effects of rainfall factors, which can cause the construction moisture content to be underestimated to varying degrees (Fig. 2.13).

The comparison between the HAM_0 and HAM model groups shows significant influence of the liquid water to the simulation results of HVAC demand, indoor hygrothermal environment and building component hygrothermal performance. For the L-type construction, due to the large moisture transport resistance of the exterior board, the influence of liquid water on the building component and the enclosed space is effectively blocked. At this time, the simulation results ratio of each indicator between the HAM_0 and HAM model groups is in the range 97.68–100.88%. For H-type construction, on the contrary, since that the core layer (masonry layer) is in direct contact with the external rainfall, it is most affected by the driving rain. Among the simulation results ratio of each indicator between the HAM_0 and HAM model groups, the HVAC's demand-related values are correspondingly P: 78.07–100.00%, $P_{\text{heating.max}}$: 78.77–100.00%, $P_{\text{cooling.max}}$: 95.02–125.95%; H: 94.94–103.51%, $H_{\text{humid.max}}$: 99.90–112.65%, and $H_{\text{dehumid.max}}$: 51.91–106.83%. For M-type construction, the impact of liquid water is between that of the L-type and H-type construction groups, and the ratio of the corresponding indicators is in the range 59.78–108.19% (Fig. 2.14).

Among the simulation results ratio between the HAM_0 and HAM model groups, the indoor hygrothermal-environment-related values for H-type construction are correspondingly $T_{\text{i.mean}}$: 100.00–104.92%, $T_{\text{i.amp}}$: 96.43–102.75%, $RH_{\text{i.mean}}$: 94.01–100.00%, and $RH_{\text{i.amp}}$: 91.61–106.57%; and the exterior walls' hygrothermal-performance-related values are correspondingly H_{flow}: 94.95–100.07% and M_{flow}:

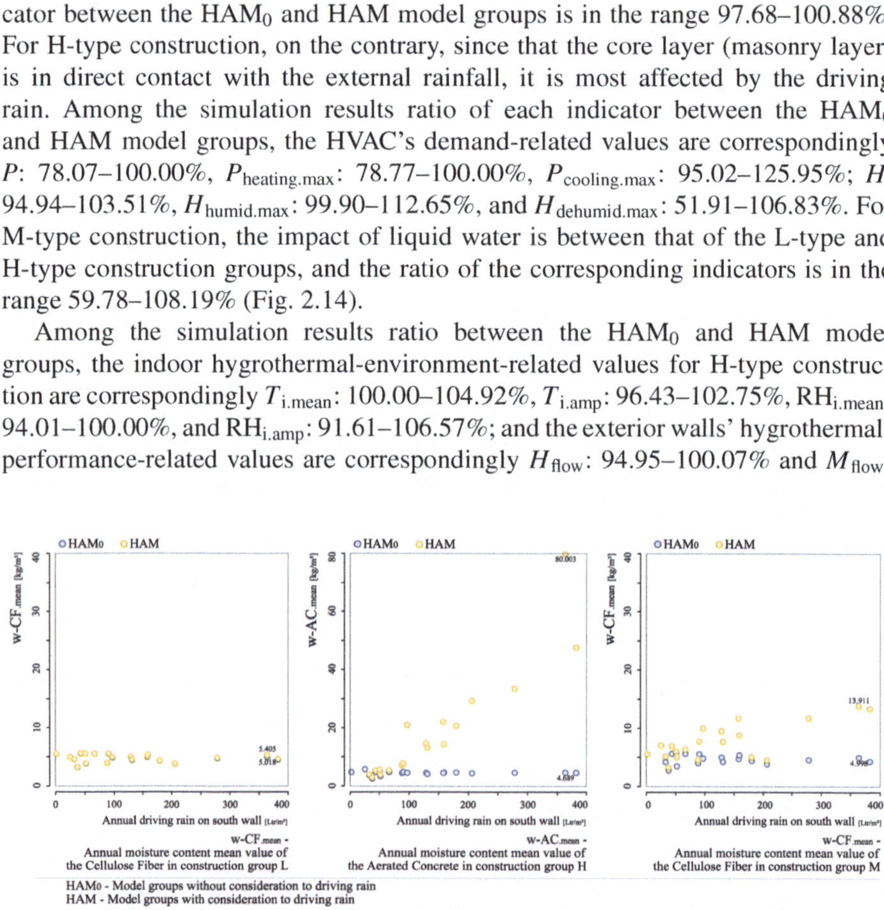

Fig. 2.13 Simulation results comparison between HAM_0 model group (without driving rain) and the HAM model group (with driving rain)—exterior walls main layer annual mean moisture content

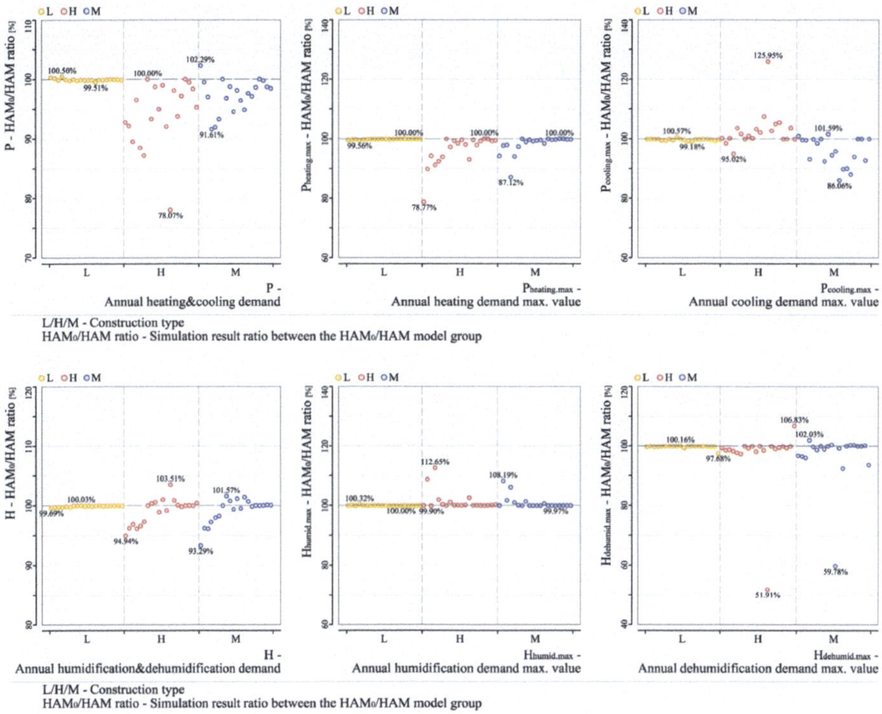

Fig. 2.14 Simulation results ratio of the HAM_0 model group (without driving rain) to HAM model group (with driving rain)—HVAC demand

92.08–100.00%. The impact of liquid water for M-type construction is between that of the L-type and H-type construction groups, and the ratio of the corresponding indicators is in the range 93.25–103.38% (Fig. 2.15).

(2) The moisture-content-dependent water vapor diffusion resistance factor

The simulation results of the annual humidification and dehumidification demand H and its peak values $H_{\text{humid.max}}$ and $H_{\text{dehumid.max}}$, annual relative humidity mean value $RH_{\text{i.mean}}$ and the amplitude $RH_{\text{i.amp}}$, as well as the moisture flow through the exterior walls M_{flow} are compared between the Glaser and HAM_0 model groups. The simulation results ratio of the Glaser model to the HAM_0 model groups show the impact of using constant value or moisture-content-dependent value of the water vapor diffusion resistance factor on the simulation results of the moisture process. Setting the HAM_0 model groups as 100%, the H, $H_{\text{humid.max}}$, $H_{\text{dehumid.max}}$, $RH_{\text{i.mean}}$, $RH_{\text{i.amp}}$, and M_{flow} of the Glaser groups are, respectively, in the range 59.79–121.91, 24.61–199.37, 0.00–115.78, 92.78–100.45, 76.86–115.70 and 81.72–143.71% (Fig. 2.16).

The comparison among the three Glaser groups is carried out, showing that the M_{flow} increases with the reducing of the water vapor diffusion resistance factor caused by the increase of moisture content. The simulation results ratio of M_{flow}

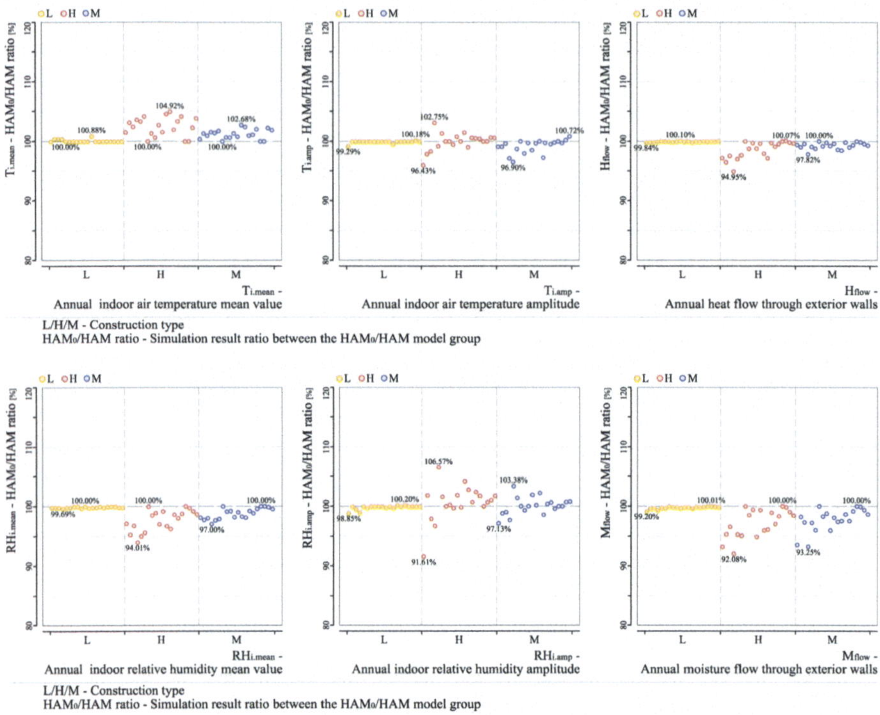

Fig. 2.15 Simulation results ratio of the HAM_0 model group (without driving rain) to HAM model group (with driving rain)—Indoor hygrothermal environment, the heat and moisture flow through exterior walls

between $Glaser_{75}$ model and $Glaser_0$ model groups is in the range 97.50–139.58%. As described above, the $RH_{i.mean}$ depends mainly on the outdoor air relative humidity, so that it is not significantly affected by the moisture transport performance of the building envelope itself. The simulation results ratio of $RH_{i.mean}$ between the $Glaser_{50/75}$ model and $Glaser_0$ model groups is 94.56–100.60%. The H value is affected not only by the moisture transport performance of the building envelope, but also by the comprehensive effects of the indoor moisture load and the direction of the moisture flow. Therefore, the impact of using constant value or moisture-content-dependent value of the water vapor diffusion resistance factor on the simulation results of H is obvious on the one hand, but shows uncertainty in the trend of deviation on the other hand.

(3) The moisture-content-dependent thermal conductivity

The simulation results of the annual heating and cooling demand P and its peak values $P_{heating.max}$ and $P_{cooling.max}$, annual indoor air temperature mean value $T_{i.mean}$ and the amplitude $T_{i.amp}$, as well as the heat flow through the exterior walls H_{flow} are compared between the Glaser and HAM_0 model groups. The simulation results ratio of the Glaser model to the HAM_0 model groups shows the impact of using constant

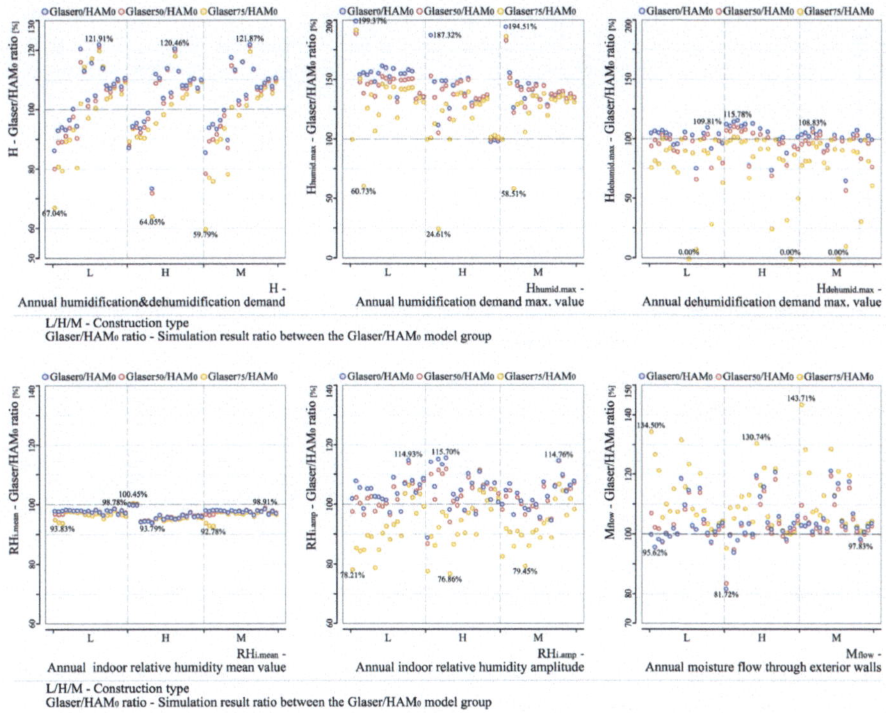

Fig. 2.16 Simulation results ratio of the Glaser model groups to HAM_0 model group—moisture-process-related indicators

value or moisture-content-dependent value of the thermal conductivity on the simulation results of the heat process. Setting the HAM_0 model groups as 100%, the P, $P_{\text{heating.max}}$, $P_{\text{cooling.max}}$, $T_{\text{i.mean}}$, $T_{\text{i.amp}}$, and H_{flow} of the Glaser groups are, respectively, in the range 78.10–116.35, 44.02–106.75, 52.73–146.66, 100.00–162.50, 78.52–102.06, and 97.06–117.38%. The deviation is different among the three types of construction, and it is the largest in H-type construction, except the $P_{\text{cooling.max}}$ value (Fig. 2.17).

The comparison among the three Glaser groups is carried out, showing that the P and H_{flow} increase with the enlarging of the thermal conductivity caused by the increase of moisture content. It is worth mentioning that, as the main indicator in a conventional building thermal calculation, the P value simulation results of the HAM_0 model groups distribute largely in the interval between the Glaser_{50} and Glaser_{75} groups, except certain cases of the H-type construction. Therefore, it can be inferred that the simplified calculation with the thermal conductivity corresponding to the equilibrium parameters in indoor and outdoor average air relative humidity would have reliable reference value for the estimation of the P value.

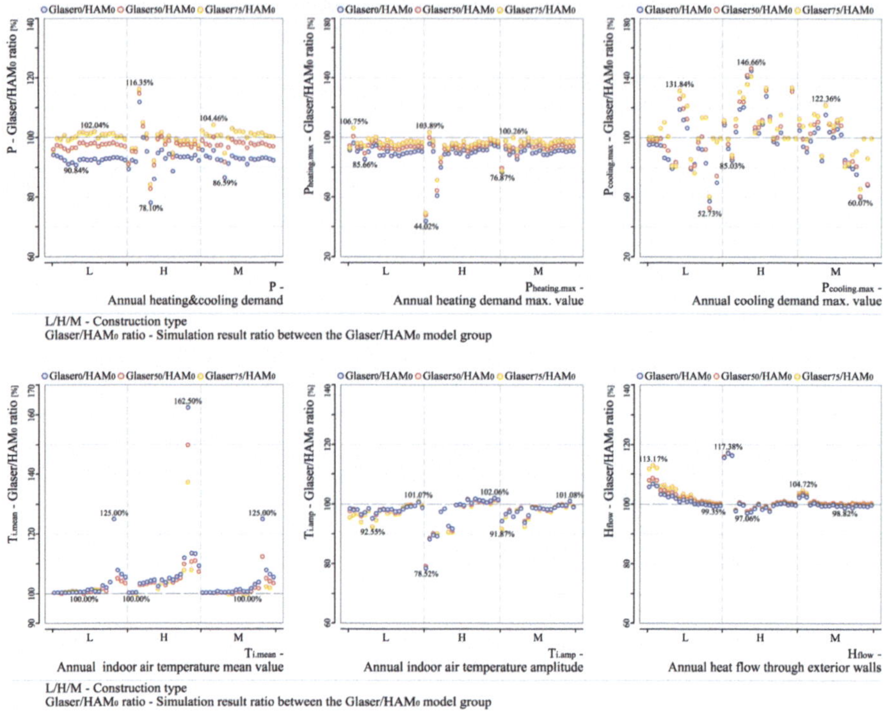

Fig. 2.17 Simulation results ratio of the Glaser model groups to HAM_0 model group—heat-process-related indicators

Compared with the bias caused by ignoring liquid water process, the deviation caused by the constant valuing of the thermal conductivity and water vapor diffusion resistance factor shows substantially smaller in H-type construction. However, the corresponding deviation could not be ignored in L-type and M-type constructions.

2.3.2.3 Difference Analysis Between Annual Dynamic Simulation and the Steady-State Evaluation

The Glaser model adopts static method for hygric performance assessment of building envelope, which assumes that there is a constant water vapor pressure difference between the inner and outer surfaces of the building envelope to drive a stable moisture flow through the construction. In the simulation of HAM model, affected by the external climate condition, indoor heat and moisture load and HVAC conditions, the annual dynamic simulation shows different laws with the results from the static assessment. The indicators from the static evaluation at building component level can not directly and accurately describe the annual HVAC demand, indoor hygric environment and the building component hygric performance. Compared with the heat process, the correlations between the hygric process simulation results and the external, internal, boundary conditions are weaker, so that the hygric performance is more difficult to predict. Therefore, the long-period, dynamic assessment on building envelope with practical conditions is of necessity.

2.4 Summary

Compared with Glaser model, the HAM model can better describe the moisture process in building envelope and its impact on the heat process for hygroscopic materials such as bamboo. The heat–air–moisture transfer model (HAM model) based on Künzel equations is analyzed and compared with the widely used Glaser model. A total of 20 representative cities from the North America typical climate zones are selected as external conditions, and 15 groups of exterior walls are constructed in WUFI Plus and simulated for annual hygrothermal performance of the exterior walls, indoor hygrothermal environment, and HVAC demand of the enclosed space units. The impact of meteorological data and material parameters on the coupled heat and moisture process simulation results are investigated, which could support the determination of theoretical model for the study on bamboo building envelope, and give reference of research methods to similar hygroscopic construction systems.

(1) Correlation analyses are carried out between the simulation results and the climate parameters, which show that the external relative humidity and driving rain affect the related results significantly; the simulation without consideration to driving rain results in significant underestimation of moisture content, heat and moisture flow, and evidently error of HVAC demand; for the constructions with core layer exposed, the influence of liquid water is far greater than the gaseous moisture.
(2) The assignment of constant value instead of moisture-content-dependent thermal conductivity leads to error of heat flow through building component, indoor thermal environment, and heating and cooling demand; on the other hand, the valuation of constant water vapor transfer coefficient, compared with moisture-content-dependent values, causes, respectively, errors to the moisture flow, hygric environment and humidification and dehumidification demand; normally, the deviation in moisture process shows greater uncertainty.
(3) The Glaser model is still widely used for simplified steady assessment on hygric performance of certain constructions. However, affected by factors such as climate conditions and construction types, the steady-state evaluation cannot accurately describe the dynamic performance of the building envelope, so that a long period dynamic assessment on building envelope with practical conditions is of necessity.
(4) The studies of building envelope have a strong bias towards heat process, while the moisture process is simply evaluated by Glaser model; the hygric environmental parameters such as driving rain are not available, and the existing

parameters for most building materials can not support the full HAM model in many cases; therefore, the promotion of the coupled heat and moisture process model requires us to carry out long-term and extensive accumulation of climate and material data.

References

1. Künzel HM (1995) Simultaneous heat and moisture transport in building components. Fraunhofer IRB Verlag, Stuttgart
2. Sedlbauer K (2001) Vorhersage von Schimmelpilzbildung auf und in Bauteilen. Dissertation, Universität Stuttgart
3. Krus M, Seidler M, Sedlbauer K (2010) Comparative evaluation of the predictions of two established mould growth models. In: Buildings XI Conference, pp 1–9
4. Urlaub S, Grün G (2016) Mould and dampness in European homes and their impact on health. Fraunhofer IBP, Holzkirchen
5. Pazold M, Antretter F, Radon J (2014) HVAC Models coupled with hygrothermal building simulation software. In: 10th Nordic Symposium on Building Physics, Lund University, pp 854–863
6. Pazold M, Antretter F, Radon J (2014) Anbindung von detaillierten Anlagetechnik an hygrothermische Gebäudesimulation. In: The 5th German-Austrian IBPSA Conference—BauSim, RWTH Aachen University, pp 263–271
7. Pazold M, Antretter F, Hermes M (2014) Coupling hygrothermal whole building simulation and air-flow modelling to determine strategies for optimized natural ventilation. In: 35th AIVC Conference, pp 537–546
8. Ge H, Baba F (2015) Dynamic effect of thermal bridges on the energy performance of a low-rise residential building. Energy Build 105:106–118
9. Kölsch P (2015) Hygrothermal simulation of ventilated pitched roofs with effective transfer parameters. Fraunhofer IBP, Holzkirchen
10. Winkler M, Nore K, Antretter F (2014) Impact of the moisture buffering effect of wooden materials on energy demand and comfort conditions. In: 10th Nordic Symposium on Building Physics, Lund University, pp 483–492
11. Künzel HM (2014) Accounting for unintended moisture sources in hygrothermal building analysis. In: 10th Nordic Symposium on Building Physics, Lund University, pp 947–952
12. Schöner T, Zirkelbach D (2016) Development of hygrothermal reference years. In: Proceedings of the CESBP Central European Symposium on Building Physics and BauSIM, pp 133–140
13. Karagiozis A, Hadjisophocleous G, Cao S (1997) Wind-driven rain distributions on two buildings. J Wind Eng Ind Aerodyn 67:559–572
14. Draft (2006) ASHRAE Standard 160P: Design criteria for moisture control in buildings
15. Krus M (1995) Moisture transport and storage coefficients of porous mineral building materials—theoretical principles and new test methods. Dissertation, Universität Stuttgart, English version published in 1996

Chapter 3
Hygrothermal Properties' Test on Bamboo

In this chapter, the material parameters for a HAM model are firstly analyzed, and then, the typical bamboo products are systematically tested according to the corresponding international specifications to obtain the targeted material parameters. The test objects include raw bamboo, bamboo-based panels, natural bamboo fiber, and bamboo charcoal, of which the former two are classified as partition boards, and the latter two as construction infills, according to their role in a layered construction. The tests on the partition boards and construction infill are based on the same specifications, but with different operation methods. When the test results are obtained, correlation analyses are carried out on the material parameters; meanwhile, the relationship of the heat and moisture transport properties with the hygric environment and the moisture content is fitted, and the diversified products produced by the industrial utilization of bamboo resources are evaluated (Fig. 3.1).

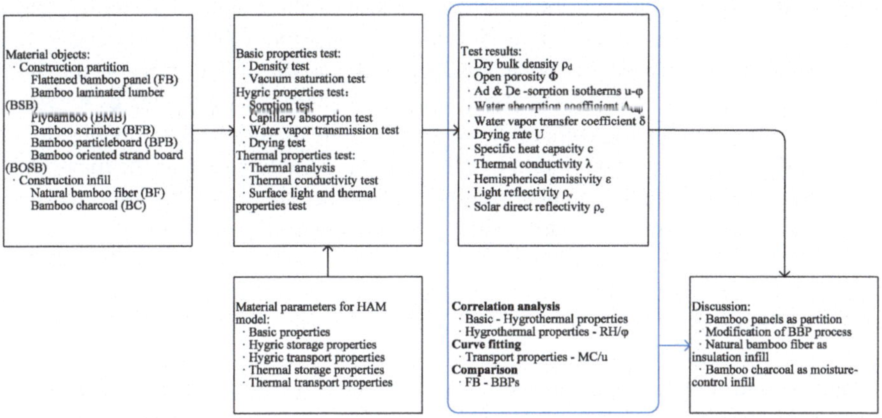

Fig. 3.1 Framework of Chap. 3

Z. Huang, *Application of Bamboo in Building Envelope*,
Green Energy and Technology, https://doi.org/10.1007/978-3-030-12032-0_3

3.1 Building Envelope-Related Properties and the Definitions

HAM model usually uses highly coupled nonlinear partial differential equations to simultaneously describe the heat and moisture process in building envelope and then solve it by numerical methods [1]. Different scholars have different views on the mechanism, cause and driving potential of the heat and moisture process and form their own description equations, which require various material parameters and climatic conditions as inputs.

Fraunhofer IBP classifies the mechanisms, formation causes, and driving potentials of heat and moisture process in building envelopes into three types, namely the heat, gaseous moisture, and liquid water. The driving potential and its selection of the corresponding material parameters are the basis for describing the heat and moisture process. Temperature and vapor pressure are commonly used as driving potentials for the heat and gaseous moisture process, which is well recognized. In contrast, for the liquid water process, mainly capillary conduction and surface diffusion, there is controversy about the selection of driving potential, which results in the situation that different HAM models have no exactly the same requirements for material parameters (Table 3.1).

3.1.1 Basic Properties

3.1.1.1 Bulk Density

Apparent density/bulk density is one of the most important physical properties of a material. In certain industry fields such as the wood-based panel technology, bulk density is often used as one of the performance indicators of the product due to the strong correlation between the mechanical strength and the bulk density. In the field of building physics, in addition to the correlation of various significances between the bulk density and the hygrothermal properties of the materials, the bulk density is also directly used for the calculation of some other parameters such as the mass–volume ratio moisture content and heat storage coefficient. Therefore, the hygrothermal properties can be predicted by this easily available material parameter.

3.1.1.2 Porosity

Compared with the skeleton structure, the pore structure is another general index in the discipline of material science, which describes the pore characteristics of the material in a microscopic scale by the pore size, quantity, specific surface area, etc. The interior pores of the material form multiple cavities that reduce the heat transport rate. The cavities are also locations where moisture is stored and therefore have a

Table 3.1 List of heat and moisture transport mechanisms occurring in building envelope, their causes, and driving potentials

Transport mechanism		Cause and potential of transport
Heat transport	• Heat conduction	• Temperature
	• Heat radiation	• Temperature in 4th power
	• Air flow	• Total pressure, density differentials
	• Enthalpy flows through moisture movement	• Vapor diffusion with phase change and liquid transport flows in temperature field
Vapor transport	• Gas diffusion	• Vapor pressure (temperature, total pressure)
	• Molecular transport (effusion)	• Vapor pressure
	• Solution diffusion	• Vapor pressure
	• Convection	• Total pressure gradient
Liquid transport	• Capillary conduction	• Capillary suction stress
	• Surface diffusion	• Relative humidity
	• Seepage flow	• Gravitation
	• Hydraulic flow	• Total pressure differentials
	• Electrokinesis	• Electrical fields
	• Osmosis	• Ion concentration

Source Künzel. Simultaneous heat and moisture transport in building components [R]. Fraunhofer IRB Verlag Suttgart, 1995
Note Fraunhofer, full name Fraunhofer-Gesellschaft, is the largest application-oriented research organization in Germany and Europe as well; IBP—Institut für Bauphysik (Institute for Building Physics) is one of the 67 research institutes of Fraunhofer, which is mainly responsible for research, development, testing, demonstration, and consulting in the fields of building physics. Its areas of expertise include the acoustics, building chemistry, building biology and hygiene, hygrothermics, life-cycle engineering, energy efficiency, indoor climate, etc.

direct impact on the moisture storage properties. The smaller pores form a capillary system that affects the capillary suction stress, while the larger pores are channels for the transport of gaseous moisture. Therefore, the pore structure also has an important influence on the transport capacity of the liquid and gaseous moisture.

Inferring the properties of a certain material from its microscopic pore structures is one of the generally accepted methods in the discipline of material science. However, this will show inapplicability to building materials with larger scales and stronger inhomogeneities, which will be cited in detail below with the results and analyses of the test. This study describes the interior porosity of a building material by ‘open porosity’ from the perspective of macroscopic phenomenology.

3.1.2 *Hygric Properties*

The state of a building material with no moisture or only bound water is called dry. When exposed to the wet air, the hygroscopic material draws moisture molecules onto the inner surface of the material pores until it is in equilibrium with the relative humidity of the ambient air, while the non-hygroscopic material remains dry. When the material is exposed to liquid water, the capillary-active material draws liquid water through capillary effect, while the hydrophobic material does not. The capillary-active material absorbs water until it reaches its saturation, which is a state known as free water saturation or capillary saturation. Higher moisture content until pore saturation or maximum water saturation can only be achieved by applying pressure or by slow diffusion of water vapor [2].

3.1.2.1 Moisture Storage

The moisture contained inside a building material can be a mixture of gaseous moisture, liquid water, or even solid ice. Generally, it is difficult to measure the moisture of each phase separately, and they can be converted into each other under certain conditions, so they are usually characterized by the overall moisture content. The relevant physical quantities are the mass–volume ratio moisture content [kg/m^3] and the mass ratio moisture content u [%]. In order to associate the moisture content of the building material to the relevant environmental parameters, Künzel divides it into three regions [2]:

(1) Sorption moisture/hygroscopic region

This region is an equilibrium moisture content from the dry state up to the ambient air relative humidity of about 95%, generally covering all the gaseous moisture regions, which are commonly described as the sorption isotherms in building physics.

The equilibrium moisture content of the hygroscopic region is affected by the ambient air temperature and relative humidity, but in the general temperature range of the built environment, the measurement carried by Kast and Jokisch at 20–70 °C, and Künzel at 5 and 15 °C indicated that the influence of temperature can be ignored [3, 4], and therefore, the hygroscopic equilibrium moisture content of the building materials is also called the moisture adsorption isotherm.

(2) Capillary water region, also known as the superhygroscopic region

This area follows the hygroscopic region until the capillary saturation, and the larger pores inside the material are filled with moisture until it reaches the state of equilibrium with the contacted liquid water.

For the capillary model, the suction stress is inversely proportional to the capillary radius. Therefore, it is generally considered that under an ideal condition, the small-sized pores inside the building material will absorb moisture from the large-sized pores until a balanced state that a certain size of pore is filled fully with moisture.

However, through observation on the microscopic morphology of some porous concrete samples, Künzel realizes that it is difficult to use the capillary model to describe the pore system in different display scales. Therefore, it is more feasible to estimate the suction stress directly instead of establishing a capillary model for pore size. The Kelvin formula describes the relationship between the relative humidity and capillary stress [5]. Based on this, as well as further test and calculation, Künzel extends the use of relative humidity as the driving potential from the hygroscopic region to the capillary water region. The method for measuring the moisture content in the capillary water area is the centrifugal test [6] and the pressure plate test [7]. Künzel's test on lime silica bricks shows that for pores with a radius of $r \geq 0.1$ mm, it would usually not be filled up by capillary suction, and for the moisture adsorption of large pores, gravity and wind pressure play a greater role than the capillary suction stress.

(3) Supersaturated region

This area is from the state of saturation with free water until all the cavities inside the material are filled with moisture, which cannot be achieved by a common suction process and is normally achieved in the laboratory by the use of pressure suction. The relative humidity in this area is usually 100% or higher, which is classified as a transient process, and there is no such stable moisture equilibrium state under natural conditions.

3.1.2.2 Moisture Transport

The building physical calculations-related moisture transport process is primarily the water vapor diffusion and the liquid water transport caused by capillary stress. Generally, water vapor diffusion occurs in larger pores of the material, while liquid water transport occurs on the smaller pores or pore walls. There is variable relationship between water vapor diffusion and liquid water transport in building components, which can result in superposition or reduction effects when they are in the same or opposite transmission directions, respectively.

Assuming that a wall is placed at the boundary between an indoor and an outdoor environment, and the outdoor side has an average relative humidity higher than the indoor, while to the contrary, the indoor vapor pressure is higher than the outdoor. When the wall is in a dry state, the moisture transport in the capillary of the material can be considered as a diffusion of pure gaseous moisture from indoor to outdoor; when the moisture content of the wall rises, the inner wall of the capillary would form a water film, of which the thickness decreases from the outside to the inside because the outdoor relative humidity is higher than the indoor; when the moisture content of the wall continues to rise, the greater the thickness of the water film, the more movable the water molecules can be, and the moving direction is from the outside to the inside. Surface diffusion is a liquid transport, and the driving potential is the relative humidity or suction stress [2].

(1) Water vapor diffusion

The diffusion due to the difference in mass fraction is the 'Fick diffusion,' and the diffusion caused by the temperature gradient is the 'Soret effect' that is usually negligible for building components. The water vapor diffusion is comparable to the diffusion in the air only when it occurs in the macropores of the porous material, and when the pores are too small and cause frequent collisions between molecules, it is called 'Knudsen transport.' Fick diffusion dominates in pores with radius larger than 10^{-6} m, and Knudsen transport dominates in pores with radius smaller than 10^{-9} m. There is a mixed transport when the pore size is between the two. As far as building physics is concerned, it is sufficient to simplify the description with a water vapor diffusion resistance factor [2].

(2) Liquid water transport

Liquid water transport involves surface diffusion and capillary conduction. The former is the moisture transport of the water molecules layer on the pore walls of hygroscopic material and microcapillary. The latter occurs when the adhesion between the moisture and the capillary wall is greater than the cohesion within the water itself, and the water is drawn into the porous material.

Through test and calculation, Künzel believes that it is feasible to calculate the surface diffusion and the capillary conduction independently from the water vapor diffusion [2]. Similar to the uniform use of vapor diffusion resistance factor to describe the Fick diffusion and Knudsen transport of the water vapor diffusion, the surface diffusion and capillary conduction also occur simultaneously in the liquid water transport process, so they can also be measured for the overall value by experiment.

The water absorption coefficient and liquid water transport coefficient are important material parameters for describing the liquid water transport. The water absorption coefficient A_{cap} [kg/(m^2 s$^{0.5}$)] is a standard parameter for characterizing the capillary suction characteristics of building materials in contact with water, and typically 20 °C in a laboratory environment is set as a temperature reference. The capillary transport coefficient D_w [m^2/s] is significantly affected by the moisture content. The redistribution process of the D_w value is usually 1–2 orders of magnitude smaller than the suction process. Currently, there is no unified approach to calculate the D_w value.

3.1.3 Thermal Properties

3.1.3.1 Heat Storage

The heat content of building materials under certain conditions is called enthalpy H_s [J/m^3], and in the temperature range of building physics, there is approximately a linear relationship between the material enthalpy and the temperature. The enthalpy of a dry building material is the product of the bulk density, specific heat capacity, and temperature. Therefore, in addition to the bulk density, the material parameter related to building heat storage is the specific heat capacity c [J/(kg K)].

For wet building materials, the enthalpy of the moisture content needs to be considered. The enthalpy of the moisture is affected by its content, physical state, and the latent heat generated during the phase change. The enthalpy of the wet material can be calculated by weighting the enthalpy of the dry material and the enthalpy of the moisture content inside.

3.1.3.2 Heat Transport

The thermal conductivity of wet building materials can be described by a moisture content-related thermal conductivity supplement a_w [(W/m K)/u(−)] value, which refers to the increase in thermal conductivity corresponding to the increase of moisture content per mass ratio. For hygroscopic materials, this value is influenced by the skeleton of the building material and its affinity with water [8].

In addition, moisture evaporation and condensation in wet building materials also affect the heat transport, but this cannot be described with heat transfer equations. The test by Künzel shows that the heat flow caused by the liquid transport is negligible compared to other heat flows, but the water vapor flow resulted from the phase change, such as the rainwater evaporation and drying, has an important effect on the heat balance [2].

3.1.4 Material Parameters Targeted in the Tests

To support the complete HAM model, this chapter obtains the storage and transport properties describing the heat and moisture processes of the material. The bulk density, porosity, surface morphology, etc., belong to the category of basic physical properties, but considering the close relationship between them and the physical properties of heat and moisture, they are also included in the following discussion (Table 3.2).

In addition, the relevant parameters are a water retention curve describing the storage property of liquid water, and a liquid water permeability coefficient describing the transport property of liquid water, of which the latter can be estimated from the former value. The measurement of water retention curve needs a pressure plate test, which cannot be carried out due to the limitations of the test condition in China on the one hand. On the other hand, it is in the capillary water area, beyond the range of moisture hygroscopic region which is commonly used, so that most of the materials in the Fraunhofer IBP have not been tested for this curve either.

Table 3.2 List of thermal and hygric properties targeted in the test items

Catalog	Describing the storage ability	Describing the transport ability	Describing both the storage and the transport abilities
Thermal properties	• Specific heat capacity	• Thermal conductivity • Surface emissivity/reflectance/absorption rate	• Thermal effusivity
Hygric properties—gaseous moisture	• Isothermal adsorption and desorption curve	• Water vapor transfer coefficient • Drying rate	
Hygric properties—liquid water	• Capillary saturation moisture content • Vacuum saturation moisture content	• Water absorption coefficient	• Liquid water transport coefficient

3.2 Test on Construction Partition Material

3.2.1 Test Objects

3.2.1.1 Material Classification Rule in This Study

People's utility of bamboo has experienced a process from raw bamboo to modified bamboo. In raw bamboo period, bamboo was commonly classified from the botanical science, e.g., Phyllostachys pubescen, Guadua angustifolia, and Dendrocalamus latiflorus. After simple manual processing, they were classified to bamboo pole, bamboo strip, and bamboo sliver, which could be further operated to bamboo weaving/net/curtain.

With reference to the glulam technology, the first bamboo-based panel was produced during the World War II. Thereafter, the modified bamboo industry had experienced a slow development period for 30–40 years. Since 1980s, Chinese scholars like Zhang have commenced with the industrialization of bamboo, mainly forcing on the bamboo-based panels, which has achieved breakthrough in the adoption of wood-based panel technology into bamboo industry. Variety of bamboo-based panels emerge, whether in exploratory production in laboratories or in market promotion. In this period, Zhang promotes several rules to the classification of modified bamboo products [9]:

Classification in respect of manufacturing technology, e.g. products made of bamboo strips, products made of bamboo chips, and products of composite materials;

Classification in respect of product structure, e.g. plybamboo products, laminated products, chipboard products and composite board products;

Classification in respect of uses, e.g. plybamboo used in vehicle making, plybamboo used in making concrete form, plybamboo used in boats, bamboo flooring, plybamboo used for making furniture and other articles.

At present, China takes the lead in terms of the production and industrialization of bamboo-based panels, which are the main products of modified bamboo. However, in contrast, the research on the technical standards of the bamboo industry is in quite backward status, and the modified bamboo even lacks uniform classification and terminology.

The classification by manufacturing process is adopted in the study, and the following factors are considered in the selection of the classification rule:

(1) The association with material technology. Variety of the modified bamboos result from the introduction of glulam manufacturing process into bamboo industry; therefore, to classify and understand the materials in manufacturing process is logical. Many steps in manufacturing process can affect or even determine the final quality of the materials; therefore, in application practice, this knowledge is necessary to match and explore products of suitable quality.
(2) The uniform of terminology. Apart from raw bamboo, there are two basic kinds of modified bamboo: homogeneous bamboo and composite bamboo. The latter is derived from the former, by combining one or several complementary materials, through auxiliary adhesive processing. All the existing and possible future material products could be uniformed with manufacturing process.

Bamboo-based panels are the main products of homogeneous bamboo. The fundamental principle of bamboo-based panels is similar to the wood panels, which mainly includes two steps: firstly to decompose the raw bamboo into basic standard units of small size, and then recombine them by means of gluing and hot pressing, to form standard panels or lumbers of larger scale and meanwhile improve utilization efficiency and strengthen the physical and chemical performance. The homogeneous bamboo can be defined into several types according to the different, respective treatments during the process of decomposition and recombination. The process mainly includes the following steps:

1. raw bamboo, 2. mechanical decomposition, 3. constituent units, 4. protective treatment, bleaching, color darkening, drying, 5. units standardization, 6. assembly, 7. gluing, 8. hot pressing, 9. post-processing (Fig. 3.2).

Common steps in all manufacturing processes include steps 1, 2, 4, 5, 7, 8, and 9, which affect the quality of the final products and do not cause essential difference. Steps 3 and 6 decide corresponding constituent units and the combination mode and finally distinguish the product types. In the long-term production practice, step 3 and step 6 have formed a relatively fixed combination.

Referring to the '*LY/T 1660-2006 Standard terminology for bamboo-based panel*' [10], typical bamboo-based panel products are classified as follows:

(1) Bamboo laminated lumber (abbreviated as BSB). The constituent unit is regular bamboo strips, which are prepared by longitudinally splitting the raw bamboo,

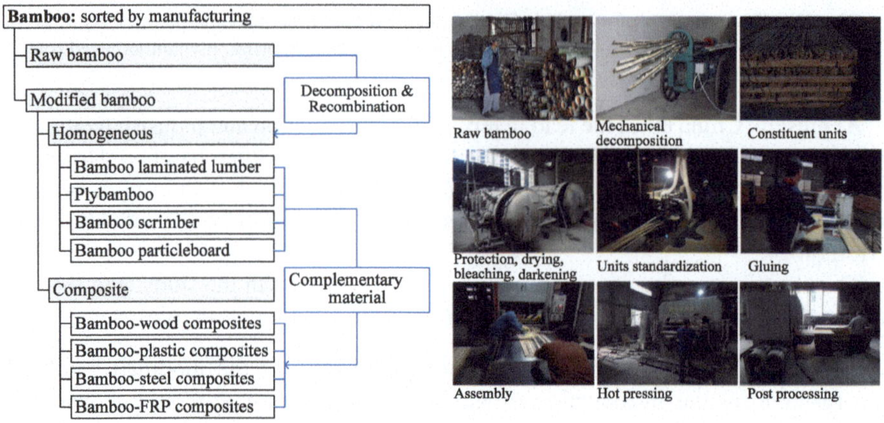

Fig. 3.2 Relation between raw bamboo and modified bamboo, between homogeneous bamboo and composite bamboo (photographs on right side are taken in Taohuajiang Industrial Co., Ltd., Hunan, China)

and then flattening or cutting to form regular units that have a thickness of 0.5–1.0 cm and a width of 1.5–2.5 cm. The regular units are assembled by flat pressing or side pressing to form the slabs, which are finally pressed in parallel or orthogonally into a plate or a square material. The bamboo laminated lumber retains the bamboo tissue and appearance features to a certain extent.

(2) Plybamboo (abbreviated as BMB). The constituent unit is bamboo sliver, or the bamboo net, bamboo mat, and bamboo curtain obtained by weaving from it. The preparation of the bamboo mat is mostly by manual operation or with simple mechanical assistance. The thickness of the bamboo sliver is 0.8–2.0 mm, and the width is 10–30 mm; the length depends on the requirement of the final product. The gluing needs to be carried out on both sides of the bamboo mat, so that large amount is required. Since the surface of the bamboo mat is uneven, the molding operation is the simplest, which can be performed by overlapping and pressing multiple layers of the mats. As the earliest product variety that has been developed and successfully promoted, it possesses the advantages of wide application, simple production process, and high utilization rate of raw bamboo.

(3) Bamboo scrimber (abbreviated as BFB). The constituent unit is bamboo fiber bundle. In the manufacturing process, the raw bamboo is firstly disintegrated into the loose but not completely separated fiber bundles, which are parallel to each other and keep the original direction of the fiber, and then formed into a plate or a square material by high pressure after drying, gluing, and assembling. There is basically no cutting process in the production of bamboo scrimber, and the bamboo fiber characteristics are retained, so that the original mechanical advantages of the bamboo fiber are fully and rationally utilized. The raw material for producing bamboo scrimber can be taken from a large number of those small-diameter,

herb, and mixed bamboo species that have not been reasonably utilized, and meanwhile the utilization rate of raw materials is effectively improved.

(4) Bamboo particleboard (abbreviated as BPB). The constituent unit is the fine bamboo shaving. According to the geometry of the bamboo shavings and the assembling method, it can be divided into the particleboard made of raw bamboo shaving and the particleboard made of bamboo filament. The latter mainly uses the residue produced in the processing of bamboo mats, bamboo toothpicks, bamboo chopsticks, and the like as raw materials. During the processing, the residue is separated into needle-shaped bamboo rods, and due to its large slenderness ratio, the thin filamentous scraps, the uniform cross section of the sheet, and the high surface quality, it can be made into thin products. Due to the wide source of raw materials, bamboo particleboard is one of the ideal ways to improve the comprehensive utilization rate of bamboo resources.

(5) Bamboo oriented strand board (abbreviated as BOSB), which is derived from bamboo particleboard technology, and the constituent unit is the larger piece of bamboo shavings, which are obtained by radial chipping of the raw bamboo, and the unit size is length × width × height = (50–100) × (3–40) × (0.3–0.8) mm. Due to the large ratio of the longitudinal strength to the lateral strength of the bamboo fiber, the oriented assembling of bamboo shavings can greatly improve the mechanical strength of the product in a single direction (Fig. 3.3).

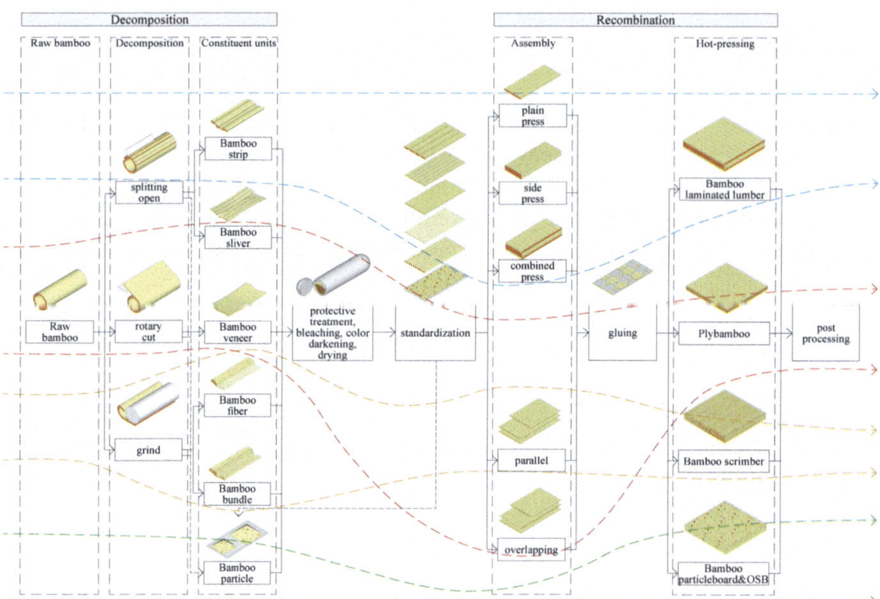

Fig. 3.3 'Decomposition–recombination' bamboo-based panel process and the corresponding products (in which bamboo OSB is derived from bamboo particleboard)

Based on the investigation of literature, market and material producers in bamboo industry, six typical bamboo products are chosen as the test objects, which are distinguished and named according to the terminology standard [10]. Flattened bamboo panel (FB) is composed of flattened bamboo culm wall; bamboo laminated lumber (BSB) is composed of bamboo strips; plybamboo (BMB) is composed of bamboo mats; bamboo scrimber (BFB) is composed of bamboo fiber bundle; bamboo particleboard (BPB) is composed of fine bamboo particles; and bamboo oriented strand board (BOSB) is composed of oriented coarse particles. Among them, the FB is flattened from round bamboo wall without gluing, which could represent the raw bamboo, while the others are bamboo-based panels (BBP) (Tables 3.3, 3.4).

3.2.1.2 Raw Bamboo (FB)

See Table 3.3.

3.2.1.3 Bamboo-Based Panels (BBP)

See Table 3.4.

3.2.2 *Test Items*

In practice, the width and height of a panel would always be much larger than its thickness. Therefore, it is reasonable to assume that the transfer of heat and moisture mainly occurs along the direction of the panels' thickness, and this is the only direction that is taken into consideration in this study. Also, the studies have shown that errors in the repeatability of these experiments are usually much smaller

Table 3.3 Sample information of flattened bamboo panel

FB (Flattened bamboo panel)	Constituent unit	Assembly method	Thickness (mm)	Main application	Sample sources
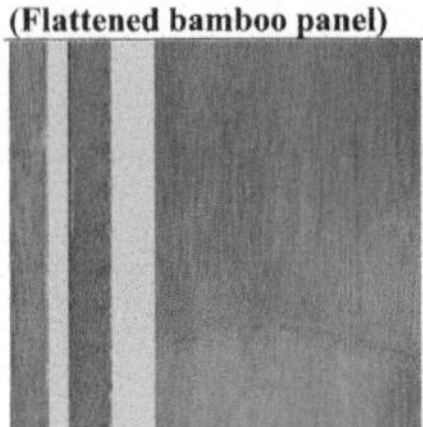	Raw bamboo wall (thickness: ca. 8 mm)	–	8.0	Indoor flooring, finishes	Dasso Industrial Group Co. Ltd. Hangzhou, China

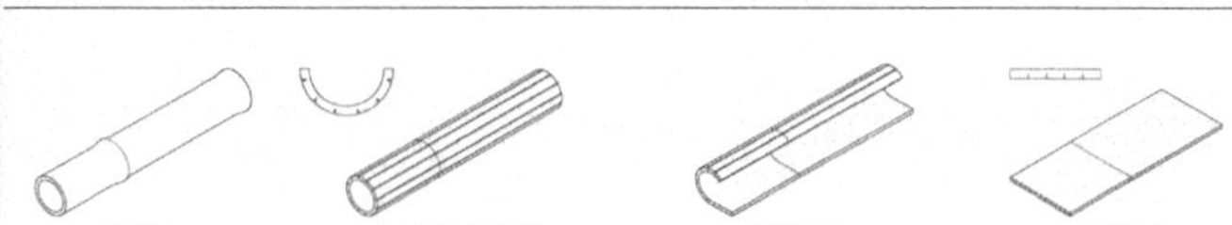

FB is flattened from round bamboo wall without gluing, which could represent the raw bamboo

Table 3.4 Sample information of the bamboo-based panels

BSB (Bamboo laminated lumber)	Constituent unit	Assembly method	Thickness (mm)	Main application	Sample sources
	Bamboo strip (width:15–20 mm, thickness: 5.0–8.0 mm)	Plain-side-plain pressed, orthogonal 3 layers	30.0	Indoor flooring, finishes, furniture	Dasso Industrial Group Co. Ltd. Hangzhou, China

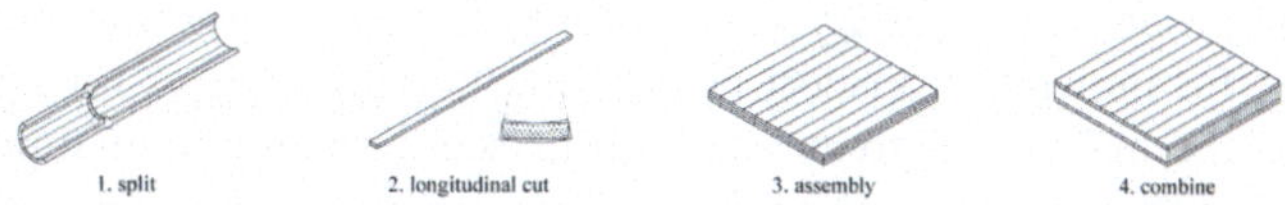

BMB (Plybamboo)	Constituent unit	Assembly method	Thickness (mm)	Main application	Sample sources
	Bamboo sliver (width:15–30 mm, thickness: 0.8–2.0 mm)	Weave sliver into mat, orthogonal 15 layers	27.8	Concrete formwork, truck and bus bottom board, load-bearing component, wallboard	Institute of Modern Bamboo, Timber and Composite Structures (IBTCS). Changsha, China

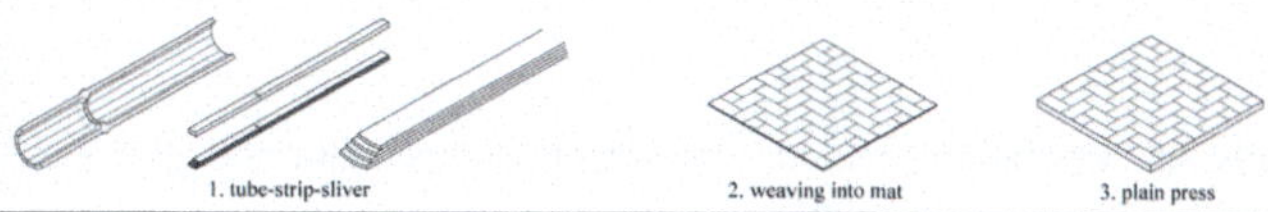

BFB (Bamboo scrimber)	Constituent unit	Assembly method	Thickness (mm)	Main application	Sample sources
	Bamboo bundle (width:10-30 mm)	Parallel	30.0	Load-bearing component, indoor/outdoor flooring, finishes, furniture	Dasso Industrial Group Co. Ltd. Hangzhou, China

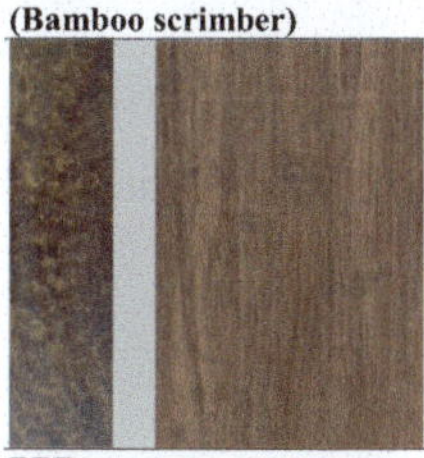

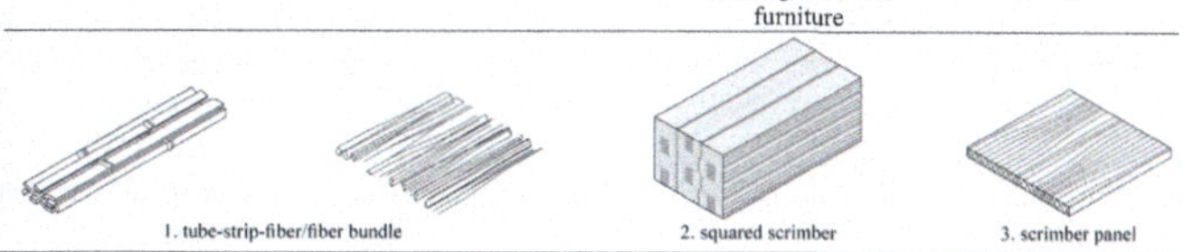

BPB (Bamboo particleboard)	Constituent unit	Assembly method	Thickness (mm)	Main application	Sample sources
	Bamboo particle (length: 20–30 mm, width: 1.0–5.0 mm, thickness: 0.1–0.5 mm)	Non-oriented, 3 layers	18.4	Furniture, packaging material	Southeast Forest Co., Ltd. Ningguo, China

BOSB (Bamboo OSB)	Constituent unit	Assembly method	Thickness (mm)	Main application	Sample sources
	Long and flat bamboo fragment (length: 50–90 mm, width: 5–20 mm, thickness: 0.3–0.7 mm)	Oriented, orthogonal 5 layers	27.6	Concrete formwork, load-bearing component, wallboard	Yunnan Yonglifa Forest Co., Ltd. Mangshi, China

Thickness specifications of bamboo panels are normally 18 and 28 mm. Using thicker samples would be conducive to the test stability. However, there are only 8 mm FB and 18.4 mm BPB available on the market than deviations induced by material inhomogeneity [11]. Hence, the tests of this study are performed in parallel on multiple test samples, instead of being repeatedly performed on a single sample.

The test items can be divided into 3 groups of 9 items, including bulk density test and vacuum saturation test for basic properties; sorption test for moisture storage properties; capillary absorption test, water vapor transmission test, and drying test for moisture transport properties; thermal analysis for heat storage properties; and thermal conductivity test and surface light and thermal properties' test for heat transport properties. The test results initially form a bamboo properties parameters' database that could support its practical application. In addition, a surface morphology analysis by scanning electron microscope is carried out, which can give reference for better analyzing the basic properties and their correlation with the hygrothermal properties (Tables 3.4, 3.5, Fig. 3.4).

Table 3.5 Operation method, equipment arrangement, and specimen treatment of the bamboo hygrothermal properties' test

Category	Item	Operation method	Equipment arrangement	Specimens treatment
Basic properties	1. Bulk density test Target value: dry bulk density	–	Drying oven: digital stainless steel electric blast oven 101A-2S, accuracy ± 1 °C Balance: SHIMADZU UX6200H, accuracy 0.01 g	Quantity: 9 copies for 1 type[a] Size: 10 × 20 cm
	2. Vacuum saturation test Target value: open porosity, vacuum saturation moisture content	Refer to the American standard ASTM D7370-2009, and the European standard DIN EN 1936-2007 [12, 13]	Drying oven: same as item 1 Balance: same as item 1 Vacuum chamber and pump: Sliding vane rotary vacuum pump Fopump-2X-8, speed 8 L/s, Extreme pressure 600 Pa	Quantity: 3 copies for 1 type Size: 10 × 10 cm
Hygric properties	3. Sorption test Target value: isothermal adsorption and desorption curve	Refer to the international standard ISO 12571:2012 and the American standard ASTM C1498-04a(2016) [14, 15]	Vacuum dryer (9 copies): inner diameter = 30 cm Balance: same as item 1 T&RH recorder: TH10R-EX, with external sensor (9 copies), accuracy T ± 0.2 °C, RH ± 2%	Quantity: 3 copies for 1 type, including 1 large sample and 2 small samples Size: large sample 10 × 20 cm[b], small samples 4 × 5 cm

(continued)

Table 3.5 (continued)

Category	Item	Operation method	Equipment arrangement	Specimens treatment
	4. Capillary absorption test Target value: water absorption coefficient; capillary saturation moisture content	Refer to the international standard ISO 15148:2002(E) [16]. With the results of water absorption coefficient and capillary saturation moisture content, liquid transport coefficient D_l [m^2/s] is then calculated [17]	Water sink: length × width × height = 40 × 30 × 10 cm; stainless steel nut supports of equal height are fixed on the bottom; distilled water is injected in the sink and keep the water level 3 mm above the supports Balance: same as item 1	Quantity: 3 copies for 1 type, seal on top and side surfaces, while keep 4 air vents of 5 × 5 mm on top surface and 6 mm gaps close to the bottom edges on side surfaces open
	5. Water vapor transmission test Target value: water vapor transfer coefficient	Refer to the international standard ISO 12572:2001(E) and the American standard ASTM E 96/E 96 M-2005 [18, 19], in which the water vapor transfer coefficient of still air δ_{air} [kg/(m s Pa)] under certain temperature T and pressure p is calculated according to the method [20]	Dry cups (3 copies), wet cups (3 copies) Balance: same as item 1 Constant T&RH curing box: HWS-250B, accuracy $T \pm 0.5$ °C, RH ± 5%	Quantity: 6 copies for 1 type Size: 10 × 10 cm
	6. Drying test Target value: drying curve; drying rate curve	Refer to the test method of Fraunhofer IBP (accredited according to DIN EN ISO/IEC 17025)	Balance: same as item 1 Constant T&RH curing box: same as item 5	Quantity: 3 copies for 1 type, seal on top and side surfaces Size: 10 × 10 cm[c]
Thermal properties	7. Thermal analysis Target value: specific heat capacity	Refer to the international standard ISO 11357-4-2005 [21]	Differential thermal scan-ning/thermogravimetric analyzer: TA4000/2910MDSC Reference material: sapphire	Quantity: 3 copies for 1 type Mass: 20 mg
	8. Thermal conductivity test Target value: thermal conductivity	Refer to the international standard ISO 8302-1991 [22]	Guarded hot plate apparatus: CD-DR3030, hot plate $T = 35$ °C, cold plate $T = 15$ °C, size of central heat transfer area 15 × 15 cm, accuracy ± 2%	Quantity: 4 copies for 1 type, including 3 large samples and 1 small sample Size: large samples 30 × 30 cm, small sample 10 × 20 cm[d]

(continued)

Table 3.5 (continued)

Category	Item	Operation method	Equipment arrangement	Specimens treatment
	9. Surface light and thermal properties' test Target value: hemispherical emissivity; light reflectivity; solar direct reflectivity; solar direct absorptivity	Refer to the European standard DIN EN 16012-2012 [23] and the Chinese standard GB/T 25968-2010 [24]	D&S digital hemisphere emissivity tester: AE1/RD1, Linear relationship ± 0.01 units Ultraviolet, visible, near infrared spectrophotometer U-4100: measurement wavelength range 200–2600 nm	Quantity: 4 copies for 1 type, including 1 large sample and 3 small samples Size: large sample 10 × 20 cm[d], small samples 10 × 10 cm

[a]Specimens after being dried in item 1 are put in the vacuum dryers for test item 3
[b]Specimens after reaching the equilibrium moisture content are carried on in test item 8 and 9
[c]Use the specimens after vacuum saturation test in test item 2
[d]Use the specimens after reaching the equilibrium moisture content in test item 3

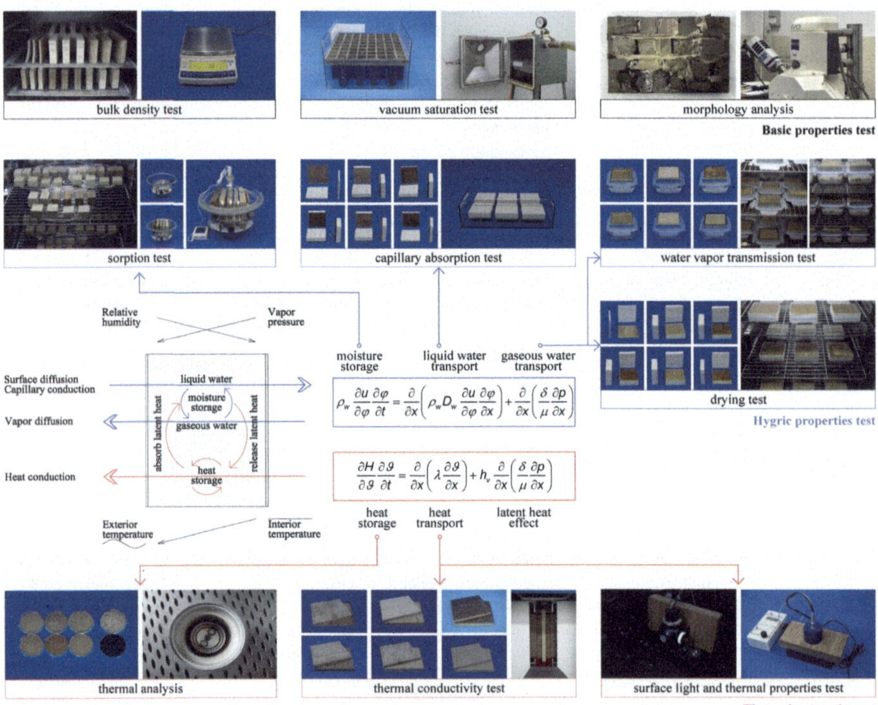

Fig. 3.4 Building envelope heat and moisture transport mechanism and the bamboo test items (the equations in middle are from Künzel)

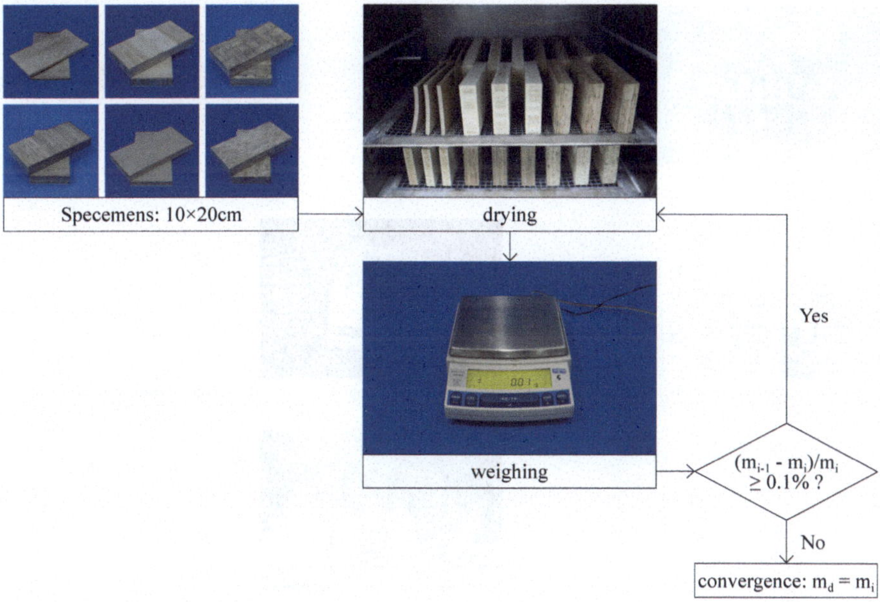

Fig. 3.5 Photograph program of the bulk density test

3.2.2.1 Bulk Density Test

Place the specimens in the drying oven, where the temperature is controlled at 105 °C. After 1 week, take the specimens out and weigh every 24 h, and then return them to the drying oven. Repeat the above operation until the difference of three consecutive weighing results does not exceed 0.1%, when the specimens are considered to have reached the dry state, and the dry mass m_d [kg] is recorded (Fig. 3.5).

According to

$$\rho_\mathrm{d} = \frac{m_\mathrm{d}}{V}$$

Calculate the bulk density ρ_d [kg/m^3] of the dried material, where V [m^3] is the volume of the specimens.

3.2.2.2 Vacuum Saturation Test

The test is performed referring to the methods recommended by the American standard ASTM D7370-2009 and the European standard DIN EN 1936–2007 [12, 13]. Stand the specimens sideways on the support at the bottom of the sink, and press them with a stainless steel frame of sufficient weight to ensure that the specimens cannot rise to the water surface. Move the specimens and the sink together into the

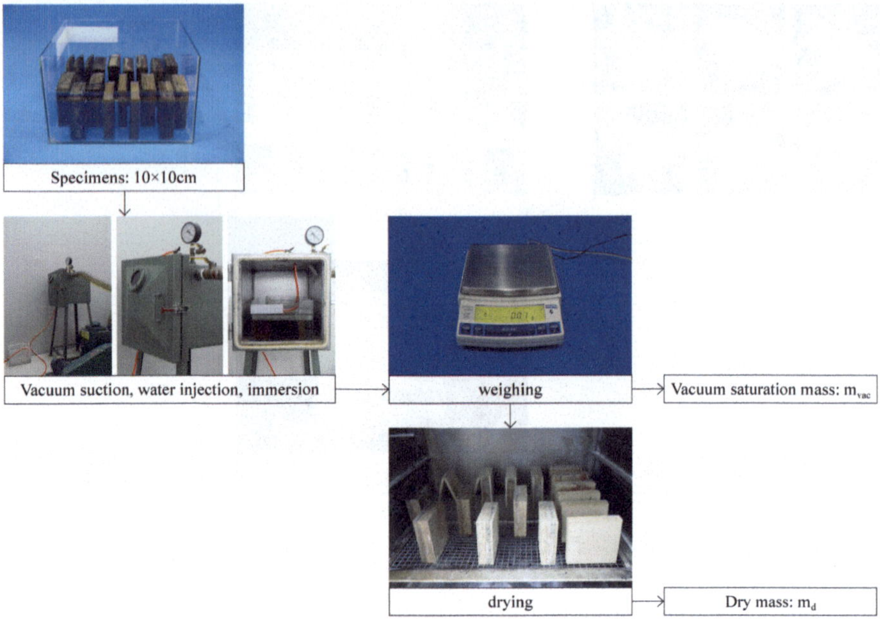

Fig. 3.6 Photograph program of the vacuum saturation test

vacuum chamber, close the chamber door, start the vacuum pump, reduce the air pressure in the chamber to below 20 mbar, and stabilize this for 6 h, so as to let the air inside the pore of the material discharge. Keep the air pressure in the vacuum chamber, open the water inlet to slowly inject the distilled water into the water sink, and keep the liquid surface rising speed about 5 cm/h until the liquid rises to about 5 cm above the specimens, then stop the water injection, and close the vacuum pump. After the specimens are immersed in water for 3 d, take them out, and remove the adhered water on the surface with a soft damp cloth, then weigh the specimens and record the vacuum saturated mass m_{vac} [kg]. Dry the specimens and then record the dry mass m_{d} [kg] (Fig. 3.6).

According to

$$w_{\mathrm{vac}} = \frac{(m_{\mathrm{vac}} - m_{\mathrm{d}})}{V}$$

calculate the vacuum saturation moisture content w_{vac} [kg/m^3] of the specimens. According to

$$\Phi = \frac{w_{vac}}{\rho_l}$$

calculate the porosity Φ [–] of the material, where ρ_l [kg/m^3] is the density of water at room temperature.

3.2.2.3 Sorption Test

The test is performed referring to the methods recommended by the international standard ISO 12571:2012 and the American standard ASTM-04a (2016) [14, 15]. Dry the specimens and record the mass m_d [kg]. Move the specimens into a drying box to cool them down to 20 °C, and then place them in the vacuum dryers for moisture adsorption process. The vacuum dryers are equipped with different saturated salt solutions to control the predetermined relative humidity. Move the vacuum dryers together with the T&RH recorders into a constant temperature and humidity chamber ($T = 20$ °C, RH = 50%) for curing, and the temperature and relative humidity changes in the vacuum dryers are monitored daily until they are stabilized at the predetermined values. After 8 weeks, open the vacuum dryers and weigh the specimens every 3 d. After the weighing is completed, return the specimens and the close the dryers. The above operation is repeated until the difference of three consecutive weighing results does not exceed 0.1%, when the specimens are considered to have reached the moisture adsorption equilibrium state. Record the mass of the wet specimens m_w [kg] (Fig. 3.7).

According to

$$u = \frac{m_w - m_d}{m_d}$$

calculate the mass ratio moisture content u [kg/kg]. According to

$$w = u \cdot \rho_d$$

calculate the mass–volume ratio equilibrium moisture content w [kg/m^3] to obtain the isothermal adsorption curve of the material. Then, move the small samples, which have been cured in the constant T&RH curing box (T = 20 °C, RH = 90%) for 12 weeks until they have reached the moisture adsorption equilibrium state, to the vacuum dryers, and repeat the above operation to obtain the isothermal desorption curve. In the experiment, it is not able to provide a higher curing relative humidity than that in the K_2SO_4 vacuum dryer (reference RH control value RH = 97.6 ± 0.6%) to form the initial moisture content of the wet specimens, and the desorption process is only performed in the remaining 8 relative humidity gradients except the K_2SO_4 (Table 3.6).

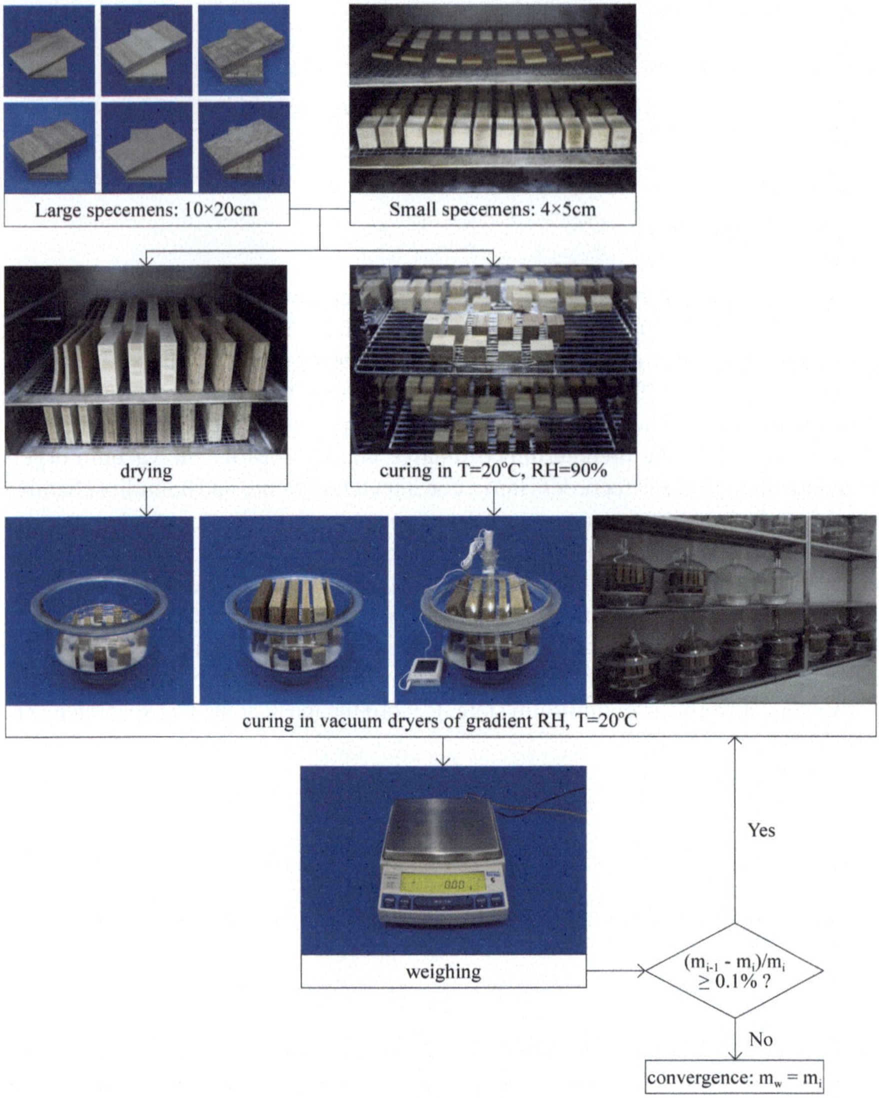

Fig. 3.7 Photograph program of the sorption test (construction partition group)

Table 3.6 Temperature and RH of ambient air control by different saturated salt solutions in vacuum dryers

No.	Saturated salt solutions	Reference RH value (%) at T = 20 °C	Measured RH value (%)	Measured T value (°C)
1	LiCl	11.31 ± 0.31	11.2	20.2
2	CH_3COOK	23.11 ± 0.25	24.4	20.2
3	$MgCl_2$	33.07 ± 0.18	33.4	20.2
4	K_2CO_3	43.16 ± 0.33	43.5	20.4
5	$Mg(NO_3)_2$	55.87 ± 0.27	55.0	20.3
6	NaBr	59.14 ± 0.44	59.7	20.3
7	NaCl	75.47 ± 0.14	77.2	20.1
8	KCl	85.11 ± 0.29	85.4	20.1
9	K_2SO_4	97.6 ± 0.6	96.3	20.3

The T&RH detectors are connected into the vacuum dryers through the opening of the glass cover, to monitor the actual ambient air temperature and relative humidity of the curing environment

3.2.2.4 Capillary Absorption Test

The test is performed referring to the methods recommended by the international standard ISO 15148:2002(E) [16]. Capillary absorption test can determine the water absorption coefficient and capillary saturation moisture content of the material, and further calculate the liquid water transport coefficient based on these two values [17, 25–28]. In the test, dry the specimens and record the mass m_d [kg]. After cooling down to 23 °C in a drying box, wrap the side and top surface of the dried specimens with hydrophobic tape, leaving 4 air vents of 5 × 5 mm on top surface and 6 mm gaps close to the bottom edges on side surfaces open, and record the total mass $m_{d,total}$ [kg]. After stainless steel nut supports of equal height are fixed on the bottom of the sink, the water of room temperature (25 °C) is poured into the sink until the liquid level is about 3 mm higher than the upper surface of the supports. Place the specimens on the supports of the water sink, take the specimens out at intervals, wipe off the attached water with a soft damp cloth, weigh the specimens to get the $m_{w,total}$ [kg], and put them back into the water to continue the water absorption (Fig. 3.8).

The water absorption process can be divided into two stages. During the experiment, the data are synchronously input into the computer for tracking and determining the stage of the water absorption process. Excluding some initial data, when the recorded water absorption per unit area is linear with the square root of time, it is considered to be in the first stage of capillary absorption. Then, linearly fit the data of the first stage. According to

$$A_{cap} = \frac{\Delta m_{moisture}}{A \cdot \sqrt{t}}$$

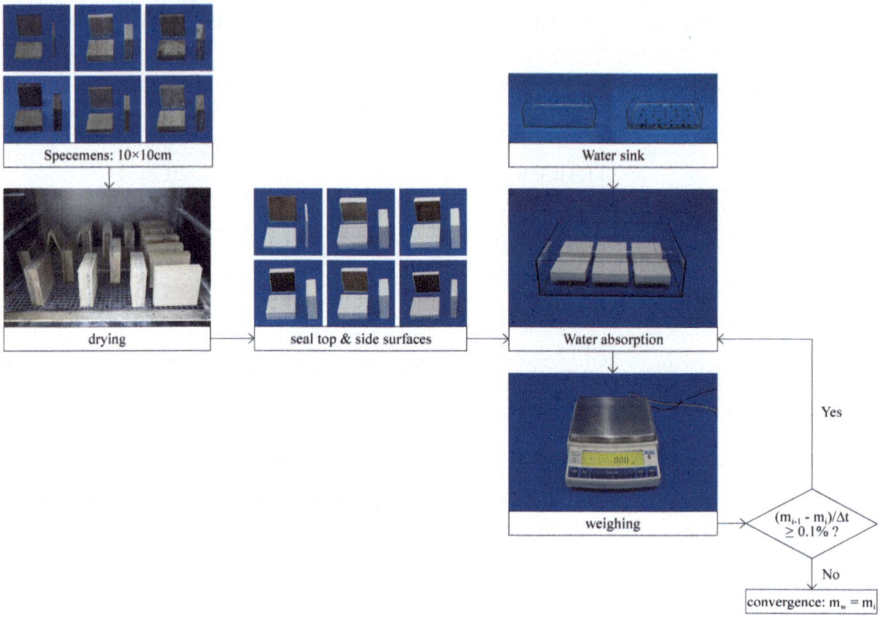

Fig. 3.8 Photograph program of the capillary absorption test

calculate the water absorption coefficient of the material A_{cap} [kg/(m^2 s$^{0.5}$)], where A [m^2] is the water absorption area of the specimens. After the specimens have reached the capillary saturation, they continue to absorb water due to the dissolution of air inside the pores of the specimens, but the rate of mass increase is obviously reduced at this period, which is considered to be the second stage of the water absorption process. Stop the process when at least five measurement points have been taken in the second stage, and then, linearly fit the data of the second stage.

Move the function fitted in the first stage to the origin, and then, calculate the ordinate of the intersection between the linear function obtained in the second stage and this moved function $\frac{\Delta m_{\text{moisture}}}{A}_{\text{cr}}$ [kg/m^2]. According to

$$w_{\text{cap}} = \frac{1}{d} \cdot \frac{\Delta m_{\text{moisture}}}{A}_{\text{cr}}$$

calculate the capillary saturation moisture content of the material w_{cap} [kg/m^3], where d [m] is the thickness of the specimens. According to

$$D_{\text{l}} = \frac{\pi}{4} \cdot \left(\frac{A_{\text{cap}}}{w_{\text{cap}}}\right)^2$$

calculate the liquid transport coefficient D_{l} [m^2/s].

3.2.2.5 Water Vapor Transmission Test

The test is performed referring to the methods recommended by the international standard ISO 12572:2001 (E) and the American standard ASTM E 96/E 96 M-2005 [18, 19]. Considering that it would test the water vapor transfer coefficient of the materials with different moisture content, the experiment device of the dry cups and wet cups is assembled from the glass bowls and their plastic covers, wherein the hydrophobic covers can be separately taken out for weighing. The airtightness of the cups is examined before the experiment, and meanwhile the quality records of three dry cups and three wet cups for modification are synchronously recorded during the experiment. Results show that the error caused by the airtightness of the cups is negligible.

Weigh the dried specimens and record the mass m_d [kg]. Seal the specimens on the side, and then, fix them on the plastic covers where there is an opening of a precise size 9.5 × 9.5 cm, and record the overall dry mass of the specimens and the plastic covers $m_{d,lid}$ [kg]. Move them together into the constant T&RH curing box, and set the predetermined relative humidity for curing. After they have reached a mass balance, seal the specimens and the plastic covers integrally to the mouth of the glass bowls, in which there are desiccant or saturated salt solution at the bottom to control the relative humidity under the specimens. Move the dry and wet cups into the constant T&RH curing box for curing, and weigh the cups every 3 d. When the mass change is linear and no less than five pieces of data are recorded, stop the curing, and remove the specimens and the plastic covers from the bowls for weighing. Record the mass $m_{w,lid}$ [kg] (Fig. 3.9).

According to

$$w = \frac{m_{w,lid} - m_{d,lid}}{V}$$

calculate the moisture content of the specimens w [kg/m^3]. Fit linearly the recorded mass and time data of each group, and the obtained mass change rate is the water vapor transfer rate G [kg/s]. According to

$$g_v = \frac{G}{A}$$

calculate the water vapor transfer rate per unit area g_v [kg/(m^2 s)]. According to

$$\Delta p_v = \Delta\varphi \cdot 610.5 \cdot e^{\frac{17.269(T_k - 273.15)}{T_k - 35.85}}$$

calculate the water vapor pressure difference Δp_v[Pa] between the two sides of the specimen. According to

$$R_{v,total} = \frac{\Delta p_v}{g_v}$$

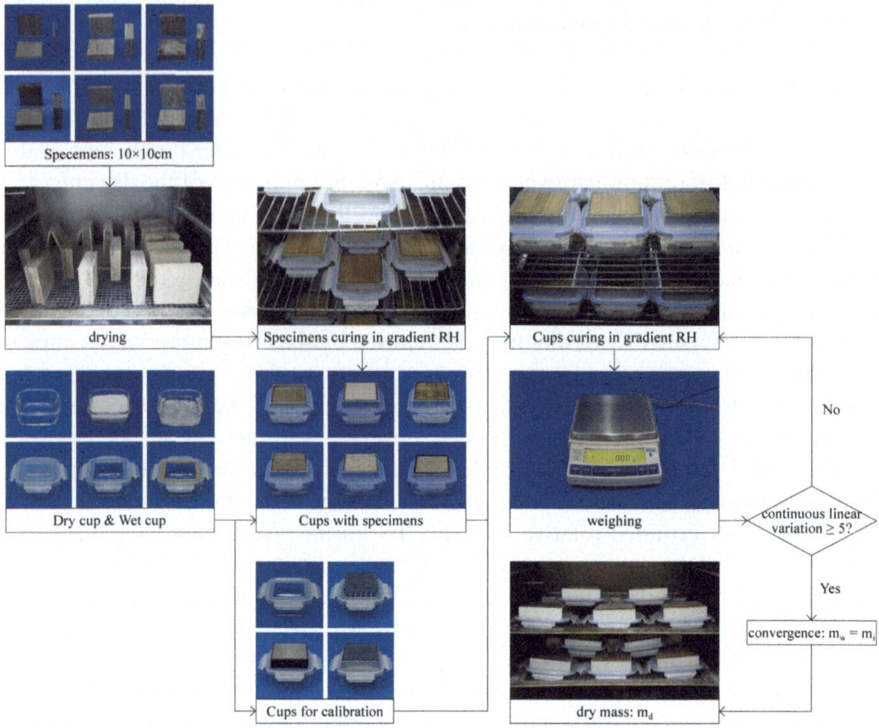

Fig. 3.9 Photograph program of the water vapor transmission test (construction partition group)

calculate the total water vapor transfer resistance $R_{\text{v,total}}$ [m^2 s Pa/kg]. According to

$$\delta_{\text{air}} = \frac{2.306 \times 10^{-5} p_{\text{o}}}{R_o T_{\text{k}} p} \left(\frac{T_{\text{k}}}{273.15} \right)^{1.81}$$

calculate the water vapor permeability of air δ_{air} [kg/(m s Pa)] with respect to a certain temperature T_{k} and barometric pressure p [20], where the standard barometric pressure $p_0 = 101{,}325$ Pa and the gas constant for water vapor $R_{\text{o}} = 461.5$ N m/(kg K). According to

$$R_{\text{v}} = R_{\text{v,total}} - \frac{h_{\text{air}}}{\delta_{\text{air}}}$$

calculate the corrected water vapor resistance R_{v} [m^2 s Pa/kg], where h_{air} [m] is the water vapor diffusion-equivalent air layer thickness, which is equal to the thickness of the air layer between the lower surface of the specimen and the desiccant/liquid level. According to

Table 3.7 RH control inside dry and wet cups and the ambient T&RH

No.	Equipment	Desiccant/saturated salt solution	Reference RH value
Condition 1 Environmental control in the curing box: $T = 23$ °C, RH = 40%			
1–3	Dry cups	Desiccant $CaCl_2$ ($d = 3$ mm)	0
4–6	Wet cups	K_2SO_4 saturated salt solution	96.3%
Condition 2 Environmental control in the curing box: $T = 23$ °C, RH = 50%			
1–3	Dry cups	Desiccant $CaCl_2$ ($d = 3$ mm)	0
4–6	Wet cups	K_2SO_4 saturated salt solution	96.3%
Condition 3 Environmental control in the curing box: $T = 23$ °C, RH = 70%			
1–3	Dry cups	Desiccant $CaCl_2$ ($d = 3$ mm)	0
4–6	Wet cups	K_2SO_4 saturated salt solution	96.3%
Condition 4 Environmental control in the curing box: $T = 23$ °C, RH = 90%			
1–3	Dry cups	Desiccant $CaCl_2$ ($d = 3$ mm)	0
4–6	Wet cups	K_2SO_4 saturated salt solution	96.3%

$$\delta = \frac{d}{R_v}$$

calculate the water vapor transfer coefficient δ [kg/(m s Pa)] of the material corresponding to the state when the moisture content is w. Then, move the specimens and the plastic covers together into the constant T&RH curing box, set the next set of relative humidity for curing, replace the desiccant/saturated salt solution in the dry and wet cups, and repeat the above operation until all the working conditions are completed (Table 3.7).

3.2.2.6 Drying Test

The test is performed referring to the methods recommended by the Fraunhofer IBP (DIN EN ISO/IEC 17025). Following the test item 2, when the specimens have reached the state of vacuum saturation, record the initial mass m_{vac} [kg]. Wipe off the adhered water and wrap the side and bottom surfaces of the specimens with a hydrophobic waterproof tape, and then weigh the total mass $m_{w,total}$ [kg]. Move the specimens in a constant T&RH curing box ($T = 23$ °C, RH = 50%) for curing. Weigh the specimens every 12 h until the differences between three successive measurements do not exceed 0.1%, when it is considered that the drying process is completed. Stop the experiment, and record the total mass $m_{d,total}$ [kg]. Remove the sealing tape, dry the specimens, and weigh the drying mass m_d [kg] (Fig. 3.10).

Calculate the instantaneous moisture content w [kg/m^3] of the specimens in turn, and differentiate the instantaneous moisture content (w) to the drying time (t) to obtain the w–t drying curve, and then according to

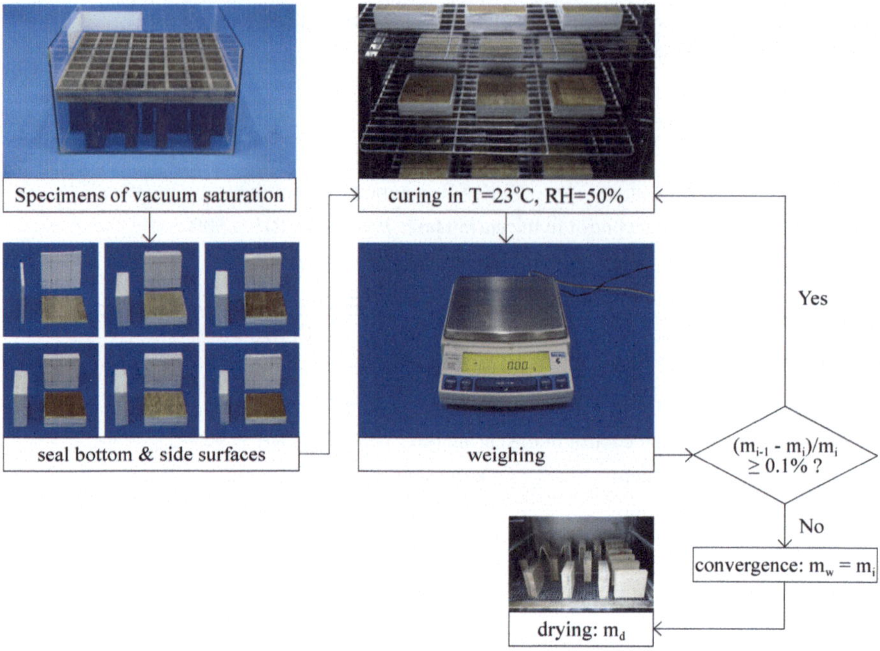

Fig. 3.10 Photograph program of the water drying test (construction partition group)

$$U = -\frac{m_\mathrm{d}}{A} \cdot \frac{\mathrm{d}w}{\mathrm{d}t}$$

calculate the drying rate U [kg/(m^2 s)] of the material corresponding to the state when the moisture content is w, and thus obtain the U–w drying rate curve.

3.2.2.7 Thermal Analysis

The test is performed referring to the methods recommended by the international standard ISO 11357-4-2005 [21]. Place the specimens in a hot pot, put the pot in a heat-flux differential scanning calorimeter (DSC), and then start the calorimeter. At the controlled temperature, measure the relationship between the temperature and the power difference input to the specimens and the reference material (sapphire). Calculate the specific heat capacity of the material c_p [kJ/(kg K)] from the recorded DSC curve (Fig. 3.11).

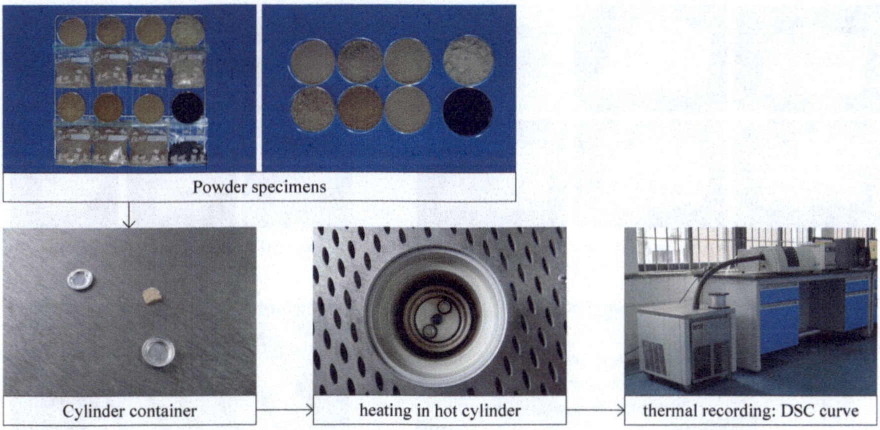

Fig. 3.11 Photograph program of the thermal analysis

3.2.2.8 Thermal Conductivity Test

The test is performed referring to the methods recommended by the international standard ISO 8302-1991 [22]. Measure the thermal conductivity of the large specimens (30 × 30 cm) of the initial moisture content λ_w [W/(m K)], using the CD-DR3030 guarded hot plate apparatus (hot plate $T = 35$ °C, cold plate $T = 15$ °C, central heat transfer area size: 15 × 15 cm). Then, dry the specimens, calculate the initial moisture content w [kg/m^3] after cooling down to 20 °C, and measure the thermal conductivity of the material in the dry state λ_d [W/(m K)]. In addition, hot wire method is adopted as a reference in this test. Take the small specimens (10 × 20 cm) after reaching their equilibrium moisture content in test item 3, and carry out the thermal conductivity test with QTM-500 hot wire tester. Repeat five times for each specimen, select the middle three values, and take their average as the thermal conductivity, which is auxiliary to display the impact of moisture content on the thermal conductivity of the material (Fig. 3.12).

Based on the test results of specific heat capacity and thermal conductivity, according to

$$S = \sqrt{\frac{2\pi\lambda \cdot c_d \cdot \rho_d}{Z}}$$

calculate the heat storage coefficient S [W/(m^2 K)] of the dry material. Taking the calculation period Z as 24 h, the 24 h heat storage coefficient S_{24h} can be obtained.

$$S_{24h} = 0.51\sqrt{\lambda \cdot c_d \cdot \rho_d}$$

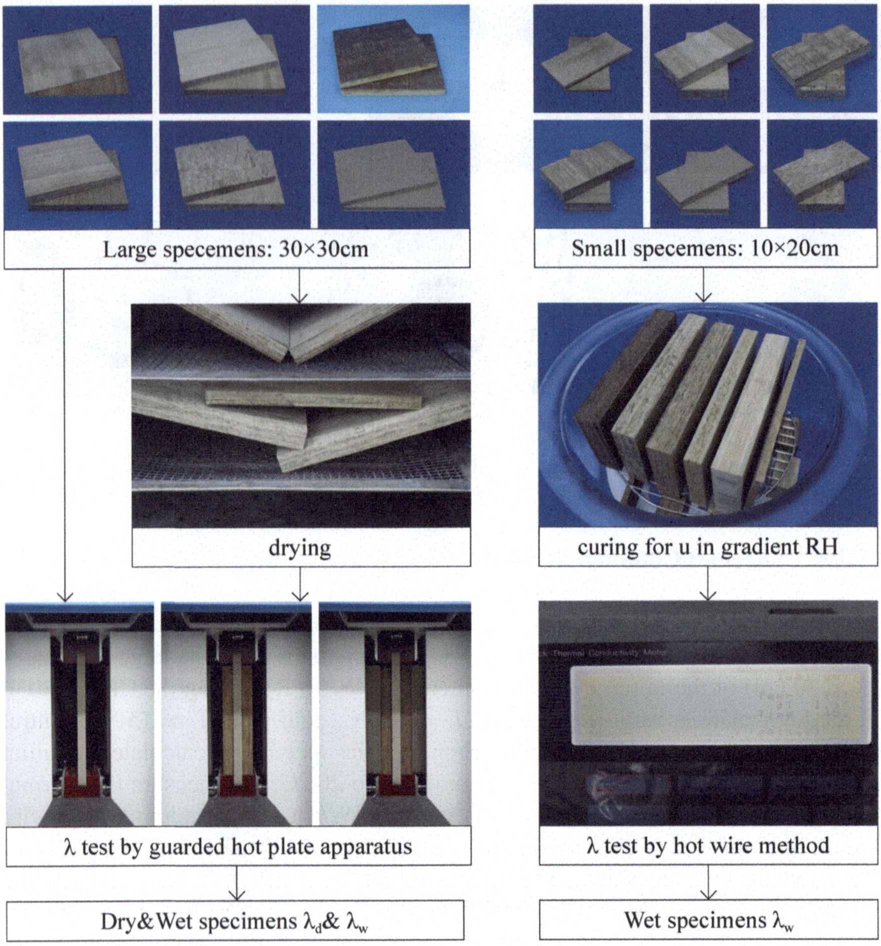

Fig. 3.12 Photograph program of the thermal conductivity test (construction partition group)

3.2.2.9 Surface Light and Thermal Properties' Test

The test is performed referring to the methods recommended by the international standard DIN EN 16012-2012 and Chinese specification GB/T 25968-2010 [23, 24]. Dry the small specimens (10 × 10 cm), and cool them down to 25 °C. Measure the hemispherical emissivity ε [–] by the D&S digital hemisphere emissivity tester, and meanwhile the light reflectivity ρ_v [%], solar direct reflectivity ρ_e [%], and solar direct absorptivity α_e [%] by the ultraviolet, visible, near infrared spectrophotometer U-4100, resulting the surface light and thermal properties of the material in dry state. Then, take the large specimens (10 × 20 cm) from the test item 3, after they have

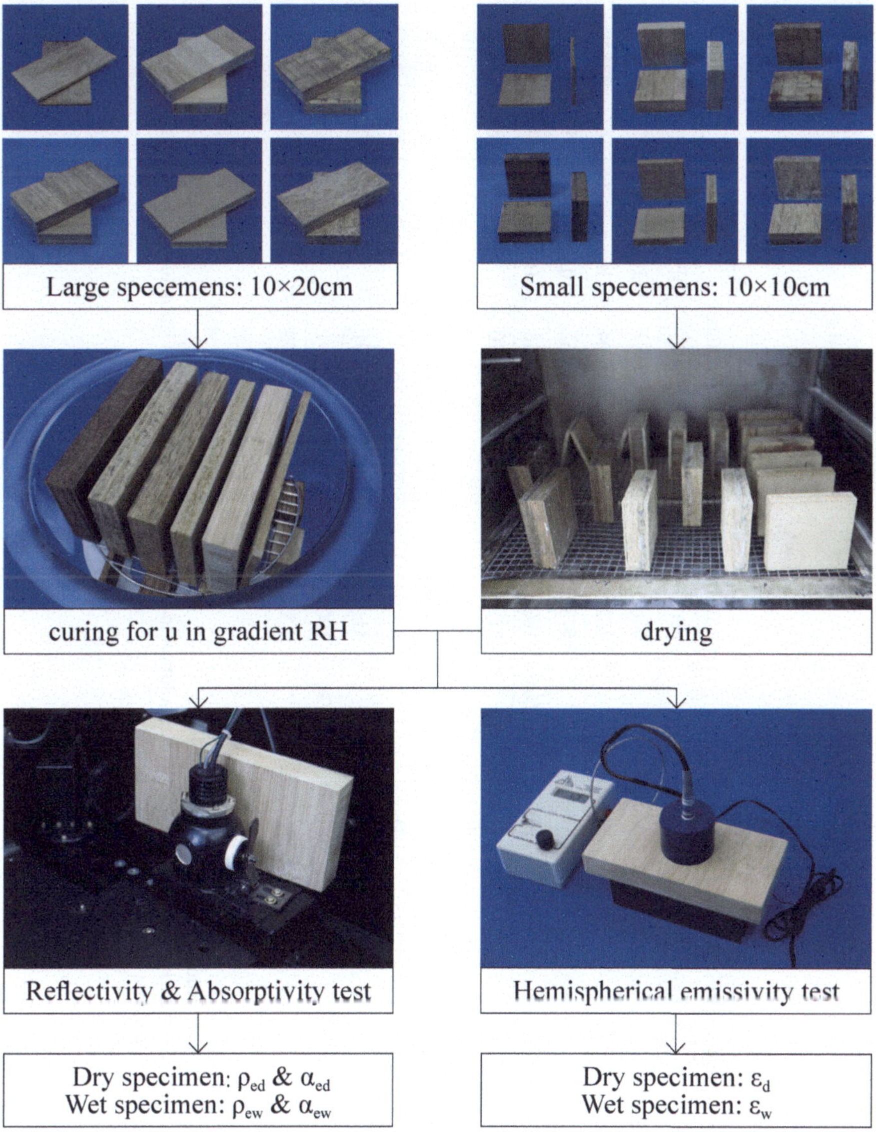

Fig. 3.13 Photograph program of the surface light and thermal properties' test

reached the equilibrium moisture content, and repeat the above operation to obtain the relationship between the surface light and thermal properties and the moisture content of the material (laboratory environment $T = 25$ °C, RH = 55%) (Fig. 3.13).

Table 3.8 Test results of the basic properties

Test item	Test and calculation results			Raw bamboo	Bamboo-based panel				
	Target value	Notation	Value	FB	BSB	BMB	BFB	BPB	BOSB
1. Bulk density test	Dry bulk density	ρ_d [kg/m^3]	Average	666.38	563.81	776.21	1108.77	623.32	895.02
			Max	722.50	587.83	815.04	1156.53	664.46	935.05
			Min	578.69	533.88	743.62	1030.37	600.41	872.68
			FB reference (%)	100	84.61	116.48	166.39	93.54	134.31
			Extreme value deviation (%)	21.58	9.57	9.20	11.38	10.28	6.97
2. Vacuum saturation test	Open porosity; vacuum saturation moisture content	Φ [–] w_{vac} [kg/m^3]	Average	52.24	53.97	49.58	17.36	63.17	42.32
			Max	53.41	57.12	50.35	18.09	63.79	42.63
			Min	51.16	49.84	48.91	16.11	62.77	41.73
			FB reference (%)	100	103.31	94.91	33.23	120.92	81.01
			Extreme value deviation (%)	4.31	13.49	2.90	11.41	1.61	2.13

w_{vac} is considered to be equivalent to Φ
Extreme value deviation = (Max. value – Min. value)/Average value × 100%
During the vacuum saturation test, BSB slabs partly slip off due to adhesive dissolution in water, which enlarges the deviation of the test results; on the other side, the smaller base radices of BFB also lead to a larger deviation

3.2.3 Results Analysis

3.2.3.1 Basic Properties

The test results show that the bulk density and open porosity of bamboo-based panels change significantly. Compared with FB (100%), the dry bulk density ρ_d [kg/m^3] of bamboo-based panels is in the range BSB 84.61%—BFB 166.39%, while the open porosity Φ [%] is in the range BFB 33.23%—BPB 120.92%.

The bulk density test results show strong inhomogeneity in bamboos, of which the extreme value deviations can reach approximately 10–20%. The extreme value deviation of FB is far larger than that of the bamboo-based panels, indicating the improved effect on material homogeneity during the decomposition and recombination process.

The open porosity test results show better homogeneity in bamboos. The open porosity of other bamboo-based panels, except BFB, changes within approximately 20% compared with FB, which means that the pore structure inside bamboos does not substantially change during the decomposition and recombination process. The Φ of BFB is reduced to only 33.23% of FB owing to the waterproofing agent applied to the constituent units and the high-pressure forming process that blocks and compresses the pore structure, which is conductive to the application in outdoor environment (Table 3.8, Fig. 3.14).

The surface morphology analysis with a scanning electron microscope (ZEISS, EVO18) shows the different microstructures caused by the assembly method and forming pressure. Comparison with FB shows that the BSB changes minimally; the BMB contains large holes (≥1 mm) owing to the overlapping of orthogonally weaved bamboo slivers; the BFB becomes the most compact as a result of the highest forming pressure; and the BPB has disorder fibers in different directions and complex pores

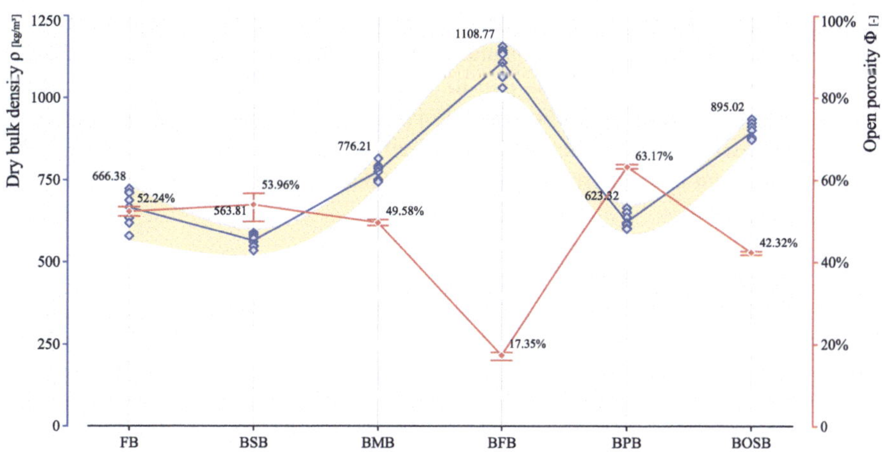

Fig. 3.14 Bamboo basic properties (dry bulk density ρ_d, open porosity Φ)

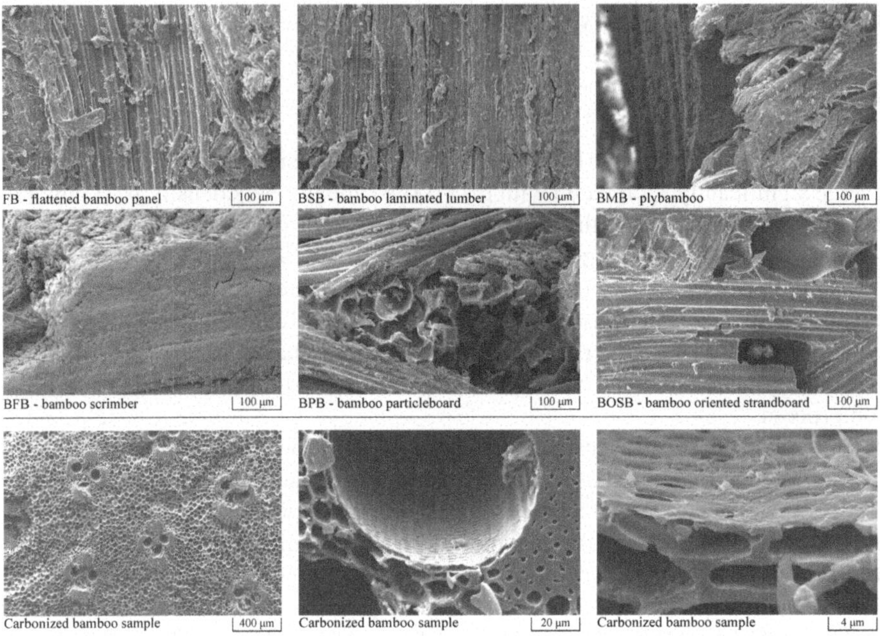

Fig. 3.15 Bamboo surface morphology analysis (the upper six photographs showed different morphology of the six bamboo samples at the same scale, while the three photographs below showed different and more complex microscopic pores at different scales)

of different sizes. On the other hand, the BOSB consists of fibers in the oriented direction and relatively underdeveloped pores. The morphology on this scale shows some characteristics of the pore systems inside the materials, which to a certain extent can explain the significant hygric and thermal differences even when the Φ values of different samples are close to each other. However, if we zoom in on the micrographs, as the carbonized bamboo sample sections show below, more complicated microscopic pores from hundreds μm to several μm can be found, which means that it is difficult to describe and define them precisely at the microscale. Therefore, the method based on macrophenomenology will be more feasible for analyzing the regularity between hygrothermal properties and basic properties (Fig. 3.15).

3.2.3.2 Hygric Properties

Moisture storage properties, characterized by the isothermal adsorption and desorption curve, are described as three stages according to Künzel. They are the sorption moisture region (or hygroscopic region, corresponding to equilibrium moisture content of RH = 0–95%), capillary water region (following the sorption moisture region, continuing to the state of free water saturation or capillary saturation), and supersaturated region (following the capillary water region, until all the air in close pore is dissolved) [2]. The former region is mainly related to gaseous moisture, while the latter two regions consider mostly the liquid water. In this study, the equilibrium moisture content of RH = 50% in gaseous water region $w_{RH=50\%}$ [kg/m^3], the capillary saturation moisture content w_{cap} [kg/m^3], and the vacuum saturation moisture content w_{vac} [kg/m^3] are chosen for analysis (Table 3.9).

Moisture storage test results show that, at FB 100%, the $w_{RH=50\%}$ of bamboo-based panels are in the range BSB 85.50%—BFB 128.41%. The w_{cap} and w_{vac} of bamboo-based panels are in the range BFB 50.83%—BPB 159.90% and BFB 33.23%—BPB 120.92%. Because of the capillary hysteresis effect, the desorption equilibrium moisture content will be larger than the sorption values, which is approximately 7.16–12.47% in this test, but the isothermal adsorption curve is sufficient in practical application. On the other hand, the size of samples we use in sorption test is close to that in practical application, which is larger than some previous tests (see Sect. 1.3.1.1 in Chap. 1); therefore, the EMC values we obtain are smaller than the previous works.

Moisture transport properties are characterized with water vapor transfer coefficient δ [kg/(m s Pa)] describing the gaseous moisture transport ability; water absorption coefficient A_{cap} [kg/(m^2 s$^{0.5}$)] describing the liquid water transport ability; and drying rate U [kg/(m^2 s)] describing both the gaseous and liquid water transport abilities. The δ value is affected by the moisture content, of which the $\delta_{RH=50\%}$ corresponding to the equilibrium moisture content of RH = 50% is chosen for analysis. The U value is also affected by the moisture content, of which the average U value corresponding to the equilibrium moisture content from $u = 12\%$ to $u = 6\%$ is chosen for analysis.

The moisture transport test results show that, at FB 100%, the $\delta_{RH=50\%}$ of bamboo-based panels are in the range BFB 15.71%—BPB 335.41%, A_{cap} are BFB 11.79%—BPB 604.62%, and $Uu_{12\%}–u_{6\%}$ are BFB 7.50%—BPB 93.57%. Except BPB (604.62%), the remaining BBPs are generally reduced in moisture transport properties (Figs. 3.16, 3.17).

3.2.3.3 Thermal Properties

Heat storage properties are characterized by specific heat capacity c [J/(kg K)]. The test results show that the c values among different bamboos vary within the small range BFB 86.30%—BMB 112.47% compared with FB (100%).

Heat transport properties are characterized by thermal conductivity λ [W/(m K)] describing the heat flow transport ability through the material; and hemispherical emissivity ε [–], light reflectivity (380–780 nm) ρ_v [%], solar direct reflectivity (200–2600 nm) ρ_e [%], and solar direct absorptivity (200–2600 nm) α_e [%] describing the surface light and thermal exchange abilities.

The heat transport property test results show that, at FB 100%, the λ values of the bamboo-based panels are in the range BPB 73.62%—BMB 159.28%. The ε values of the bamboo-based panels are in the range BSB 81.16%—BFB 95.65%, which

Table 3.9 Test results of the hygric properties

Test item	Test and calculation results			Raw bamboo	Bamboo-based panel				
	Target value	Notation	Value	FB	BSB	BMB	BFB	BPB	BOSB
3. Sorption test	20 °C isothermal adsorption and desorption curve	$w_{RH=50\%}$ [kg/m^3]	Average	31.07	30.44	39.20	28.17	36.89	46.07
			FB reference (%)	100	97.97	126.17	90.69	118.74	148.28
		$w_{RH=11.2\%}$	Average	8.58	9.08	11.58	6.79	10.25	13.63
		$w_{RH=24.4\%}$		13.91	15.26	19.36	14.05	17.92	23.00
		$w_{RH=33.4\%}$		19.99	21.46	25.48	20.26	23.81	31.50
		$w_{RH=43.5\%}$		26.03	26.27	34.59	25.55	32.31	40.86
		$w_{RH=55.0\%}$		34.94	33.64	42.74	30.19	40.41	50.07
		$w_{RH=59.7\%}$		38.65	36.57	46.38	32.24	44.42	55.50
		$w_{RH=77.2\%}$		54.53	53.60	63.18	53.75	61.07	75.83
		$w_{RH=85.4\%}$		69.88	64.35	81.59	69.58	74.17	91.74
		$w_{RH=96.3\%}$		118.61	100.78	120.55	155.04	105.83	137.53

(continued)

Table 3.9 (continued)

Test item	Test and calculation results			Raw bamboo	Bamboo-based panel				
	Target value	Notation	Value	FB	BSB	BMB	BFB	BPB	BOSB
4. Capillary absorption test	Water absorption coefficient	A_{cap} [E^{-04} kg/(m^2 s$^{0.5}$)]	Average	74.06	78.74	38.71	8.73	447.78	59.15
			FB reference (%)	100	106.32	52.27	11.79	604.62	79.87
	Capillary saturation moisture content	w_{cap} [kg/m^3]	Average	326.41	317.21	221.44	165.93	521.95	342.68
			FB reference (%)	100	97.18	67.83	50.83	159.90	105.00
5. Water vapor transmission test	Water vapor transfer coefficient	$\delta_{RH=50\%}$ [E^{-13} kg/(m s Pa)]	Average	27.81	27.99	19.28	4.37	93.28	7.23
			FB reference (%)	100	100.63	69.33	15.71	335.41	26.01
		$w_{RH=20.0\%}$	Average	10.69	21.66	9.10	2.12	82.27	3.96
		$w_{RH=25.0\%}$		11.46	22.72	10.79	2.50	84.11	4.51
		$w_{RH=35.0\%}$		13.00	24.83	14.19	3.25	87.78	5.60
		$w_{RH=45.0\%}$		22.87	26.94	17.58	3.99	91.44	6.69
		$w_{RH=50.0\%}$		27.81	27.99	19.28	4.37	93.27	7.23
		$w_{RH=73.0\%}$		41.03	47.98	37.61	18.40	140.61	25.26
		$w_{RH=83.0\%}$		128.32	188.16	68.89	26.47	234.47	38.16
		$w_{RH=93.0\%}$		337.32	522.22	242.92	40.97	754.10	77.50

(continued)

Table 3.9 (continued)

Test item	Test and calculation results			Raw bamboo	Bamboo-based panel				
	Target value	Notation	Value	FB	BSB	BMB	BFB	BPB	BOSB
6. Drying test	Drying rate ($T = 23$ °C, RH = 50%)	$U_{u12\%\text{–}u6\%}$ [E^{-07} kg/(m^2 s)]	Average	135.15	60.20	18.32	10.14	126.46	38.01
			FB reference (%)	100	44.54	13.56	7.50	93.57	28.12
		$U_{u12\%\text{–}u11\%}$	Average	266.32	74.22	22.32	12.58	179.53	48.35
		$U_{u11\%\text{–}u10\%}$		187.78	66.64	21.15	11.46	163.16	45.26
		$U_{u10\%\text{–}u9\%}$		166.94	58.87	19.52	10.38	156.75	41.36
		$U_{u9\%\text{–}u8\%}$		108.52	60.01	17.57	9.71	123.94	37.13
		$U_{u8\%\text{–}u7\%}$		65.45	52.75	15.96	8.98	83.92	31.20
		$U_{u7\%\text{–}u6\%}$		15.92	48.72	13.43	7.74	51.43	24.76

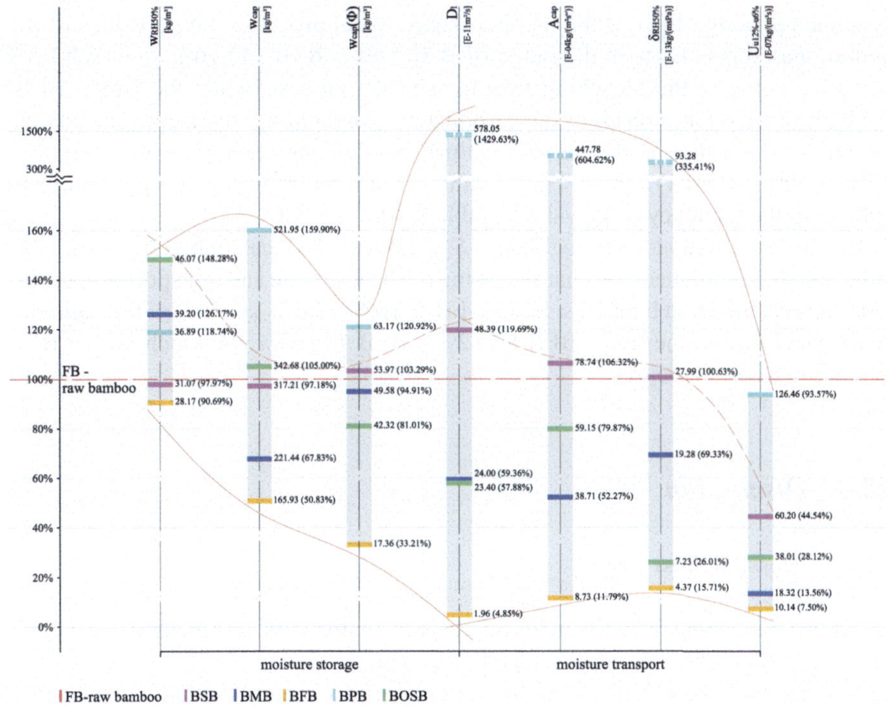

Fig. 3.16 Bamboo hygric properties (moisture storage, moisture transport) test results; the values in brackets are the parameters ratio between the five bamboo-based panels and the raw bamboo

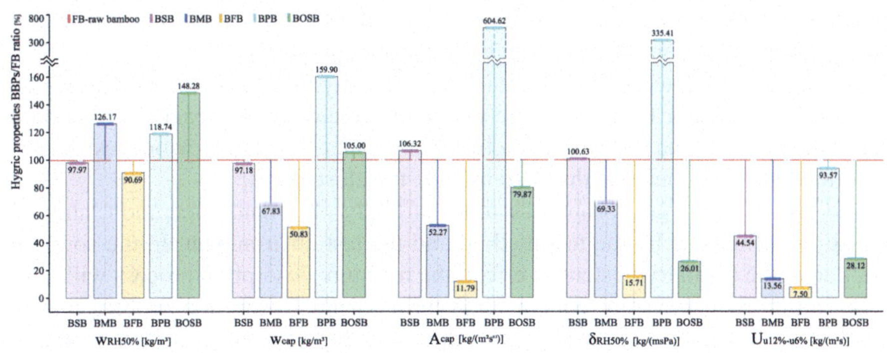

Fig. 3.17 Hygric properties' test results bamboo-based panels/raw bamboo ratio

show the reducing effect of the bamboo-based panel process. The ρ_v values of the bamboo-based panels are in the range BFB 50.88%—BSB 132.76%, in which BSB increases owing to the bleaching treatment in the process, while the BMB, BFB, and BPB decrease as a result of the carbonization and gluing treatment. The change regularity of ρ_e values is similar to ρ_v values, but the magnitude is smaller, which is BFB 74.49%—BSB 112.90%. The α_e values, resulting from $\alpha_e = 1 - \rho_e$, change in contrast to the tendency of ρ_e values (Table 3.10, Figs. 3.18, 3.19).

The 24 h thermal storage coefficient S_{24h} [W/(m^2 K)], calculated from the c, λ, and ρ values, simultaneously characterizing the heat storage and transport properties, varies among different bamboos. Setting FB as 100%, the S_{24h} values of the bamboo-based panels are in the range BPB 82.12%—BFB 146.37%, of which all but BPB increase to a certain degree.

3.2.4 Discussion

3.2.4.1 Possible Test Error

Test errors may come from the material's representativeness, test methods, test environment, equipment, data analysis, and other aspects.

As to the basic properties' tests, the operation process and calculation method are simple; the equipment is of high precision; and there are no special requirements for the test environment; hence, the test results are reliable. However, the unignorable inhomogeneity of the materials themselves, such as the 6.97–21.58% extreme value deviation in bulk density, can cause deviations to the results of the rest test items, as basic properties have great impact on hygrothermal properties.

The porosity targeted in this study is open porosity that can be filled with water, which is the indicator to describe the maximum liquid water capacity of a certain material; therefore, the test result is a little bit smaller than the previous test results performed with CT methods that also include the close pores.

The test of the hygric properties relies strongly on manual operation, and the results are influenced by the test method and the test environment. In the sorption test, due to the capillary hysteresis effect, the moisture desorption process will have larger equilibrium moisture content than the moisture adsorption process, but in practical applications, it is usually sufficient to use the moisture adsorption isotherm for characterizing the equilibrium moisture content [2]. In this test, the obtained desorption isotherm is 7.16–12.47% higher than the adsorption isotherm. In the water vapor transmission test, the water vapor transfer coefficient is not obviously affected by the temperature, but significantly influenced by the moisture content, which is generally described with a function [29]. During the water vapor transmission test, the wet cups with K_2SO_4 saturated salt solution cause high moisture content to the specimens above, which leads to liquid water transfer and results in values higher than the actual. In addition, the capillary water absorption experiment is affected by the water temperature [30], and the drying curve test is influenced by the ambient

Table 3.10 Test results of the thermal properties

Test item	Test and calculation results			Raw bamboo	Bamboo-based panel				
	Target value	Notation	Value	FB	BSB	BMB	BFB	BPB	BOSB
7. Thermal analysis	Specific heat capacity	c [J/(kg K)]	Average:	1796	1960	2020	1550	1760	1663
			FB reference (%)	100	109.13	112.47	86.30	98.00	92.62
	24 h heat storage coefficient	S_{24h} [W/(m^2 K)]	Average	5.93	6.63	8.57	8.68	4.87	6.94
8. Thermal conductivity test			FB reference (%)	100	111.80	144.52	146.37	82.12	117.03
	thermal conductivity (dry)	λ [W/(m K)]	Average	0.1088	0.1475	0.1733	0.1625	0.0801	0.1197
			FB reference (%)	100	135.57	159.28	149.36	73.62	110.02
9. Surface light and thermal properties' test	Hemispherical emissivity	ε [–]	Average	0.69	0.56	0.60	0.66	0.59	0.59
			FB reference (%)	100	81.16	86.96	95.65	85.51	89.86
	Light reflectivity (380–780 nm)	ρ_v [%]	Average	38.80	51.51	26.40	19.74	31.59	31.59
			FB reference (%)	100	132.76	68.04	50.88	81.42	100.80
	Solar direct reflectivity (200–2600 nm)	ρ_e [%]	Average	59.75	67.46	49.51	44.51	46.39	46.39
			FB reference (%)	100	112.90	82.86	74.49	77.64	97.29
	Solar direct absorptivity (200–2600 nm)	α_e [%]	Average	40.25	32.54	50.40	55.49	53.61	53.61
			FB reference (%)	100	80.84	125.44	137.86	133.19	104.02

Note The thermal property values shown in the table correspond to the dry state of the material

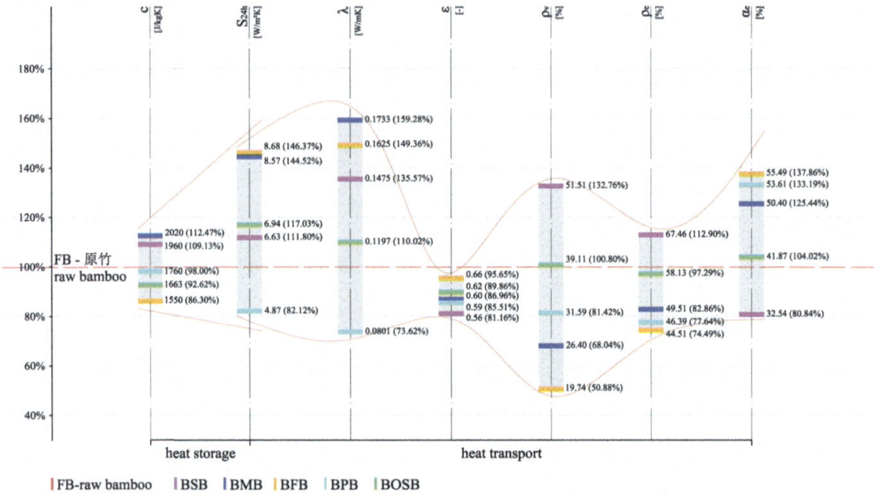

Fig. 3.18 Bamboo thermal properties (heat storage, heat transport) test results; the values in brackets are the parameters ratio between the five bamboo-based panels and the raw bamboo. *Note* The thermal property values shown in the figure correspond to the dry state of the material

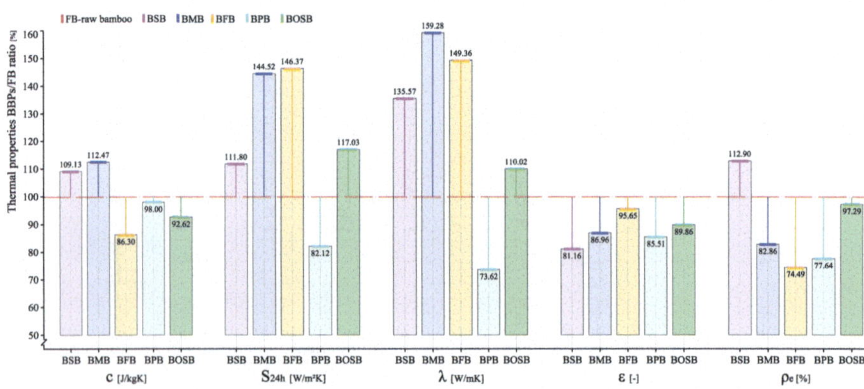

Fig. 3.19 Thermal properties' test and calculation results bamboo-based panels/raw bamboo ratio. *Note* The thermal property values shown in the figure correspond to the dry state of the material

temperature, relative humidity and wind speed. The two experiments are carried out in a laboratory environment of $T = 23\ °C$, $RH = 50\%$, following the recommendations of Fraunhofer IBP.

The test of thermal properties is largely determined by the equipments, the data is output by computer, and generally there is no special requirement for the test environment, so that the difference in test results is mainly due to the material itself. In addition to the non-homogeneity, it is also particularly affected by the moisture content of the material. This effect on the heat storage properties (specific heat capacity) can be obtained by the weighted calculation of the dry material and the moisture contained [8], and the effect on the heat transport properties requires actual measurement. In this work, the thermal properties of different moisture contents in the moisture hygroscopic region are tested. The results indicate that the thermal conductivity is greatly affected by the moisture content of the material, showing a nonlinear relationship, which is decided by the material itself and the affinity of the contained moisture. In practical applications, the moisture-related thermal conductivity supplement a_w is usually used for an approximate correction, and the thermal conductivity of the wet building material is regarded as a linear function of the moisture content. The a_w value obtained in this test is in the range BSB 0.2137—BOSB 0.6890.

The surface light and thermal properties are affected by the moisture content of the material, which normally would show a nonlinear relationship. However, in this test, it is found that the surface light and thermal properties of bamboo in the moisture hygroscopic region are not obviously affected by the moisture content, and the difference between the test samples of various moisture contents (corresponding to the three groups of vacuum dryers in which RH = 11.2/43.5/77.2%, RH = 24.4/55.0/85.4%, and RH = 33.4/59.7/96.3%) is similar to the difference resulted from their own non-homogeneity of the dry samples. For the hemispherical emissivity ε [–], the extreme value difference D_{ev} [(max. value − min. value)/average value] of the dry samples/wet samples test results is: FB—7.21/6.61%, BSB—3.57/5.31%, BMB—1.66/2.46%, BFB—1.51/3.13%, BPB—6.82/7.83%, and BOSB—4.81/4.87%; for the solar direct reflectivity ρ_e [%], the corresponding value is: FB—4.79/7.04%, BSB—5.40/9.67%, BMB—6.09/5.50%, BFB—1.10/1.29%, BPB—3.49/2.30%, and BOSB—9.72/8.29%. Therefore, it is considered that in the moisture hygroscopic region, the influence of gaseous moisture content on the surface light and thermal properties of the bamboo is negligible. However, the relationship between the properties and the moisture content of liquid water in the capillary water region and the supersaturated region needs further study.

3.2.4.2 Correlation Between Basic Properties and Hygrothermal Properties

As explained above, the characteristics of bamboo at a microscale and the relation between hygrothermal properties are difficult to describe and define quantitatively and accurately. Therefore, a macrophenomenological approach for correlation analysis to explain the relation among basic, hygric, and thermal properties is adopted. Basic properties such as bulk density are commonly used as quality control indicators during the BBP process, as they have decisive effects on mechanical properties. The correlation analysis between hygrothermal properties and basic properties can help to explain the difference between various bamboos and prejudge the hygrothermal performance of certain BBPs by simple indicators such as ρ and Φ. All hygrothermal properties show larger r values and smaller sig values when correlation analysis is carried out with Φ than that with ρ.

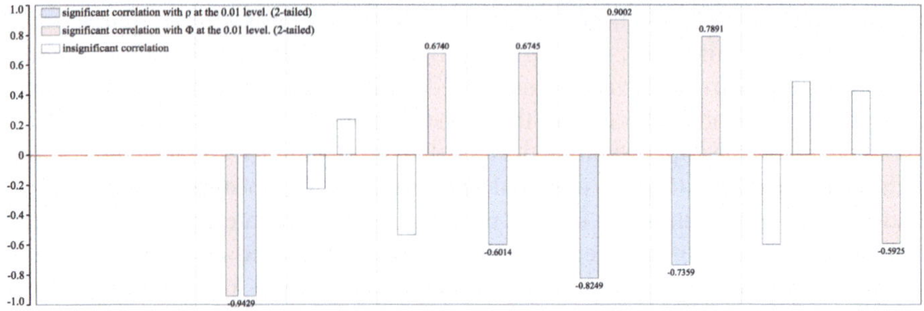

	Correlation indicators	ρ/Φ	u	W_{cap}	δ	A_{cap}	U	c	λ
	Correlation coefficient-r:	-0.9429	-0.2272	-0.5361	-0.6014	-0.8249	-0.7359	-0.6000	0.4232
ρ	Statistical significance-sig:	4.7743E-9	0.098568	0.021829	0.000006	0.000025	3.1545E-7	0.208000	0.080111
	Number-N:	18	54	18	48	18	36	6	18
	Correlation coefficient-r:	-0.9429	0.2348	0.6740	0.6745	0.9002	0.7891	0.4857	-0.5925
Φ	Statistical significance-sig:	4.7743E-9	0.087438	0.002159,	1.4681E-7	3.5973E-7	1.0776E-8	0.328723	0.009565
	Number-N:	18	54	18	48	18	36	6	18

Fig. 3.20 Correlation between bulk density ρ and open porosity Φ, between basic properties and hygrothermal properties

The Spearman correlation analysis results between ρ and Φ are as follows: $r = -0.9429$, sig $= 4.7743\text{E}^{-09}$, which indicates a significant negative correlation between ρ and Φ. Gaseous moisture storage properties, characterized by mass rate water content u values, show no significant correlation with basic properties; however, the liquid water storage properties, characterized by w_{cap}, have a significant positive correlation with Φ. All moisture transport properties, δ, A_{cap}, and U values have significant positive correlation with Φ and negative correlation with ρ. Heat storage properties, characterized by specific heat capacity c values, show no significant correlation with basic properties. On the other hand, the heat transport properties, characterized by λ values, show significant negative correlation with Φ.

Although the moisture storage properties show no significant correlation with basic properties, the moisture content of the materials affects the transport properties of both moisture and heat, which is always ignored in non-HAM models. The impact of gaseous moisture content on δ, U, λ, and surface light and thermal transport properties would be discussed in Sect. 3.2.4.3. The water vapor transmission resistance factor and thermal conductivity in previous works are carried out in steady state, without consideration to the effect caused by moisture content. In this study, the impacts of moisture content on moisture and heat transport properties are tested and quantitatively analyzed. The results show that the increase of moisture content will increase the transport properties to varying degree and therefore cannot be avoided (Fig. 3.20).

3.2.4.3 Impact of Moisture Content on Heat and Moisture Transport Properties

The vapor transmission coefficients with different moisture content are tested, and the results show that they are significantly affected by moisture content. The Spearman correlation analysis results between δ and u are as follows: correlation coefficient (r) $= 0.814870$ and statistical significance (sig) $= 3.1586\text{E}^{-12}$ ($N = 47$). Curve fitting results show that there is an exponential function relation between δ and u values that can be defined as $\delta = a \times \text{EXP}(b \times u)$, with R^2 in the range 0.9083–0.9972, and sig < 0.01. Owing to the surface liquid water accumulation, the BFB specimens corresponding to $u_{\text{RH} = 93\%}$ are too high and removed from this analysis.

The impact of moisture content on drying rate is tested, and the results show that they are significantly affected by moisture content. The Spearman correlation analysis results between U and u are $r = 0.651771$ and sig $= 7.2251\text{E}^{-09}$ ($N = 63$). Curve fitting results show that there are logarithmic function relations between U and u values that can be defined as $U = a \times \ln(u) + b$, with R^2 in the range 0.9379–0.9904, and sig < 0.01 (Fig. 3.21).

The impact of moisture content in the gaseous moisture region on heat transport properties is tested, and the results show that the λ values have a nonlinear relation with moisture content, which is determined by the affinity between the material framework and moisture inside [8]. Generally, a thermal conductivity supplement,

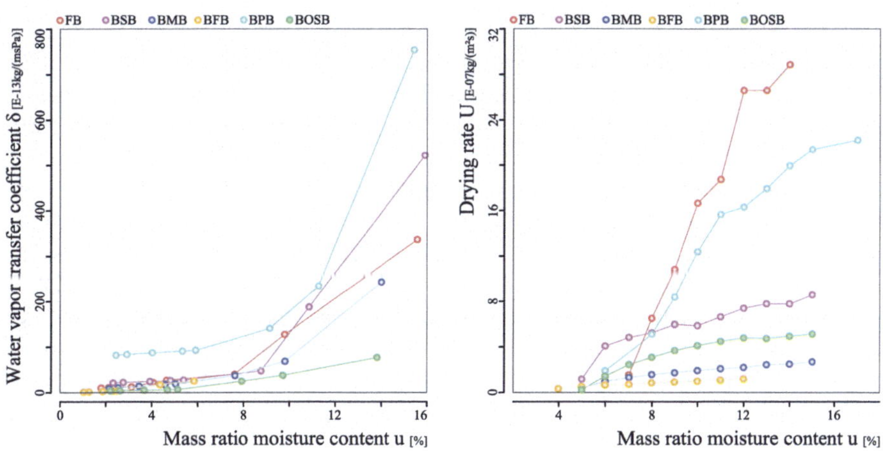

δ=a×EXP(b×u)				
Material	a	b	R²	sig.
FB	6.9969E-13	25.901810	0.972738	0.000006
BSB	9.4511E-13	24.486998	0.941027	0.000066
BMB	5.1624E-13	26.974973	0.997189	6.9458E-9
BFB	1.1541E-13	56.169734	0.976291	0.000030
BPB	4.3498E-12	16.285983	0.908253	0.000250
BOSB	2.1042E-13	27.584925	0.973023	0.000006

U=a×ln(u)+b				
Material	a	b	R²	sig.
FB	36.653893	101.056969	0.959219	0.000004
BSB	5.676184	19.350068	0.937934	9.8166E-7
BMB	1.952089	6.431779	0.987276	7.6886E-10
BFB	0.748983	2.796362	0.990416	2.5176E-8
BPB	20.134181	58.648931	0.958443	3.1046E-8
BOSB	4.377563	13.909484	0.958203	1.6425E-7

Fig. 3.21 Impact of moisture content on water vapor transmission coefficient (left) and drying rate (right), and the curve fitting

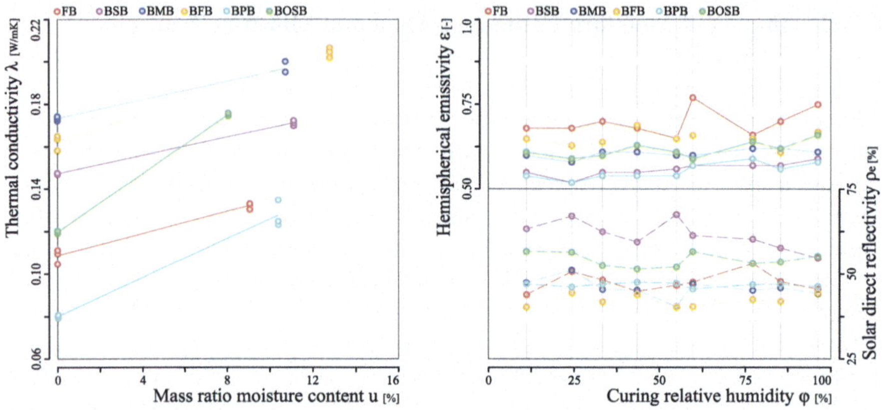

λ=aw×(u)+b				
Material	a_w	b	R^2	sig.
FB	0.258728	0.108800	0.969227	0.000359
BSB	0.213729	0.147500	0.996019	0.000006
BMB	0.218498	0.173300	0.977901	0.000184
BFB	0.328903	0.162500	0.986426	0.000069
BPB	0.458253	0.080033	0.976924	0.000201
BOSB	0.688971	0.119667	0.999661	4.3146E-8

Correlation indicators		ε	ρ_e
φ	Correlation coefficient-r:	0.195052	-0.059824
	Statistical significance-sig:	0.157534	0.667404
	Number-N:	54	54

Fig. 3.22 Impact of moisture content on thermal conductivity (left) and surface light and thermal properties (right)

a_w [(W/m K)/u(–)] value, is used to approximate the linear function between λ value [W/(m K)] and the moisture content u value [–]. The test results show that a_w values are 0.2587, 0.2137, 0.2185, 0.3289, 0.4583, and 0.6890 for FB, BSB, BMB, BFB, BPB, and BOSB, respectively.

On the other hand, the impact of gaseous moisture content on the surface light and thermal transport properties is uncertain in this test. Spearman correlation analysis shows that the r and sig values are 0.1951 and 0.1575, respectively, between ε and φ; and −0.0598 and 0.6674, respectively, between ρ_e and φ, which means that there is no significant correlation between the surface light and thermal transport properties and the gaseous moisture content. Therefore, the impact of moisture content in the gaseous water region on the surface light and thermal transport properties can be ignored, but it requires further study in the liquid water region (Fig. 3.22).

3.2.4.4 Assessment on the Bamboo-Based Panel Process

In the test, the flattened bamboo panel (FB) represents the raw bamboo, while the remaining five are modified bamboo through 'decomposition' and 'recombination' manufacturing process. The comparison between five BBPs and the FB shows that the

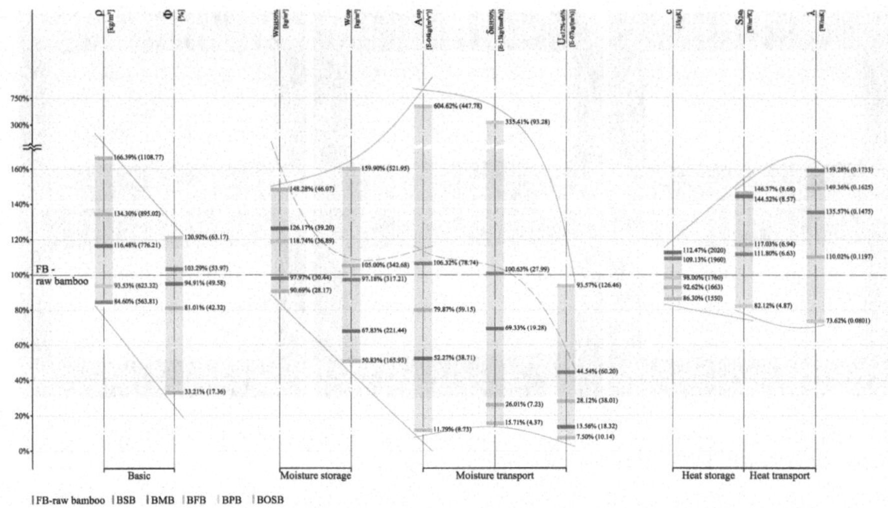

Fig. 3.23 Material properties relative position of bamboo-based panel (BBPs) and raw bamboo (FB) (the percentage values outside the brackets are BBPs/FB ratios and the values inside the brackets are the corresponding measured/calculated results). *Note* The calculation method of the liquid water transport coefficient in Sect. 3.2.2.4 is controversial, so it is not analyzed in this figure

BBP process improves homogeneity and broadens the material properties spectrum, which is beneficial for diversified use. Except BPB, the distinction between bamboo and timber is enhanced, which is manifested by the increase in bulk density, decrease in moisture storage and transport properties, and increase in heat storage properties, as well as an increase in thermal conductivity.

Considering the most commonly used hygrothermal parameters, the moisture properties of BFB are significantly reduced to 7.50–15.71%, making it suitable for outdoor use, while its heat transport properties increase to 146.37–149.36%. Changes in BMB and BOSB are similar to those of BFB, in the range 13.56–79.87% for moisture transport properties and 110.02–159.28% for heat transport properties. The BPB's heat transport properties decrease to 73.62–82.12%, while its moisture transport properties increase to 335.41–604.62%, making it a better interlayer material for thermal insulation. The BSB changes slightly in terms of both hygric and thermal aspects (Fig. 3.23).

3.2.4.5 Mold Growth on Specimens in Vacuum Dryer of High Humidity

In the sorption test, mold does not appear in the specimen group of the 8 relative humidity gradient of RH = 11.3–85.4%, but occurs in the vacuum dryer of RH = 96.3% (K_2SO_4). The mildews grow first on the BPB and then gradually appear on the remaining specimens. Further research on the growth of bamboo molds needs

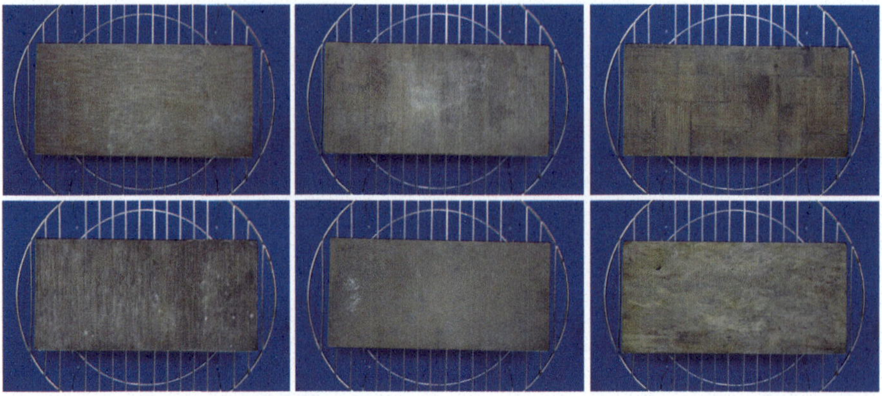

Fig. 3.24 Mold growth on surface of the specimens in vacuum dryer of RH = 96.3% (photographs are FB, BSB, BMB, BFB, BPB, and BOSB successively)

to be carried out, including a necessary test of the isopleth range for the growth of certain mildews at the material level [31]. The test results can be compared with the temperature and relative humidity environment simulated by the HAM model to assess the risk of mold growth. The WUFI Bio computer program of Fraunhofer IBP has the function of performing such analysis on the inner surface of building components (Fig. 3.24).

3.3 Test on Construction Infill Material

Bamboo fiber can be divided into natural bamboo fiber (hereinafter referred to as BF) and regenerated bamboo fiber. The latter is chemically treated on the basis of the former and is mainly used in the textile and the paper industry (bamboo pulp fiber). The crude bamboo fiber obtained by mechanical rolling or flash explosion process has the potential to become a mechanical reinforcement in composite materials, a fiber thermal insulation or a sound absorbing material. Plant fiber is widely used in low-rise civil buildings in Europe and the North America. The enriched bamboo resources make bamboo fiber a preferred target for the development of plant fiber construction infills in the bamboo producing area.

Bamboo charcoal (hereinafter referred to as BC) is formed by carbonization of raw bamboo with high temperature through isolation of oxygen. According to the carbonization temperature, it can be divided into three types: High-temperature carbonized bamboo charcoal is generated with carbonization temperature higher than 1000 °C, and the product has higher conductivity; medium-temperature carbonized bamboo charcoal is produced with carbonization temperature between 700–800 °C, and the product has good adsorption performance; low-temperature carbonized bamboo charcoal is prepared with carbonization temperature of about 600 °C, and the product has better moisture-control ability. Bamboo charcoal and its activated charcoal have excellent adsorption performance due to its unique microstructure. In the process of preparing bamboo charcoal, bamboo vinegar is also produced as a by-product. Compared with wood charcoal, bamboo charcoal has advantages in its tissue structure and function. The outer surface of the bamboo wall is dense, and the specific surface area (BET) is greatly increased during the carbonization and activation process with high temperature. The vascular bundle tissues per unit area of bamboo are about twice that of wood, and the BET is about three times as much as wood, making bamboo charcoal have higher adsorption capacity than wood charcoal and other materials. The adsorption capacity of bamboo charcoal is eight times, and meanwhile the anion is five times that of wood charcoal. Anions have strong deodorizing and purifying abilities, so that they have a special effect on water and air purification. Wood charcoal has a carbon content of about 87–93%, so it can be used as an ideal fuel, while bamboo charcoal has about 75–86%, so in comparison it is not suitable to be a fuel. Bamboo charcoal after activation has better performance, in terms of adsorption, filtration, far infrared ray, and the electromagnetic wave shielding.

Bamboo resources are widely distributed in temperate, subtropical, and tropical regions. These areas are hot and humid, with strong solar radiation and abundant rainfall, resulting in a large heat and moisture load on the building façade. The BF and BC resources that are nearby available may become certain kinds of climate-adapted and local building materials; the study on the hygrothermal properties of BF and BC as climate-adaptive building infills has the dual significance of effective utilization of bamboo forest resources and improvement of building physical performance.

The aim of this test is to obtain the hygrothermal properties of BF as a thermal insulation infill and discuss the effect of hygric environment on the hygrothermal properties of BF; the BC is tested for the hygric properties parameters as a moisture-control infill, and its control capacity is evaluated. The material parameters obtained provide a basis for further performance studies at the building component level.

3.3.1 Test Objects

3.3.1.1 Natural Bamboo Fiber (BF)

The BF samples are natural bamboo fibers extracted from Phyllostachys pubescens (Phyllostachys Sieb. et Zucc., the main industrial utilized bamboo species in China) by rolling technology and are mainly used as raw materials for regenerated bamboo fiber products. The unit length is 2.0–10.0 cm (Fig. 3.25).

3.3.1.2 Bamboo Charcoal (BC)

The BC samples are moisture-control bamboo charcoals that are produced through 600 °C carbonization process and are mainly used as raw materials for infills of

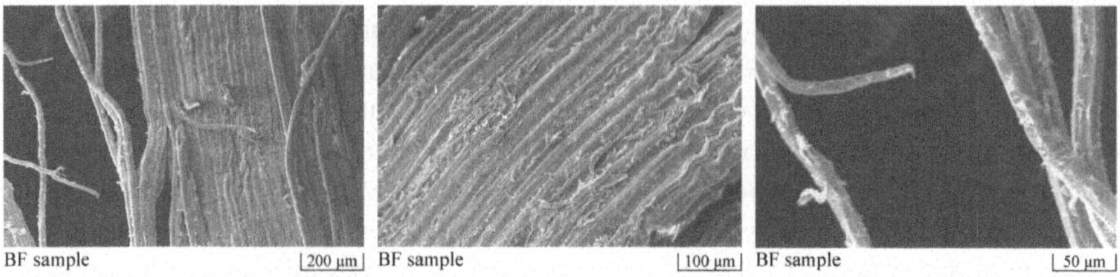

Fig. 3.25 Natural bamboo fiber (BF) sample surface morphology (images are taken by Zeiss EVO18 scanning electron microscopy)

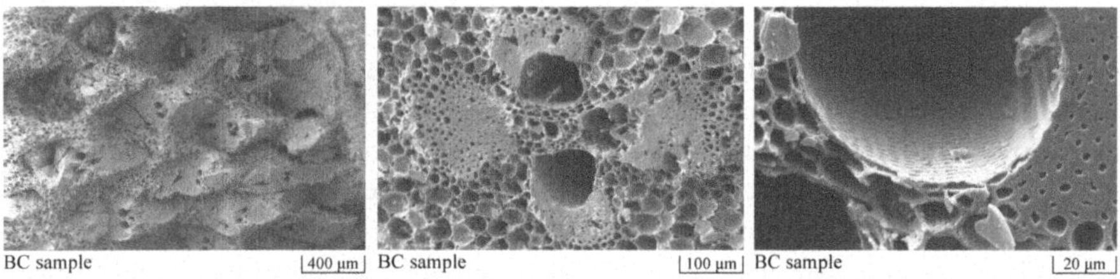

Fig. 3.26 Bamboo charcoal (BC) sample surface morphology (images are taken by Zeiss EVO18 scanning electron microscopy)

moisture-control products. The unit diameter is 1.0–1.5 cm, and bulk density is 270 kg/m^3 (Fig. 3.26).

3.3.2 Test Items

The test items include the sorption test for moisture storage properties, the water vapor transmission test and drying test for moisture transport properties, and the thermal conductivity test for heat transport properties (Table 3.11).

3.3.2.1 Sorption Test

The sorption test is used to test the isothermal adsorption and desorption curve of the material to obtain the equilibrium moisture content u values that characterize the moisture storage property. In the experiment, a predetermined mass of dry BF is weighed and filled in a glass dish to obtain a BF specimen with a filling density of $\rho = 110$ kg/m^3, which is prepared 6 copies for 1 type, including 3 copies for each of the moisture adsorption and desorption processes; the BC which is dried and calculated in predetermined quality is filled in a glass dish to obtain a BC specimen with a filling density of $\rho = 270$ kg/m^3, which is, in the same way, prepared 6 copies for 1 type, including 3 copies for each of the moisture adsorption and desorption

Table 3.11 Operation method, equipment arrangement, and specimen treatment of the hygrothermal properties' test

Target value	Operation method	Equipment arrangement	Specimens treatment
Item 1: Isothermal adsorption and desorption curve w [kg/m^3]	Refer to the international standard ISO 12571:2012 and the American standard ASTM-04a (2016)[a] [14, 15]	Vacuum dryer (9 copies): $d = 30$ cm Drying oven: digital stainless steel electric blast oven 101A-2S, accuracy ± 1 °C Balance: SHIMADZU UX6200H, accuracy 0.01 g T&RH recorder: TH10R-EX, with exterior sensor (9 copies), accuracy $T \pm 0.2$ °C, RH ± 2% Glass dish (102 copies): $d =$ 100 mm, $h = 20$ mm	Bulk density: $\rho_{BF} = 110$ kg/m^3, $\rho_{BC} = 270$ kg/m^3 Size: filled in the glass dishes Quantity: 6 copies for 1 type, of which 3 copies for both adsorption and desorption process
Item 2: Water vapor permeability coefficient δ [kg/(m s Pa)]	Refer to the international standard ISO 12572:2001(E) and the American standard ASTM E 96/E 96 M-2005[b] [18, 19]	Dry cups (6 copies), wet cups (6 copies): 12 × 12 cm, $h = 5$ cm Acrylic container tube (12 copies): d = 10 cm, $h = 10$ cm Drying oven: same as item 1 Balance: same as item 1 Constant T&RH conditioning box: HWS-250B, accuracy $T \pm 0.5$ °C, RH ± 5%	Bulk density: $\rho_{BF} = 110$ kg/m^3, $\rho_{BC} = 270$ kg/m^3 Size: filled in the acrylic container tubes Quantity: 6 copies for 1 type, of which 3 copies for both dry cups and wet cups
Item 3: Drying curve; adsorption and desorption rate U [kg/(m^2 s)]	Refer to the test method of Fraunhofer IBP (accredited according to DIN EN ISO/IEC 17025)	Acrylic container tube (24 copies): d = 10 cm, $h = 4/6/8/10$ cm (sealed at bottom, reserved 5 pores, $d = 5$ mm) Drying oven: same as item 1 Balance: same as item 1 Constant T&RH conditioning box: same as item 2	Bulk density: $\rho_{BF} = 110$ kg/m^3, $\rho_{BC} = 270$ kg/m^3 Size: filled in the acrylic container tubes Quantity: 3 copies for 1 type

(continued)

Table 3.11 (continued)

Target value	Operation method	Equipment arrangement	Specimens treatment
Item 4: Thermal conductivity λ [W/(m K)]	Refer to the international standard ISO 8302-1991 [22]	Guarded hot plate apparatus: CD-DR3030, hot plate $T = 35$ °C, cold plate $T = 15$ °C, size of central heat transfer area 15 × 15 cm, accuracy ± 2% EPS template: 30 × 30 cm, $h =$ 28 mm (inside hollow 20 × 20 cm) Drying oven: same as item 1 Balance: same as item 1 Constant T&RH conditioning box: same as item 2	Bulk density: $\rho_{BF} =$ 70/90/110/130/150/170 kg/m^3 Size: 20 × 20 cm (filled in the EPS templates) Quantity: 2 copies for 1 type

[a]Temperature (T) and relative humidity (RH) of ambient air control by different saturated salt solutions in vacuum dryers are shown in Table 3.6

[b]Relative humidity (RH) control by desiccant/saturated salt solutions in cups are shown in Table 3.7; the water vapor transmission coefficient δ_{air} [kg/(m s Pa)] of stationary air under temperature T_k and barometric pressure p is calculated according to [20]

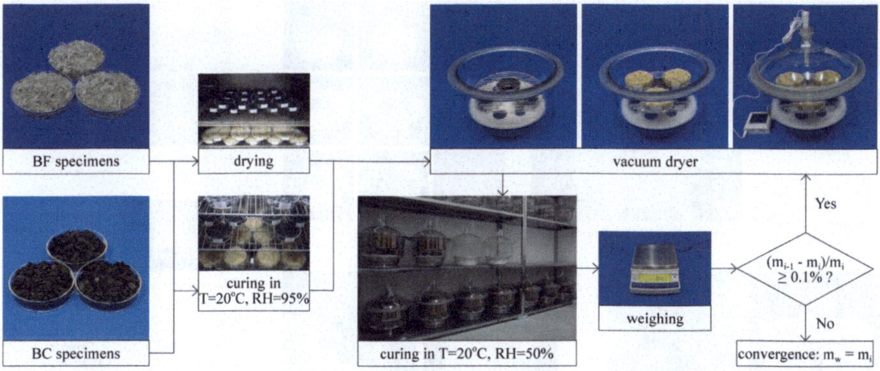

Fig. 3.27 Photograph program of the sorption test (construction infill group)

processes. The main experimental devices are: 9 vacuum dryers, inner diameter $d =$ 30 cm; 1 temperature and relative humidity recorder TH10R-EX and its 9 external sensors, accuracy $T \pm 0.2$ °C, RH $\pm$ 2%; 102 glass dishes, $d = 100$ mm, $h = 20$ mm.

The methods recommended by the international ISO 12571:2012 standard and the American ASTM-04a (2016) standard are adopted for the test [14, 15]. The dried specimens are kept standing still at 20 °C, and the dry mass m_d [kg] is recorded. Then, the dried specimens are placed in vacuum dryers, in which relative humidity is controlled using specific quantities of saturated saline solution at the bottom. The vacuum dryers are moved together with the temperature (T) and relative humidity (RH) recorders into a constant temperature and humidity chamber ($T = 20$ °C, RH = 50%) for curing, and the T and RH changes in the vacuum dryers are monitored every day until they reach the predetermined values. 4 weeks later, the dryers are opened every 3 d to weigh the specimens that are returned after they are weighed. The procedure is repeated until the differences between three successive weighting results are less than 0.1%, when the material is assumed to reach its sorption equilibrium. The mass of the wet specimens m_w [kg] at this point is recorded (Fig. 3.27).

According to

$$u = \frac{m_\mathrm{w} - m_\mathrm{d}}{m_\mathrm{d}}$$

the mass ratio moisture content u [kg/kg] is calculated and the isothermal moisture adsorption curve of the specimen is thus acquired. Based on the same principles, the specimens that are kept in a constant T&RH conditioning box ($T = 20$ °C, RH = 90%) for 2 weeks until they reach their adsorption equilibrium are placed into the vacuum dryers, and the procedures described above are repeated to obtain the isothermal moisture desorption curve of the specimens. In this experiment, it is not able to provide a higher curing hygric environment than that in the K_2SO_4 vacuum dryer (reference RH control value RH = 97.6 $\pm$ 0.6%) to form the initial moisture

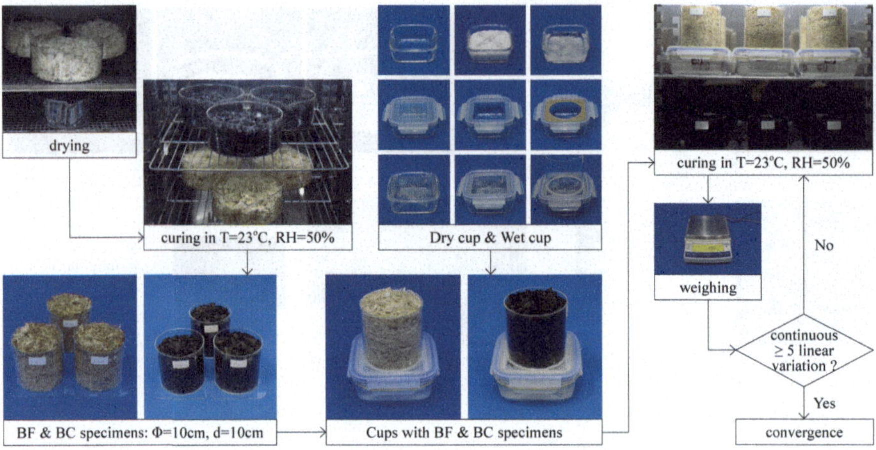

Fig. 3.28 Photograph program of the water vapor transmission test (construction infill group)

content as the wet specimens, so that the desorption process is conducted only in the remaining eight relative humidity gradients.

3.3.2.2 Water Vapor Transmission Test

The water vapor transmission test is used to test the water vapor transfer coefficient δ to characterize the moisture transport properties of the material. In the experiment, the predetermined mass of dry BF is weighed and filled in the acrylic glass tube containers to obtain the BF specimens with a filling density of $\rho = 110$ kg/m^3, which is prepared 6 copies for 1 type, including each 3 copies for the dry and wet cups; similarly, the BC which is dried and calculated for predetermined mass is filled the acrylic glass tube containers to obtain the BC specimens having a filling density of $\rho = 270$ kg/m^3, which is prepared 6 copies for 1 type, including each 3 copies for the dry and wet cups. The dry cups contain desiccant $CaCl_2$ ($d = 3$ mm), in which the reference value of relative humidity is RH = 0%, while the wet cups contain K_2SO_4 saturated salt solution, in which the reference value of relative humidity is RH = 96.3%. The main experimental devices include: balance SHIMADZU UX6200H, accuracy 0.01 g; constant T&RH conditioning box HWS-250B, accuracy $T \pm 0.5$ °C, RH ± 5%; 6 acrylic glass tube containers, $d = 100$ mm, $h = 100$ mm; 6 square glass bowls as the bottom containers of the dry and wet cups, with the interior dimensions 120 × 120 mm, $h = 50$ mm (Fig. 3.28).

The test is based on the methods recommended by the international ISO 12572:2001(E) standard and the American ASTM E 96/E 96 M-2005 standard [18, 19]. The specimen containers formed by using acrylic tubes are sealed and fixed on the plastic covers of the glass bowls. A circular opening ($d = 100$ mm) is retained in each of the bowl cover, which is used as the testing area for the specimens. The specimen and the plastic cover is weighed for the overall dry mass $m_{d,lid}$ [kg] and then moved into the constant T&RH conditioning box ($T = 23$ °C, RH = 50%) for curing. Two weeks later, when the specimens reach their adsorption equilibrium, the specimens and the plastic covers are sealed together to the opening of the glass bowls, which contain drying agents or the saturated saline solution to control the relative humidity of the environment under the specimen.

The above overall apparatuses, namely the 'dry cups' and 'wet cups,' are moved to a constant T&RH conditioning box ($T = 23$ °C, RH = 50%) for curing. The cups are weighted every 3 d, and the process is completed after linear changes are observed in the total weight over five consecutive measurements. The specimen and plastic cover are subsequently removed from the bowl and weighed to measure the wet mass $m_{w,lid}$ [kg]. Subsequently, they are dried and weighted for the dry mass m_d [kg]. According to

$$w = \frac{m_{w,lid} - m_{d,lid}}{V}$$

the moisture content of the specimens w [kg/m^3] is calculated, where V is the volume of the acrylic glass tube. The recorded mass and time data of each group are linearly fitted, and the obtained mass change rate is the water vapor transfer rate G [kg/s]. According to

$$g_v = \frac{G}{A}$$

The water vapor transfer rate per unit area g_v [kg/(m^2 s)] is obtained, where A is the opening area of the bowl cover. According to

$$\Delta p_v = \Delta \varphi \cdot 610.5 \cdot e^{\frac{17.269(T_k - 273.15)}{T_k - 35.85}}$$

the water vapor pressure difference Δp_v [Pa] between the two sides of the specimen is calculated, where T_k is the absolute temperature. According to

$$R_{v,total} = \frac{\Delta p_v}{g_v}$$

the total water vapor transfer resistance $R_{v,total}$ [m^2 s Pa/kg] is calculated. According to

$$\delta_{air} = \frac{2.306 \times 10^{-5} p_o}{R_o T_k p} \left(\frac{T_k}{273.15} \right)^{1.81}$$

the water vapor permeability of air δ_{air} [kg/(m s Pa)] with respect to a certain temperature T_k and barometric pressure p [20] is obtained, where the standard barometric pressure $p_o = 101{,}325$ Pa and the gas constant for water vapor $R_o = 461.5$ N m/(kg K). According to

$$R_v = R_{v,total} - \frac{h_{air}}{\delta_{air}}$$

the corrected water vapor resistance R_v [m^2 s Pa/kg] is calculated, where h_{air} [m] is the water vapor diffusion-equivalent air layer thickness, which is equal to the thickness of the air layer between the lower surface of the specimen and the desiccant/liquid level. According to

$$\delta = \frac{d}{R_v}$$

The water vapor transfer coefficient δ [kg/(m s Pa)] of the specimen is calculated, which corresponds to the state when the moisture content of the specimen is w.

3.3.2.3 Drying Test

The drying test is used to test the drying curve of the material and calculate the drying rate, which is regarded as the moisture adsorption and desorption rate U value that characterizes the moisture transport property. In the experiment, a predetermined mass of dry BF is weighed and filled in a transparent acrylic glass tube container to obtain a BF specimen having a filling density of $\rho = 110$ kg/m^3, which is prepared 3 copies for 1 type; similarly, the BC which is dried and calculated for predetermined mass is filled in transparent acrylic glass tube containers to obtain the BC specimens having a filling density of $\rho = 270$ kg/m^3, which is prepared 3 copies for 1 type. The main experimental devices are: balance SHIMADZU UX6200H, accuracy 0.01 g; constant T&RH conditioning box HWS-250B, accuracy $T \pm 0.5$ °C, RH $\pm$ 5%; 12 acrylic glass tubes, with the bottom sealed and five airholes of $d = 5$ mm reserved, $d = 100$ mm, $h = 40/60/80/100$ mm.

The test is based on the method recommended by Fraunhofer IBP (DIN EN ISO/IEC 17025 certified). The specimens with predetermined mass are placed in a glass dish to which distilled water is added to soak the specimens, and they reach a state of capillary saturation after 2 d. The specimens are taken out from the water, and the adhered waters are removed. Subsequently, they are inserted into a preweighed acrylic glass tube and transferred to a constant T&RH conditioning box ($T = 23$ °C, RH = 50%) for curing. The specimens are weighed every 12 h until the differences between three successive measurements do not exceed 0.1%, when it is assumed that the specimens are fully dried. The total mass of the specimens $m_{d,total}$ [kg] is recorded (Fig. 3.29).

After the materials are taken out from the acrylic glass tubes and dried, the dry mass m_d [kg] is recorded. Based on this, the instantaneous moisture content w [kg/m^3]

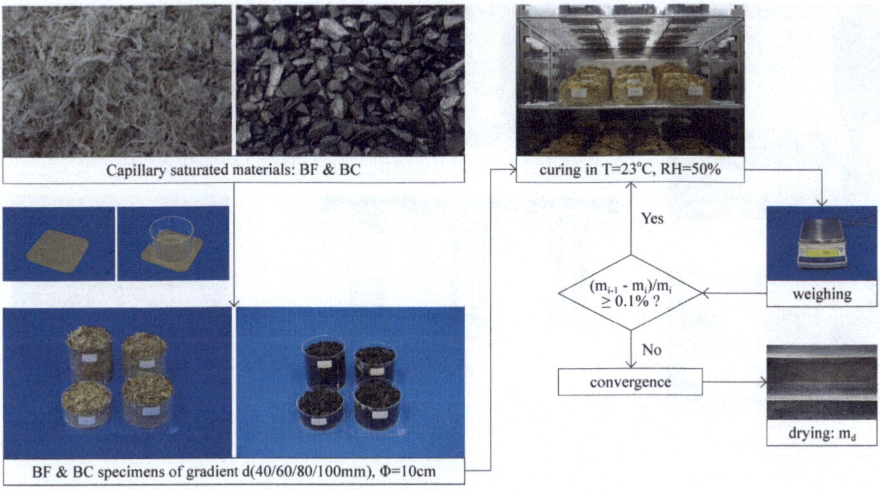

Fig. 3.29 Photograph program of the water drying test (construction infill group)

values of the specimens are estimated, and the instantaneous moisture content (w) is derivatized to the drying time (t) to obtain a w–t drying curve. Then, according to

$$U = -\frac{m_\mathrm{d}}{A} \cdot \frac{\mathrm{d}w}{\mathrm{d}t}$$

the drying rate is calculated, where A [m^2] is the drying area of the specimens, which is the open area of the acrylic glass tubes. The calculation result is regarded as the moisture adsorption and desorption rate U [kg/(m^2 s)] of the materials.

3.3.2.4 Thermal Conductivity Test

The test is used to obtain the thermal conductivity λ value that characterizes the heat transfer properties. Considering that the bulk density of the wood fiber insulation board is generally 70–170 kg/m^3, the dry BF of predetermined mass is weighed during the experiment and filled in the EPS templates of 30 $\times$ 30 cm ($d = 28$ mm, with a hollow 20 $\times$ 20 cm in the middle) to obtain the BF specimens of filling density $\rho = 70/90/110/130/150/150/170$ kg/m^3, 2 copies for 1 type. The specimens are first dried and placed in a constant T&RH conditioning box to cure for a set equilibrium moisture content, in which the relative humidity of the curing environment is set to be RH = 40/50/60/70/80/90%. The main experimental devices are: guarded hot plate apparatus CD-DR3030, hot plate $T = 35$ °C, cold plate $T = 15$ °C, size of central heat transfer area 15 $\times$ 15 cm, accuracy $\pm$ 2%; constant T&RH conditioning box HWS-250B, accuracy $T \pm 0.5$ °C, RH $\pm$ 5% (the T&RH recorder TH10R-EX with

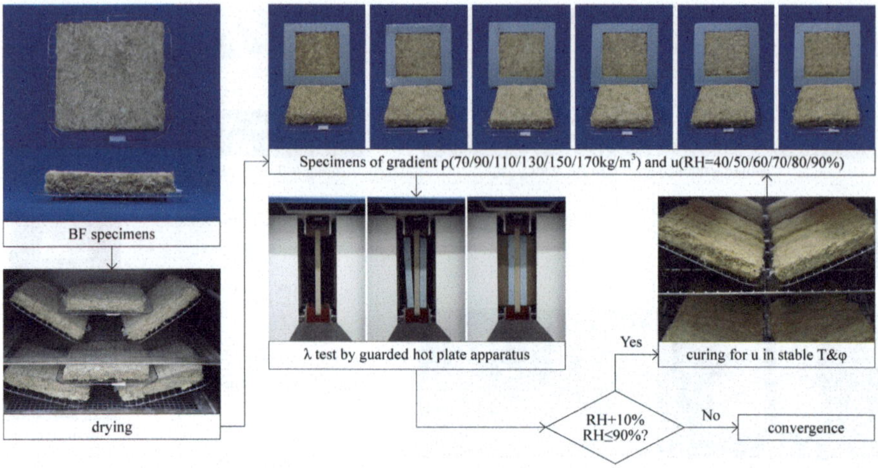

Fig. 3.30 Photograph program of the thermal conductivity test (construction infill group)

an accuracy of $T \pm 0.2$ °C and RH $\pm$ 2% is put inside for real-time monitoring of the actual temperature and relative humidity in the box).

The test is based on the method recommended by the international ISO 8302-1991 standard [22]. First, a CD-DR3030 guarded hot plate apparatus is used to conduct thermal conductivity tests on the dried specimens, thus obtaining the dry thermal conductivity λ_d [W/(m K)]. The specimens are subsequently placed in a constant T&RH conditioning box for conditioning at a specified level of relative humidity. After 2 weeks, the specimens are weighed, and the wet λ values corresponding to the moisture content of the specimens are measured. After the measurement is completed, the specimens are replaced in the constant T&RH conditioning box and conditioned at the next level of relative humidity. The operation is repeated until the λ values of the specimens that correspond to all filling density gradients and moisture content gradients are obtained (Fig. 3.30).

3.3.3 Results Analysis

3.3.3.1 Natural Bamboo Fiber

In the mid-twentieth century, Luikov et al. proposed highly coupled nonlinear partial differential equations to describe the coupled heat and moisture migration in porous materials [32, 33]. In 1995, Künzel modified the Luikov equations, considered relative humidity and water vapor pressure as the driving potentials of the moisture transport and developed the coupled heat and moisture process simulation computer program WUFI [2]. The hygrothermal properties are commonly described by using moisture content, bulk density, ambient air temperature, and relative humidity with various fitting functions. Occasionally, dozens of fitting equations are obtained for a certain property from different perspectives. In this study, exponential function relations are obtained that fitted well with the equilibrium moisture content u values, adsorption and desorption rate U values, and thermal conductivity λ values with the ambient relative humidity φ values as described with a general function:

$$Y = a \cdot e^{b \cdot \varphi} + c$$

Y referred to u, U, and λ; a, b, and c are fitting coefficients.

(1) u–φ test results

The BF isothermal adsorption and desorption u–φ curve indicates an anti-'S' curve feature that is commonly termed as the 'second type of isotherm' that is normally described with BET equation for the anterior part and Kelvin equation for the posterior. When fitting the u–φ test results with exponential function, the results are

$$u = 0.0039\mathrm{e}^{4.2827\varphi}$$

($R^2 = 0.9584$, sig $= 0.000004$) for the adsorption curve,

$$u = 0.0145\mathrm{e}^{2.8363\varphi}$$

($R^2 = 0.9274$, sig $= 0.000031$) for the desorption curve, and

$$u = 0.0088\mathrm{e}^{3.3637\varphi}$$

($R^2 = 0.9495$, sig $= 0.000009$) for the mean curve.

From the R^2 and sig values, it can be seen that exponential function describes the u–φ curve to a favorable degree (Table 3.12).

The BF indicates high moisture adsorption capacity in which the mass ratio moisture content reaches $u = 25.84\%$ when $\varphi = 96.3\%$. Within the test range, the curve slope corresponding to $\varphi = 85.4$–96.3% is considerably higher than that of the $\varphi = 11.2$–85.4%, and this is due to the higher water vapor pressure and also because the moisture content of BF reaches the fiber saturation point when $\varphi = 85.4$–96.3% and therefore corresponds to the free water adsorption stage such that the u values are significantly increased.

The isothermal adsorption and desorption curve is a collection of equilibrium moisture content u values and the corresponding ambient relative humidity. The process through which the material reaches the u values leads to different results. Typically, the u values resulting from desorption process exceed those of the adsorption process to varying degree, and this is termed as the 'hysteresis effects.' The phenomenon varies among different materials. In certain mineral building materials,

Table 3.12 Temperature (T) and relative humidity (φ) of ambient air control by different saturated salt solutions in vacuum dryers and the measured equilibrium moisture content (u value)

No.	Saturated salt solutions	T measured value [°C]	Relative humidity [%]		Measured u values [kg/kg]		
			φ Reference value	φ Measured value	Adsorption (%)	Desorption (%)	Average (%)
1	LiCl	20.2	11.31 ± 0.31	11.2	0.49	1.50	1.00
2	CH_3COOK	20.2	23.11 ± 0.25	24.4	0.84	2.67	1.75
3	$MgCl_2$	20.2	33.07 ± 0.18	33.4	2.06	4.52	3.29
4	K_2CO_3	20.4	43.16 ± 0.33	43.5	3.68	6.70	5.19
5	$Mg(NO_3)_2$	20.3	55.87 ± 0.27	55.0	5.06	7.92	6.49
6	NaBr	20.3	59.14 ± 0.44	59.7	5.55	9.05	7.30
7	NaCl	20.1	75.47 ± 0.14	77.2	8.65	10.59	9.62
8	KCl	20.1	85.11 ± 0.29	85.4	11.33	11.85	11.59
9	K_2SO_4	20.3	97.6 ± 0.6	96.3	25.84	–	25.84

T&RH sensors are inserted inside the vacuum dryers to monitor the real values of ambient air

the adsorption curve is considered as sufficient for practical applications when the hysteresis effect can be neglected [2]. Nevertheless, with respect to vegetable fibers, the hydroxyl groups of the cellulose molecule are partly connected to each other by deputy valences and mutual saturation is achieved during the drying process. Therefore, the subsequent adsorption capacity for moisture is weakened, and this results in lower u values. In the test range $\varphi = 11.2$–85.4%, an evident hysteresis effect is observed in which the u values of desorption process exceed those of the adsorption process by 0.52–3.50%, and thus, the use of the average u values is recommended (Fig. 3.31).

(2) U–φ test results

The BF drying curve indicates two different stages in which the adsorption and desorption rate U values corresponding to $\varphi = 75$–95% significantly exceed those of $\varphi = 50$–75%. The U values in former stage mainly depend on the ambient temperature and relative humidity, while the latter stage is affected by its own properties. An exponential function

$$U = a \cdot e^{b \cdot \varphi}$$

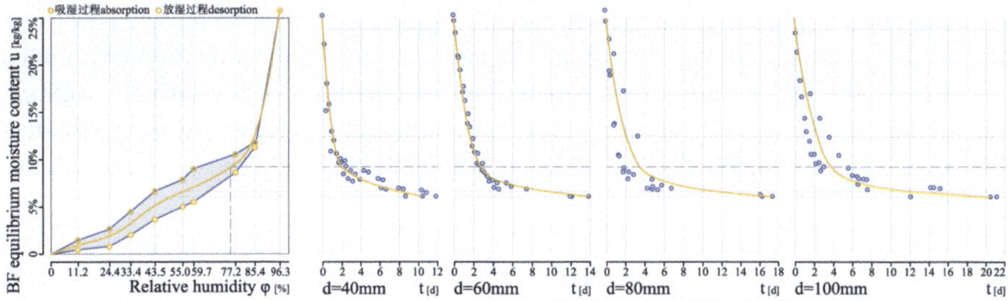

u=a×EXP(b×φ)

Process	a	b	R²	sig.
absorption	0.0039	4.2827	0.9584	0.000004
desorption	0.0145	2.8363	0.9274	0.000031
average	0.0088	3.3637	0.9495	0.000009

U=a×EXP(b×φ)

Specimens	a	b	R²	sig.
d=40mm	2.2794E-10	0.109511	0.958236	0.000004
d=60mm	2.7671E-10	0.111487	0.989946	2.9770E-8
d=80mm	3.1354E-10	0.110578	0.995909	1.2762E-9
d=100mm	2.7676E-10	0.113991	0.993365	6.9418E-9

Fig. 3.31 BF isothermal adsorption and desorption u–φ curve (left) and the drying curve of specimens with gradient thickness (right)

Table 3.13 BF moisture adsorption and desorption rate test results

φ Region corresponding to the moisture content (%)	Measured U values [kg/(m^2 s)]				
	40	60	80	100	Average
$\varphi = 95$–90	5.18488E^{-06}	5.43272E^{-06}	5.93655E^{-06}	6.46256E^{-06}	5.75418E^{-06}
$\varphi = 90$–85	3.17065E^{-06}	3.74002E^{-06}	3.67350E^{-06}	4.15984E^{-06}	3.68600E^{-06}
$\varphi = 85$–80	1.69439E^{-06}	2.22363E^{-06}	2.27235E^{-06}	2.85198E^{-06}	2.26059E^{-06}
$\varphi = 80$–75	7.46166E^{-07}	1.34196E^{-06}	1.52849E^{-06}	1.62270E^{-06}	1.35379E^{-06}
$\varphi = 75$–70	2.88939E^{-07}	8.18542E^{-07}	7.58969E^{-07}	9.66016E^{-07}	7.08116E^{-07}
$\varphi = 70$–65	1.94335E^{-07}	3.55418E^{-07}	3.61248E^{-07}	4.66671E^{-07}	3.44418E^{-07}
$\varphi = 65$–60	1.41433E^{-07}	1.72079E^{-07}	2.37770E^{-07}	2.65632E^{-07}	2.04229E^{-07}
$\varphi = 60$–55	1.06312E^{-07}	1.16062E^{-07}	1.38669E^{-07}	1.34067E^{-07}	1.23777E^{-07}
$\varphi = 55$–50	8.68895E^{-08}	8.60425E^{-08}	7.68911E^{-08}	7.36279E^{-08}	8.08628E^{-08}

is used to fit the U–φ test results and results in $R^2 = 0.958236$–0.995909 and sig $=$ 1.2762E^{-09}–0.000004, which indicates a high fitting degree. The fitting coefficients are $a = 2.2794\text{E}^{-10}$–3.1354E^{-10} and $b = 0.109511$–0.113991 and vary in a range that indicates the slight impact of thickness of the specimens in the test range of $d =$ 40–100 mm (Table 3.13).

(3) λ–φ test results

Spearman nonlinear correlation analysis for the thermal conductivity λ test results indicates that the correlation coefficients correspond to 0.8688 (sig $= 8.833\text{E}^{-14}$) between λ and φ, and 0.8406 (sig $= 3.2946\text{E}^{-12}$) between λ and u. Both correlations between λ and φ and λ–u indicate significance at 0.01 level (2-tailed), and the former is slightly higher than the latter. Conversely, 'S' curve relationship exists between λ

and u, and this implies that the growth rate of λ increases when the u increases from the dry state to the initial wet state. After the u reaches a certain value (approximately $u = 11.5$–13.0% in the study), the growth rate of λ decreases. It is relatively difficult to describe the 'S' curve by using a single function, and it is typically approximated with a linear function. When a linear equation is adopted to fit the λ–u test results, there is

$$\lambda = 0.0421 - 0.016870u$$

of which the correlation index $R^2 = 0.822019$–0.967023 and sig $= 0.000224$–0.004860.

When the exponential function is used, there is

$$\lambda = 0.000457e^{4.3974\varphi} + 0.0437$$

of which the corresponding $R^2 = 0.938057$–0.996640 and sig $= 2.2253E^{-07}$–0.000332, and this indicates a better fitting degree between thermal conductivity and hygric environment by the exponential function (Table 3.14, Fig. 3.32).

The dry λ values of BF with filling density $\rho = 70$–170 kg/m^3 are 0.0423–0.0465 W/(m K), and these are similar to other vegetable fiber insulation materials. The wet λ values increase with the increases of φ, and the growth rate is positively correlated with the ρ values. For example, the dry λ values' difference between specimens of $\rho = 170$ kg/m^3 and $\rho = 70$ kg/m^3 is 9.46%, and it increases to 20.88% with respect to the corresponding $\varphi = 90\%$. This indicates that the effects of hygric environment on λ values of BF are positively correlated with its own filling density.

3.3.3.2 Bamboo Charcoal

The mass ratio moisture content of BC reaches $u = 16.67\%$ when $\varphi = 96.3\%$, and correspondingly the mass volume ratio moisture content is $w = 45.01$ kg/m^3. In the test range $\varphi = 11.2$–85.4%, the u values of desorption process exceed that of the adsorption process by 0.25–1.75%, and this indicates that insignificant desorption hysteresis effect is conducive in playing the moisture-control capacity of BC. The water vapor transmission coefficient δ values are closer to each other with respect to the dry cups and wet cups. The dry cups' δ value ranges from $8.08130E^{-11}$ to $1.17718E^{-10}$ kg/(m s Pa), and the wet cups' δ value ranges from $7.91200E^{-11}$ to $9.03744E^{-11}$ kg/(m s Pa). The average value $\delta = 9.32E^{-11}$ kg/(m s Pa) is adopted. (Tables 3.15, 3.16, Fig. 3.33)

Five bamboo panels chosen as the reference in the study for comparison purposes including flatten bamboo board (FB), bamboo mat board (BMB), bamboo scrimber (BFB), bamboo particleboard (BPB), and bamboo oriented strand board (BOSB)

Table 3.14 BF thermal conductivity test results

Ambient conditioning relative humidity φ [%]	Specimens filling density ρ [kg/m^3]											
	70		90		110		130		150		170	
	u [kg/kg] (%)	λ [W/(m K)]	u [kg/kg] (%)	λ [W/(m K)]	u [kg/kg] (%)	λ [W/(m K)]	u [kg/kg] (%)	λ [W/(m K)]	u [kg/kg] (%)	λ [W/(m K)]	u [kg/kg] (%)	λ [W/(m K)]
0	0.00	0.0423	0.00	0.0425	0.00	0.0432	0.00	0.0447	0.00	0.0455	0.00	0.0465
40	6.92	0.0433	6.66	0.0444	6.37	0.0469	6.57	0.0480	5.86	0.0484	5.53	0.0506
50	7.14	0.0443	7.25	0.0458	6.89	0.0478	7.15	0.0497	6.29	0.0497	6.46	0.0515
60	8.63	0.0453	8.51	0.0467	9.65	0.0494	9.27	0.0527	8.42	0.0543	8.98	0.0556
70	9.95	0.0469	9.73	0.0499	10.73	0.0525	9.89	0.0546	10.04	0.0576	10.30	0.0600
80	11.46	0.0509	12.95	0.0572	12.70	0.0597	12.21	0.0637	12.21	0.0654	12.80	0.0661
90	17.55	0.0592	17.34	0.0634	18.54	0.0687	19.68	0.0696	17.09	0.0711	17.82	0.0730

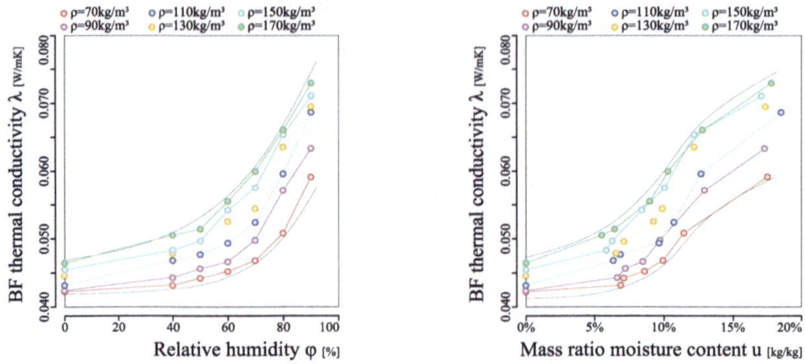

λ=a×EXP(b×φ)+c					
Specimens	a	b	c	R^2	sig.
ρ=70kg/m³	0.000101	0.056430	0.0423	0.996640	2.2253E-7
ρ=90kg/m³	0.000126	0.059320	0.0425	0.981392	0.000016
ρ=110kg/m³	0.000163	0.059683	0.0432	0.943838	0.000259
ρ=130kg/m³	0.000162	0.060797	0.0447	0.947828	0.000215
ρ=150kg/m³	0.000150	0.062387	0.0455	0.961199	0.000102
ρ=170kg/m³	0.000172	0.061625	0.0465	0.938057	0.000332

λ=a×u+b				
Specimens	a	b	R^2	sig.
ρ=70kg/m³	0.001008	0.038579	0.822019	0.004860
ρ=90kg/m³	0.001309	0.038307	0.865294	0.002379
ρ=110kg/m³	0.001437	0.039283	0.890184	0.001414
ρ=130kg/m³	0.001408	0.041685	0.894327	0.001282
ρ=150kg/m³	0.001669	0.041716	0.915370	0.000730
ρ=170kg/m³	0.001610	0.043375	0.947023	0.000224

Fig. 3.32 BF thermal conductivity test results (left: λ–φ relation, right: λ–u relation)

Table 3.15 Temperature (T) and relative humidity (φ) of ambient air control by different saturated salt solutions in vacuum dryers and the measured equilibrium moisture content (u value)

No.	Saturated salt solutions	T measured value [°C]	Relative humidity		Measured u values			
			φ Reference value [%]	φ Measured value [%]	Adsorption [%]	Desorption [%]	Average [%]	Average w [kg/m^3]
1	LiCl	20.2	11.31 ± 0.31	11.2	0.72	1.41	1.07	2.88
2	CH_3COOK	20.2	23.11 ± 0.25	24.4	1.74	2.33	2.03	5.49
3	$MgCl_2$	20.2	33.07 ± 0.18	33.4	2.72	3.85	3.29	8.87
4	K_2CO_3	20.4	43.16 ± 0.33	43.5	4.80	6.55	5.68	15.33
5	$Mg(NO_3)_2$	20.3	55.87 ± 0.27	55.0	7.74	7.99	7.87	21.24
6	NaBr	20.3	59.14 ± 0.44	59.7	8.60	9.49	9.05	24.42
7	NaCl	20.1	75.47 ± 0.14	77.2	10.37	10.82	10.59	28.60
8	KCl	20.1	85.11 ± 0.29	85.4	11.64	12.07	11.85	32.00
9	K_2SO_4	20.3	97.6 ± 0.6	96.3	16.67	–	16.67	45.01

Table 3.16 Relative humidity (φ) control by desiccant/saturated salt solutions in dry and wet cups and the measured water vapor transfer coefficient (δ value)

Group	Desiccant/saturated salt solutions	φ Reference value [%]	Measured δ values [kg/(m s Pa)]			
			No. 1	No. 2	No. 3	Average
Dry cup	$CaCl_2$ ($d = 3$ mm)	0	8.08130E^{-11}	1.04000E^{-10}	1.17718E^{-10}	1.00604E^{-10}
Wet cup	K_2SO_4	96.3	9.03744E^{-11}	7.91200E^{-11}	8.77608E^{-11}	8.57517E^{-11}

Ambient air in conditioning chamber: $T = 20$ °C, $\varphi = 50\%$

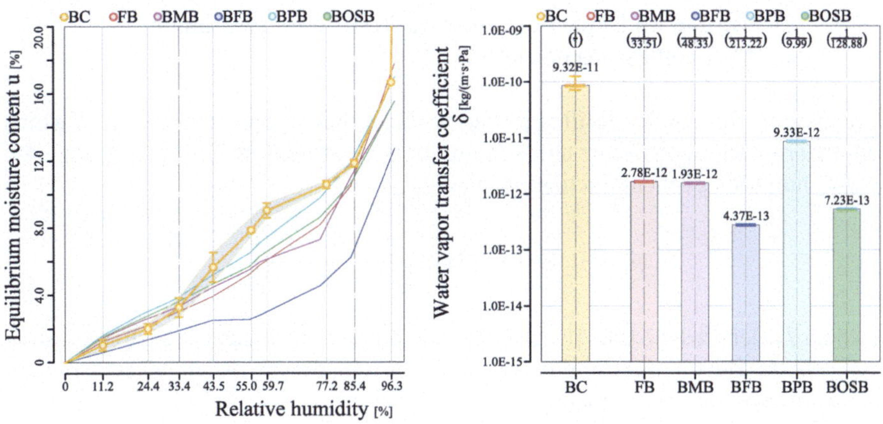

Fig. 3.33 Equilibrium moisture content (left) and water vapor transfer coefficient (right) of BC and bamboo panels (FB—flatten bamboo board, BMB—bamboo mat board, BFB—bamboo scrimber, BPB—bamboo particleboard, BOSB—bamboo oriented strand board)

are uncarbonized bamboos that are tested by the author through the same methods. The isothermal adsorption and desorption curve indicates that the moisture-control capacity of BC increases when $\varphi = 33.4$–85.4% where the curve rises upwards. Water vapor transmission coefficient of BC is also expanded to 9.99–213.2 times that of the uncarbonized bamboo panels. The higher moisture capacity and transport rate contribute to the better regulation ability with respect to BC.

3.3.4 Discussion

3.3.4.1 Evaluation on BF as a Vegetable Fiber Insulation Material

The dry thermal conductivity λ values of BF with filling density $\rho = 70$–170 kg/m^3 are 0.0423–0.0465 W/(m K), and this is comparable to those of other vegetable fiber insulation materials. However, the wet λ values of untreated BF are exponentially affected by the hygric environment (here referred to the ambient relative humidity). The relation respectively between u, U, λ, and φ can generally be defined with exponential function $Y = a \times \mathrm{EXP}(b \times \varphi) + c$. Within the test range, the u–φ curve and U–φ grow rapidly when respectively $\varphi = 85.4$–96.3% and $\varphi = 85.4$–96.3%.

The practical application of the BF required either moisture resistance treatment by reducing the U value or the control of hygric environment in contact with BF. In the following section, the improvement effect of the BF is indicated through projection and correlation analysis.

3.3.4.2 Derivation on Impact of Moisture Resistance Treatment on Thermal Properties of BF

Here, assume that the BF infill is placed in the middle of an exterior wall in which both exterior and interior air temperature are set constant as T. The outdoor 24 h relative humidity cyclical fluctuation curve is:

$$\varphi_e = 0.80 + 0.12\cos\left[\frac{\pi \cdot (T_h - 6)}{12}\right]$$

in which T_h denoted the time [h], and the indoor relative humidity is set to a constant as $\varphi_i = 50\%$. The relative humidity of the air surrounding the BF in the construction interlayer cavity is approximately:

$$\varphi = \frac{\varphi_e + \varphi_i}{2}$$

The fitted exponential functions for the test results of $\rho = 110$ kg/m^3 and $d = 60$ mm are selected for further projections. Two calculation groups for comparison are defined. The hourly thermal conductivity λ values and the index of thermal inertias D values in the 24 h denote the thermal transport and storage performance of the BF layer, respectively. In group 1, the original functions between u and φ, U–φ, and λ–φ are used to iteratively calculate the hourly λ_0 and D_0 values. In group 2, the U values are reduced to 1/2 of the original, while the u and λ values are maintained. Hourly $\lambda_{-1/2U}$ and $D_{-1/2U}$ values are calculated to evaluate the impact of U value on the thermal performance of BF.

Based on the u–φ, U–φ, and λ–φ fitted results above, there are:

$$\begin{aligned} u &= 0.0088\mathrm{e}^{3.3637\varphi} \\ U &= 2.7610E^{-10}\mathrm{e}^{0.1110\varphi} \\ \lambda &= 0.000457\mathrm{e}^{4.3974\varphi} + 0.0437 \end{aligned}$$

When the calculation period Z is set as 24 h, the heat storage coefficient is:

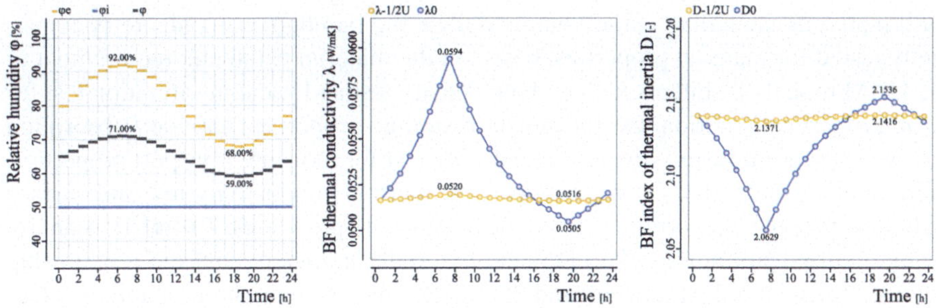

Fig. 3.34 Relative humidity φ in 24 h cycle and the BF thermal conductivity λ and index of thermal inertia D calculation results

$$S_{24\text{h}} = 0.51\sqrt{\lambda \cdot c \cdot \rho}$$

and the corresponding D value is:

$$D = R \cdot S = 0.51d \cdot \sqrt{\frac{c \cdot \rho}{\lambda}}$$

in which:

$$c = c_0 + 0.65u$$
$$\rho = \rho_0 \cdot (1 + u)$$

in which c_0 denotes the specific heat capacity of dry BF and is approximately set as 2.0 kJ/(kg K), and ρ_0 denotes the filling density of dry BF, and $\rho_0 = 110$ kg/m^3.

Calculation results indicate that the BF thermal performance fluctuates along with the periodically fluctuating hygric environment. Here, the extreme value deviance D_{ev} is used to characterize the fluctuation amplitude: D_{ev} = (Max. value–Min. value)/Mean. value. The D_{ev} values of λ_0 and D_0 are 16.60 and 4.28%, respectively.

When the U values are reduced to 1/2, the fluctuation amplitude of BF thermal performance significantly decreases, in which the D_{ev} values of $\lambda_{-1/2\text{U}}$ and $D_{-1/2\text{U}}$ are 0.80 and 0.21%, respectively. Additionally, the λ values decrease, in which the 24 h mean value of λ is reduced from 0.0536 W/(m K) to 0.0518 W/(m K) (Fig. 3.34).

3.4 Summary

Six construction partition materials, including flattened bamboo panel (FB), bamboo laminated lumber (BSB), plybamboo (BMB), bamboo scrimber (BFB), bamboo particleboard (BPB), and bamboo oriented strand board (BOSB), and two construction infill materials, including natural bamboo fiber and bamboo charcoal, are systematically tested for material properties, based on the material parameter requirement of the HAM model. Test items include bulk density test and vacuum saturation test for basic properties; sorption test for moisture storage properties; capillary absorption test, water vapor transmission test, and drying test for moisture transport properties; thermal analysis for heat storage properties; thermal conductivity test and surface light and thermal properties' test for heat transport properties. Correlation analyses are carried out among the obtained material parameters, and the relationship between the material parameters and the hygric environment, the heat and moisture transport properties, and the moisture content of the materials are fitted to quantify the relationship between them. In addition, the bamboo-based panel process, industrial utilization of natural bamboo fiber and bamboo charcoal, etc., are given modification assessment at the material level.

(1) Construction partition group

The test results initially form a properties' database, which has practical value, because the test results are obtained through the corresponding test standards specialized for building materials, and the size of the test sample is close to the actual size in practical project; the test object not only contains the raw bamboo, but also covers the representative bamboo-based panels that are made up of the typical constituent units, including the bamboo strip, bamboo sliver, bamboo fiber bundles, fine shavings and coarse shavings. The test content can support the complete building envelope HAM model. The obtained material parameters include the bulk density, specific heat capacity, thermal conductivity (constant value) and water vapor diffusion resistance factor (constant value), as well as the open porosity and isothermal adsorption and desorption curve, capillary saturated moisture content, capillary water absorption rate, water vapor transfer coefficient (moisture content-independent), thermal conductivity (moisture content-independent), and the surface light and thermal properties.

Correlation analysis between hygrothermal properties and basic properties shows that macroscopically gaseous moisture and heat storage properties show insignificant correlation with basic properties; both moisture and heat transport properties show stronger correlation with open porosity Φ than with dry bulk density ρ. Curve fitting between moisture transport properties and the moisture content shows that there is an exponential function relation between water vapor transfer coefficient δ and the mass rate moisture content u that can be defined as $\delta = a \times \text{EXP}(b \times u)$, and a logarithmic function relations between drying rate U and u that can be defined as $U = a \times \ln(u) + b$; for heat transport properties, an linear function $\lambda = a_w \times u + b$ is generally used to approximate the relation between thermal conductivity λ and u, resulting in the thermal conductivity supplement, a_w [(W/m K)/u(–)] values, are in the range 0.2137–0.6890; Spearman correlation analysis shows that the impact of moisture content in the gaseous moisture region on the surface light and thermal transport properties can be ignored, but it requires further study in the liquid water region (a, b, and c in the formulas are fitting parameters).

The comparison between the 5 BBPs (bamboo-based panels, including BSB, BMB, BFB, BPB, and BOSB), and the FB (raw bamboo) shows that the BBP process can improve homogeneity and broaden the material properties' spectrum. In particular, the dry bulk density inhomogeneity of the BBPs is narrowed down from ca. 20% to ca. 10%; at FB 100%, the dry bulk density range increases to 84.60–166.39%; open porosity is 33.21–120.92%; hygric properties, including moisture storage and transport, are 7.50–604.62%; and thermal properties, including heat storage and heat transport, are 50.88–159.28%. During the BBP process, the change magnitude of hygric and transport properties is greater than that of the thermal and storage properties, respectively.

(2) Construction infill group

For natural bamboo fiber (BF), results show that there are exponential relation respectively between equilibrium moisture content u, moisture adsorption and desorption rate U, thermal conductivity λ, and the ambient relative humidity φ, that can generally be defined with exponential function $Y = a \times \mathrm{EXP}(b \times \varphi) + c$; within the range of the test, the u–φ curve and U–φ curve grow rapidly, respectively, when $\varphi = 85.4$–96.3% and $\varphi = 75$–95%; the dry λ values of BF with bulk density $\rho =$ 70–170 kg/m^3 are 0.0423–0.0465 W/(m K), which is comparable to other vegetable fiber insulation materials; the wet λ values increase significantly with the rise of φ, and the growth rate is in positive correlation with the ρ values; the practical application of BF requires either moisture resistance treatment by reducing the U value or the control of hygric environment. For bamboo charcoal (BC), compared with the five uncarbonized bamboo boards, the moisture-control capacity of BC is expanded when $\varphi = 33.4$–85.4%; the water vapor transmission coefficient is enlarged to 9.99–213.2 times, which is 9.32E^{-11} kg/(m s Pa). The improvement of moisture-control capacity and transport rate contribute better moisture-control ability to BC (a, b, and c in the formulas are fitting parameters).

References

1. Van Belleghem M, Steeman HJ, Steeman M et al (2010) Sensitivity analysis of CFD coupled non-isothermal heat and moisture modelling. Build Environ 45(11):2485–2496
2. Künzel HM (1995) Simultaneous heat and moisture transport in building components. Fraunhofer IRB Verlag Suttgart
3. Kast W, Jokisch F (1972) Überlegungen zum Verlauf von Sorptionsisothermen und zur Sorptionskinetik an porösen Feststoffen (Considerations regarding the development of sorption isotherms and the sorption kinetics in porous solids). Chemie-Ingenieur Technik 1972(44), H. 8, S. 556–563
4. Künzel HM (1991) Feuchteeinfluß auf die Warmeleitfähigkeit bei hygroskopischen und nicht hygroskopischen Stoffen (The moisture effect on the thermal conductivity of hygroscopic and non-hygroscopic materials). WKSB (36), H. 29, S. 15–18
5. Kießl K (1983) Kapillarer und dampfförmiger Feuchtetransport in mehrschichtigen Bau-teilen (Capillary and vaporous moisture transport in multi-layered building com-ponents). Diss. Universität-Gesamthoch-schule Essen

6. Sommer E (1971) Beitrag zur Frage der kapillaren Flüssigkeitsbewegung in porigen Stoffen bei Be- und Entfeuchtungs-vorgängen (Contribution to the question of capillary liquid movement in porous materials during humidification and dehumidification processes). Diss. Technische Hochschule Darmstadt
7. Richards LA (1949) Methods of measuring soil moisture tensions. Soil Sci H. 68, S. 95–112
8. 龙激波. 基于多孔介质热质传输理论的竹材结构建筑热湿应力研究 [D]. 长沙: 湖南大学博士学位论文, 2013/Long J B. Thermal and moisture stress analysis of bamboo buildings based on heat and mass transfer method of porous medium. Changsha: Doctoral thesis of Hunan University, 2013
9. Zhang QS, Jiang SX, Tang YY (2002) INBAR Technical Report No. 26—Industrial Utilization on Bamboo. International Network for Bamboo and Rattan (INBAR), Beijing
10. LY/T 1660–2006竹材人造板术语 [S]. 北京: 国家林业局, 2006/LY/T 1660-2006 Standard Terminology for Bamboo-based Panel. The state forestry administration of the People's Republic of China, Beijing, 2006
11. 冯驰. 多孔建筑材料湿物理性质的测试方法研究 [D]. 华南理工大学博士论文, 2014/Feng C (2014) Study on the test methods for the hygric properties of porous building materials. Doctoral thesis of South China University of Technology
12. ASTM D7370–2009: standard test method for determination of relative density and absorption of fine, coarse and blended aggregate using combined vacuum saturation and rapid submersion
13. DIN EN 1936–2007: Natural stone test method—Determination of real density and apparent density, and of total and open porosity; English version of DIN EN 1936:2007-02
14. ISO 12571:2012: Hygrothermal performance of building materials and products—Determination of hygroscopic sorption properties
15. ASTM C1498—04a(2016): Standard test method for hygroscopic sorption isotherms of building materials
16. ISO 15148:2002(E): Hygrothermal performance of building materials and products—Determination of water absorption coefficient by partial immersion
17. Scheffler GA, Plagge R (2010) A whole range hygric material model: modelling liquid and vapour transport properties in porous media. Int J Heat Mass Transf 53(1–3):286–296
18. ISO 12572:2001(E): Hygrothermal performance of building materials and products - Determination of water vapour transmission properties
19. ASTM E 96/E 96 M-2005: Standard test methods for water vapor transmission of materials
20. Mukhopadhyaya P, Kumaran MK, Lackey J (2005) Use of the modified cup method to determine temperature dependency of water vapor transmission properties of building materials. J Test Eval 33(5):316–322
21. ISO 11357-4-2005: Plastics—differential scanning calorimetry (DSC)—Part 4: Determination of specific heat capacity
22. ISO 8302-1991: Thermal insulation; determination of steady-state thermal resistance and related properties; guarded hot plate apparatus
23. DIN EN 16012-2012: Thermal insulation for buildings—reflective insulation products—determination of the declared thermal performance; German version EN 16012:2012
24. GB/T 25968-2010: Test methods of solar transmittance and solar absorptance of materials by spectrophotometer
25. Roels S, Carmeliet J, Hens H et al (2004) Interlaboratory comparison of hygric properties of porous building materials. J Building Phys 27(4):307–325
26. Krus M, Künzel H (1993) Determination of D_W from A-value. IEA Annex 24 Report T3-D-93/02
27. Carmeliet J, Hens H, Roels S et al (2004) Determination of the liquid water diffusivity from transient moisture transfer experiments. J Therm Envelope Build Sci 27(4):277–305
28. Nizovtsev MI, Stankus SV, Sterlyagov AN et al (2008) Determination of moisture diffusivity in porous materials using gamma-method. Int J Heat Mass Transf 51(17–18):4161–4167
29. Pazera M, Salonvaara M (2012) Multilayer test method for water vapor transmission testing of construction materials. J Building Phys 35(3):224–237

30. Mukhopadhyaya P, Kumaran K, Normandin N et al (2002) Effect of surface temperature on water absorption coefficient of building materials. J Therm Envelope Build Sci 26(2):179–195
31. Sedlbauer K (2001) Vorhersage von Schimmelpilzbildung auf und in Bauteilen. Dissertation Universität Stuttgart
32. Luikov AV (1964) Heat and mass transfer in capillary-porous bodies. Adv Heat Transf. Elsevier, Amsterdam. 123–184
33. Luikov AV (1975) Systems of differential equations of heat and mass transfer in capillary-porous bodies (review). Int J Heat Mass Transf 18(6):1–14

Chapter 4
Hygrothermal Performance Assessment on Bamboo Building Envelope in Typical Climate Regions

Based on the material properties' test in Chap. 3, timber and timber units are set as reference models, and accordingly, bamboo and bamboo units of the same construction and space size as evaluation models. The performance on levels of material, building component, and space unit are compared, with comprehensive consideration of the representative conditions in terms of the external climate, building function, construction type, and HVAC. On the one hand, the comparison results can support the feasibility evaluation on 'substitute timber with bamboo' in building envelope industry in terms of physical performance. On the other hand, the reference timber units for comparison are set as benchmarks to provide performance assessment on the application of bamboo in building envelope (Fig. 4.1).

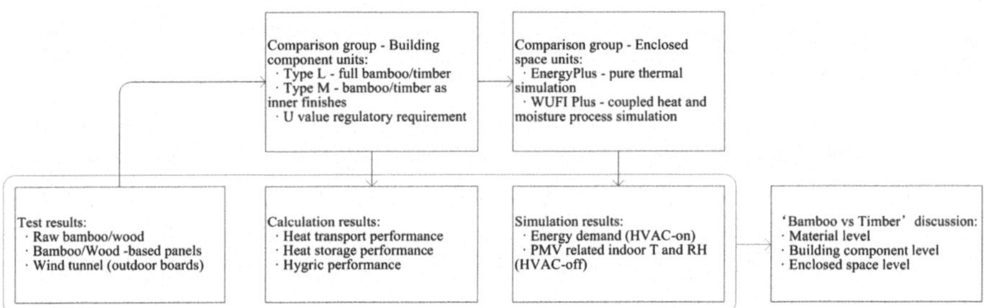

Fig. 4.1 Framework of Chap. 4

Z. Huang, *Application of Bamboo in Building Envelope*,
Green Energy and Technology, https://doi.org/10.1007/978-3-030-12032-0_4

4.1 Properties Comparison Between Corresponding Material Varieties of Bamboo and Timber

4.1.1 Collection of Reference Timber

Since the knowledge on timber is relatively sufficient and available, the comparison with timber can be helpful to the evaluation of bamboo. To set comparison reference for bamboo, related parameters from various databases are collected to form the reference timber (RT).

However, the parameters from different databases vary greatly and it is difficult to judge which are representative and authoritative, which means that to obtain precise bamboo–timber ratio is impractical. Here we define the RT range with the available maximum and minimum values, and the relative position (RP) of the bamboo parameters in RT is used to characterize the relative relation between bamboo and the corresponding timber products:

$$\mathrm{RP} = \frac{\text{Bamboo value} - \text{RT.min value}}{\text{RT.max value} - \text{RT.min value}} \times 100\%$$

when

RP < 0%: Bamboo value < RT;
0% ≤ RP ≤ 20%: Bamboo value is in a low position of RT;
20% < RP ≤ 40%: Bamboo value is in a middle to low position of RT;
40% < RP ≤ 60%: Bamboo value is in a middle position of RT;
60% < RP ≤ 80%: Bamboo value is in a middle to high position of RT;
80% < RP ≤ 100%: Bamboo value is in a high position of RT;
RP > 100%: Bamboo value > RT.

The bamboo parameters, as described in Chap. 3, are chosen for the comparison between bamboo and RT. The $w_{\mathrm{RH}=50\%}$, $w_{\mathrm{RH}=95\%}$, and w_{cap} are selected out from the isothermal adsorption and desorption curve to describe, respectively, the gaseous moisture and liquid water storage properties. The $1/\mu$ is reciprocal of the water vapor diffusion resistance factor, so as to describe the moisture transport rate in a positive correlation.

4.1.2 Comparison Between Bamboo and Reference Timber

(1) Raw material

FB is compared separately with softwood and hardwood. The bulk density ρ value of FB > Softwood and open porosity Φ value of FB < Softwood, while the ρ value of FB is in a low position of hardwood, and the Φ value of FB is in a middle position of hardwood, which means that the basic properties of bamboo are closer to hardwood, and quite different from softwood which is normally the raw material for wood-based panels. Meanwhile, as mentioned above, the bamboo fibers are almost totally longitudinally arranged, with extremely few radial fibers at the bamboo nodes, which is far different from wood which microscopically consists of multi-directional fiber. These can significantly affect the moisture transport ability.

(2) Bamboo/Wood-based panels

When made into bamboo/wood-based panels, the difference in raw material, constituent unit, assembly method and protective treatment, etc., during the manufacturing process can affect the hygrothermal properties of the final products. The bamboo/wood-based panels below are separated into three groups according to their usages, constituent units, and assembly methods for more pertinent comparison:

Group 1 Exterior board: BFB, compared RT is hardwood, plywood high, wood fiberboard hard;
Group 2 Interior board: BSB, compared RT is laminated board;
Group 3 Interlayer boards: BMB, BPB, BOSB, compared RT are, respectively, BMB—laminated veneer lumber (LVL), plywood; BPB—particleboard/chipboard, fiberboard (MDF); BOSB—wood oriented strand board (OSB).

Judging from the table below, it can be seen that except BPB, the other bamboo-based panels have higher bulk density and lower porosity than RT. Most of the hygric properties are lower than RT or distribute in a low or middle to low position of RT, which means lower moisture capacity and vapor permeation rate. The higher vapor permeation resistance can be dialectical that can help to weaken the vapor transport, but might also cause water retardation on the boundaries between two layers (as would be shown in the exterior wall condensation in Sect. 4.2.3.2). The BFB has a higher liquid water transport resistance, which is beneficial to outdoor use. Except BPB, most of the thermal properties are higher than RT or distribute in a high or middle to high position of RT, which means higher heat storage ability that is beneficial to thermal stability, and higher heat transport ability that is unfavorable to the thermal insulation (Table 4.1).

Fraunhofer IBP is one of the authoritative material testing institutions for hydrothermal properties, and the computer program WUFI it developed is the main tool for the heat and moisture process simulation in this study. The source of material database in WUFI covers the timber products from the major timber construction activity areas, including Europe, North America, Australia, and Japan. Specifically, its source consists of Fraunhofer IBP database (Holzkirchen, Germany), MASEA database (Germany), Materials for Dena Constructions (Germany Energy Agency), Materials for thermal calculations (from DIN 4108-4 or other sources), North America Database, Japan Database, University of Technology Vienna (Austria), NTNU Norwegian University of Science and Technology, LTH Lund University (Sweden), and Generic Database. The 18 timber panel products from WUFI database would be selected to form the reference timber (RT) and compared with the test results of bamboo properties (Fig. 4.2).

Table 4.1 Relative position (RP) of bamboo parameters in the range of RT

Items			Bamboo	Timber. max	Timber. min	RP	RP location < 0%	0%-20%	20%-40%	40%-60%	60%-80%	80%-100%	100% >
Raw material: FB—Softwood													
Basic properties	ρ	(kg/m^3)	666.38	530.00	350.00	175.77%							
	Φ	[–]	0.52	0.90	0.73	–122.12%							
Raw material: FB—Hardwood													
Basic properties	ρ	(kg/m^3)	666.38	740	650	18.20%							
	Φ	[–]	0.52	0.72	0.35	46.59%							
Exterior board: BFB—Hardwood, Plywood high, Wood fiberboard hard													
Basic properties	ρ	(kg/m^3)	1108.77	959.00	400.00	126.79%							
	Φ	[–]	0.17	0.96	0.35	–28.92%							
Hygric properties—gaseous moisture	$w_{RH50\%}$	(kg/m^3)	28.17	82.50	19.30	14.03%							
	$w_{RH95\%}$	(kg/m^3)	144.85	228.00	108.10	30.65%							
	$1/\mu$	[–]	0.0022	0.0277	0.0029	–2.53%							
Hygric properties—liquid water	w_{cap}	(kg/m^3)	165.93	573.00	236.00	–20.79%							
	A_{cap}	[kg/(m^2 s$^{0.5}$)]	0.000873	0.004200	0.000720	4.40%							
	D_l	[m^2/s]	1.96E^{-11}	4.61E^{-10}	1.64E^{-11}	0.73%							
Thermal properties	c	[J/(kg K)]	1550	2100	1400	21.43%							
	λ	[W/(m K)]	0.1625	0.1800	0.0680	84.38%							
	S_{24h}	[W/(m^2 K)]	8.52	7.98	3.65	112.54%							
Interior board: BSB—Laminated board													
Basic properties	ρ	(kg/m^3)	563.81	500.00	400.00	163.81%							
	Φ	[–]	0.54	0.90	0.56	–5.97%							
Hygric properties—gaseous moisture	$w_{RH50\%}$	(kg/m^3)	30.44	53.00	18.00	35.54%							
	$w_{RH95\%}$	(kg/m^3)	96.44	268.03	88.00	4.69%							
	$1/\mu$	[–]	0.0143	0.0500	0.0101	10.65%							
Thermal properties	c	[J/(kg K)]	1960	2100	1300	82.50%							
	λ	[W/(m K)]	0.1475	0.1400	0.086	113.89%							
	S_{24h}	[W/(m^2 K)]	6.51	5.65	3.69	143.65%							

(continued)

Table 4.1 (continued)

Interlayer board: BMB—Laminated veneer lumber (LVL), Plywood						
Basic properties	ρ	(kg/m^3)	776.2	708.0	400.00	122.15%
	Φ	[–]	0.50	0.96	0.50	–0.91%
Hygric properties—gaseous moisture	$w_{RH50\%}$	(kg/m^3)	39.20	62.00	19.30	46.60%
	$w_{RH95\%}$	(kg/m^3)	115.90	228.00	75.97	26.26%
	$1/\mu$	[–]	0.0099	0.1200	0.0050	4.23%
Thermal properties	c	[J/(kg K)]	2020	1880	1400	129.17%
	λ	[W/(m K)]	0.1733	0.1300	0.0680	169.84%
	S_{24h}	[W/(m^2 K)]	8.41	5.70	3.65	232.14%
Interlayer board: BPB—Particleboard/Chipboard, Fiberboard (MDF)						
Basic properties	ρ	(kg/m^3)	623.32	1000.00	500.00	24.66%
	Φ	[–]	0.63	0.80	0.50	43.90%
Hygric properties—gaseous moisture	$w_{RH50\%}$	(kg/m^3)	36.89	74.00	32.70	10.15%
	$w_{RH95\%}$	(kg/m^3)	102.05	291.00	11.80	32.32%
	$1/\mu$	[–]	0.0478	0.0311	0.0080	172.00%
Thermal properties	c	[J/(kg K)]	1760	2100	1300	57.50%
	λ	[W/(m K)]	0.0801	0.1400	0.0820	–3.28%
	S_{24h}	[W/(m^2 K)]	4.78	6.63	4.38	17.64%
Interlayer board: BOSB—Wood oriented strand board (OSB)						
Basic properties	ρ	(kg/m^3)	895.02	725.00	518.00	182.14%
	Φ	[–]	0.42	0.95	0.41	2.44%
Hygric properties—gaseous moisture	$w_{RH50\%}$	(kg/m^3)	46.07	71.00	22.10	49.02%
	$w_{RH95\%}$	(kg/m^3)	132.07	245.00	106.89	18.23%
	$1/\mu$	[–]	0.0037	0.0160	0.0016	14.35%
Thermal properties	c	[J/(kg K)]	1663	2100	1400	37.57%
	λ	[W/(m K)]	0.1197	0.1300	0.0840	77.61%
	S_{24h}	[W/(m^2 K)]	6.81	6.94	4.55	94.30%

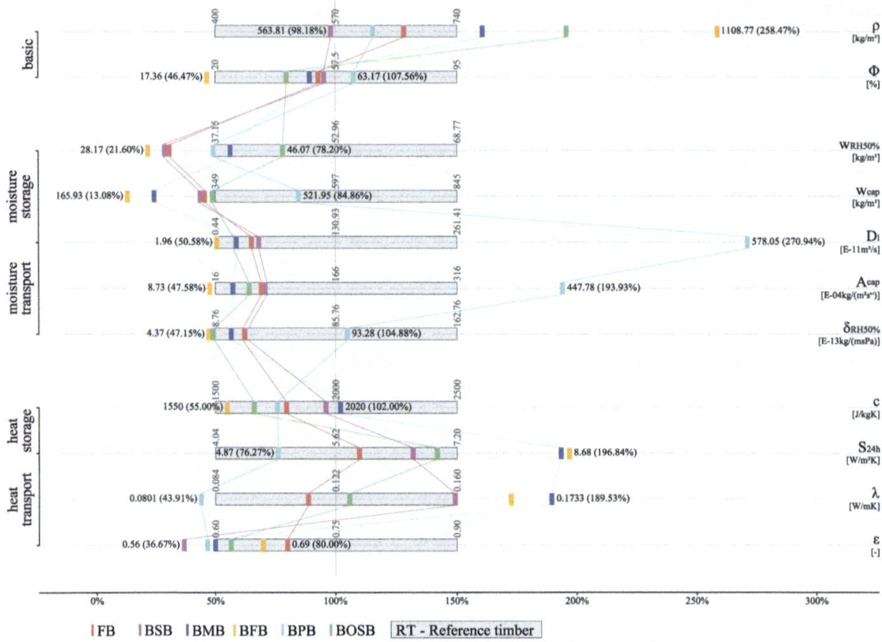

Fig. 4.2 Relative position of bamboo property parameters in the range of RT, the values in brackets are the relative position of bamboo compared with RT, which are calculated by: (Bamboo value – Average.RT)/(Max.RT – Min.RT) × 100% + 100%

(1) Basic properties. Bulk density (ρ_d) comparison shows that the ρ_d of bamboo is much higher than that of the timber. Setting the RT median value as 100%, the ρ_d of bamboo is in the range 98.18–258.47%. Both the ρ_d of BFB and BOSB (1108.77 and 895.02 kg/m^3, respectively,) exceed the maximum value of RT (740 kg/m^3). Comparison for open porosity (Φ) shows that the Φ of bamboo is generally in a middle to low position of RT, among which the Φ of BFB (17.36%) is smaller than the minimum value of RT (20%).

(2) Hygric properties. Moisture storage properties comparison results show that the equilibrium moisture content of bamboo corresponding to RH = 50% $w_{RH=50\%}$ (kg/m^3) is in the range 28.17 (BFB)—46.07 (BOSB), which is lower than the 37.15–68.77 (RT). Meanwhile, the capillary saturation moisture content w_{cap} (kg/m^3) is 165.9 (BFB)—342.7 (BOSB), also lower than the 349 (RT). Moisture transport properties comparison results show that, when compared with reference timber, the water absorption coefficient A_{cap} [E^{-04} kg/(m^2 s$^{0.5}$)] is in the range 8.73 (BFB)—78.74 (BSB), which is in the low position of 16-316 (RT). The liquid water transport coefficient D_l [E^{-11} m^2/s] is 1.96 (BFB)—48.39 (BSB), generally locating in the low position of 0.44–261.41 (RT). The water vapor transfer coefficient of bamboo corresponding to RH = 50% $\delta_{RH=50\%}$ [E^{-13}

kg/(m s Pa)] is in the range 4.37 (BFB)—27.99 (BSB), which is lower than the 8.76–162.76 (RT).

(3) Thermal properties. Heat storage properties comparison results show that the specific heat capacity c [J/(kg K)] is in the range 1550 (BFB)—1960 (BSB), which is in the middle to low position of 1500–2500 (RT). Heat transport properties comparison results show that the dry thermal conductivity λ_d [W/(m K)] is 0.1088 (FB)—0.1733 (BMB), which is slightly higher than the 0.084–0.16 (RT). The 24 h heat storage coefficient S_{24h} [W/(m^2 K)] that describes both the heat storage and transport properties is in the range 5.93 (FB)—8.68 (BFB), which is significantly higher than the 4.04–7.20 (RT).

4.1.3 Wind Tunnel Test on Bamboo Scrimber and Hardwood

As described in Sect. 1.1.2.4 in Chap. 1, the production of bamboo scrimber reaches its technical maturity in a late period. It is a kind of bamboo-based panel composed of incompletely separated fiber bundles as constituent units, which are exerted with high pressure to form standard plate or square. Bamboo scrimber technology can improve the utilization rate of the raw material, achieve stronger mechanical strength, cracking resistance and corrosion resistance, so that can be applied in the outdoor environment (Table 4.2).

4.1.3.1 Standard Test Results Comparison

As described in Sect. 4.1.2, the group 1 exterior board is taken for comparison, including bamboo scrimber (BFB) and the 28 reference timbers (RT), which consist of hardwood, plywood of high bulk density and hardwood fiberboard. The range of RT is defined by the maximum and minimum values, and the relative position of BFB in RT range is analyzed below to describe the properties relation between bamboo and timber (Table 4.3, Fig. 4.3).

(1) Basic properties. Both the bulk density and open porosity of bamboo scrimber exceed the range of the reference timber. The high bulk density, as 1108.77 kg/m^3, rarely exists in plant materials, which is mainly due to the high density of the raw material and strong pressing during the manufacturing process. The low open porosity resulted from the high pressure and waterproofing agent that compress and block the pore structure is conductive to the application in outdoor environment.

(2) Hygric properties. Compared with the reference timber, the isothermal adsorption and desorption curve of bamboo scrimber is in a lower position. The gaseous moisture transport property is greatly influenced by the moisture content. The water vapor transfer resistance factor falls from 921.28 to 47.67 when moisture content rise from $w_{RH=20.0\%}$ to $w_{RH=93.0\%}$. The relative position of bamboo

Table 4.2 Information of the bamboo scrimber sample

BFB (Bamboo scrimber)	Constituent unit	Assembly	Thickness	Main application	Sample sources
	Bamboo bundle (width:10–30 mm)	Parallel	30.0 mm	Load-bearing component, indoor/outdoor flooring, finishes, furniture	Dasso Industrial Group Co. Ltd. Hangzhou, China

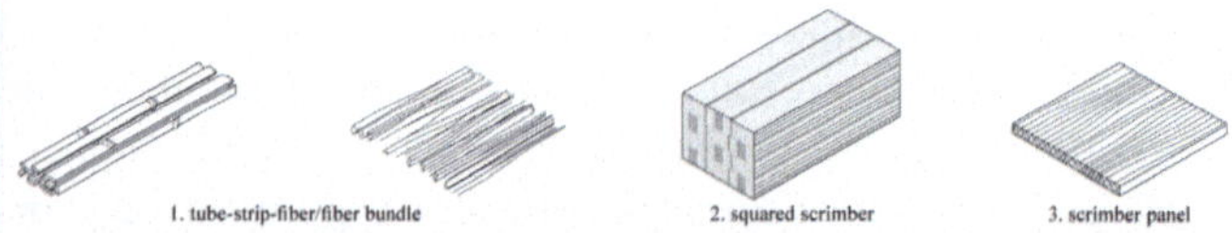

Main manufacturing processes of bamboo scrimber (Photos are taken in Yongyu Bamboo Manufacture Co., Ltd. Anji, Zhejiang, China)

scrimber to reference timber also falls with the increasing moisture content. Due to the lower open porosity, the liquid water storage and transport abilities are lower than the reference timber.

(3) Thermal properties. Both the heat storage and transport properties of BFB are higher than the RT, which is mainly caused by the higher bulk density and lower porosity. On the one hand, this gives BFB the advantage of thermal capacity, but on the other hand, the shortcomings of thermal resistance cannot be ignored.

4.1.3.2 Wind Tunnel Test Results Comparison

For examining the behavior of the materials in outdoor conditions, a wind tunnel test is carried out with a typical summer day weather data of Guangzhou, a subtropical city locating in south China. The weather data, including dynamic solar radiation, air temperature, relative humidity and constant wind speed, are repeated 72 h for

Table 4.3 Bamboo scrimber properties and the comparison with RT

Items		Unit	BFB—Bamboo scrimber		RT—Reference timber			
					Min. value		Max. value	
Basic properties	Dry bulk density ρ	[kg/m^3]		1108.77		400		708
	Open porosity Φ	[–]		17.36%		41%		90%
Hygric properties	20 °C isothermal adsorption and desorption curve	[kg/m^3]	$w_{RH} = 11.2\%$ $w_{RH} = 24.4\%$ $w_{RH} = 33.4\%$ $w_{RH} = 43.5\%$ $w_{RH} = 55.0\%$ $w_{RH} = 59.7\%$ $w_{RH} = 77.2\%$ $w_{RH} = 85.4\%$ $w_{RH} = 96.3\%$	6.79 14.05 20.26 25.55 30.19 32.24 53.75 69.58 155.04	$w_{RH} = 0\%$ $w_{RH} = 20\%$ $w_{RH} = 30\%$ $w_{RH} = 50\%$ $w_{RH} = 65\%$ $w_{RH} = 80\%$ $w_{RH} = 90\%$ $w_{RH} = 93\%$ $w_{RH} = 95\%$ $w_{RH} = 97\%$	0 5.70 9.70 22.10 31.90 45.50 62.29 80.10 88.00 94.40	$w_{RH} = 0\%$ $w_{RH} = 20\%$ $w_{RH} = 30\%$ $w_{RH} = 50\%$ $w_{RH} = 65\%$ $w_{RH} = 80\%$ $w_{RH} = 90\%$ $w_{RH} = 93\%$ $w_{RH} = 95\%$ $w_{RH} = 97\%$	0 42.00 61.00 74.00 90.00 115.00 151.00 197.00 245.00 318.00
	Water vapor transfer coefficient ($\mu_{RH} = 0.0\%$ and $\mu_{RH} = 100.0\%$ are regarded as equal respectively to $\mu_{RH} = 20.0\%$ and $\mu_{RH} = 93.0\%$)	[–]	$\mu_{RH} = 20.0\%$ $\mu_{RH} = 25.0\%$ $\mu_{RH} = 35.0\%$ $\mu_{RH} = 45.0\%$ $\mu_{RH} = 50.0\%$ $\mu_{RH} = 73.0\%$ $\mu_{RH} = 83.0\%$ $\mu_{RH} = 93.0\%$	921.28 781.24 600.96 489.50 446.94 106.15 73.79 47.67	$\mu_{RH} = 0.0\%$ $\mu_{RH} = 100\%$	9 7	$w_{RH} = 0\%$ $w_{RH} = 25\%$ $w_{RH} = 75\%$ $w_{RH} = 100\%$	845 845 376 376

(continued)

Table 4.3 (continued)

Items		Unit	BFB—Bamboo scrimber		RT—Reference timber			
					Min. value		Max. value	
	Drying rate (T = 23 °C, RH = 50%)	[E^{-07} kg/(m^2 s)]	$U_{u12-u11\%}$ $U_{u11-u10\%}$ $U_{u10-u9\%}$ $U_{u9-u8\%}$ $U_{u8-u7\%}$ $U_{u7-u6\%}$	12.58 11.46 10.38 9.71 8.98 7.74		–		–
	Capillary saturation moisture content w_{cap}	[kg/m^3]		165.93		326.00		864.50
	Water absorption coefficient A_{cap}	[kg/(m^2 s$^{0.5}$)]		0.000873		0.0004		0.0045
Thermal properties	Specific heat capacity c	[J/(kg K)]		1550		1300		2100
	Thermal conductivity λ_{dry}	[W/(m K)]		0.1625		0.09		0.13
	24 h heat storage coefficient S_{24h}	[W/(m^2 K)]		8.52		3.62		5.55

(continued)

Table 4.3 (continued)

Items		Unit	BFB—Bamboo scrimber		RT—Reference timber			
					Min. value		Max. value	
	Hemispherical emissivity ε	[–]		0.66		–		–
	Light reflectivity (380–780 nm) ρ_v	[%]		19.74				
	Solar direct reflectivity (200–2600 nm) ρ_e	[%]		44.51				
	Solar direct absorptivity (200–2600 nm) α_e	[%]		55.49				

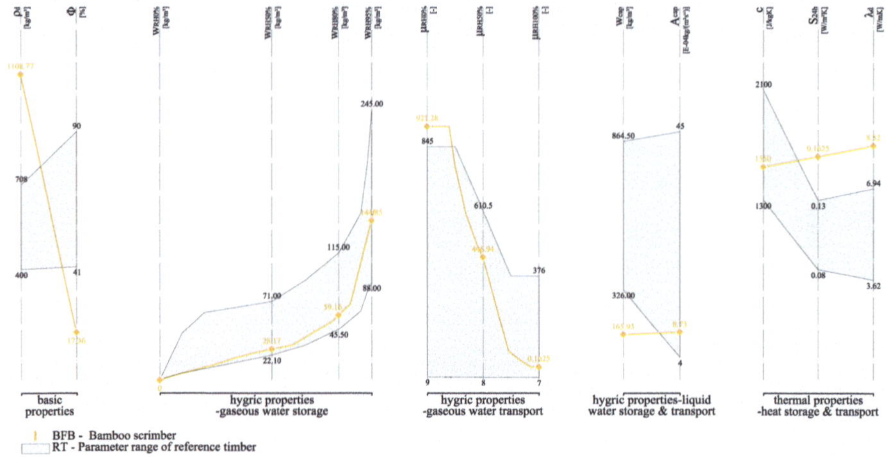

Fig. 4.3 Relative position of bamboo scrimber property parameters in the range of reference timber

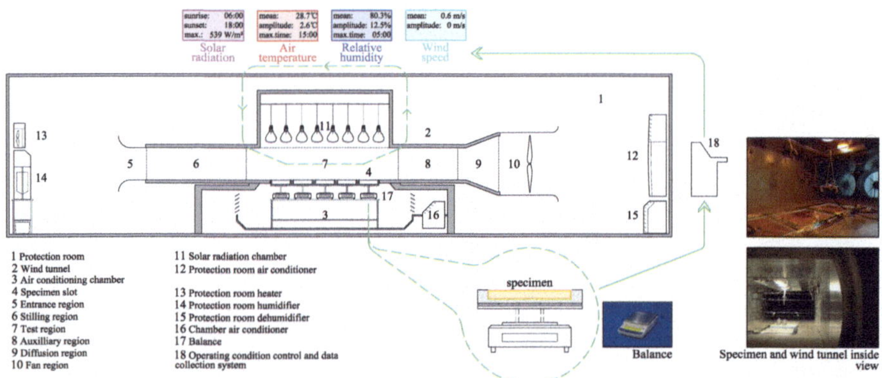

Fig. 4.4 Wind tunnel test: section of the wind tunnel and the test photograph. *Source* the section basemap is redrawn from the materials provided by the State Key Laboratory of Subtropical Building Science

the entire test, of which the recorded data in the last 48 h are selected for analysis. The reference timber in this test is Intsia.spp (Caesalpiniaceae), a kind of antiseptic hardwood for outdoor flooring (Fig. 4.4, Table 4.4).

(1) The hourly mass change rate comparison between bamboo scrimber and hardwood shows that bamboo scrimber has slower adsorption and desorption rate U_{ad} value. Except certain areas, mostly the adsorption–desorption transition stage, the hourly U_{ad} values of bamboo scrimber are up to 33.69E^{-06} kg/(m^2 s) smaller than that of the hardwood specimen.
(2) Affected by the strong outdoor solar radiation and temperature, the adsorption and desorption rate of bamboo scrimber are, respectively, up to

Table 4.4 Weather condition, operation, equipment, and specimens of the wind tunnel test

Weather condition (24 h repeated)	Operation method	Main equipment arrangement	Specimens treatment
Solar radiation: Sunrise–06:00 Sunset–18:00 Max. value—539.0 W/m^2 Air temperature: Mean value–28.7 °C Amplitude–2.6 °C Max. time–15:00 Relative humidity: Mean value–80.3% Amplitude–12.5% Max. time–05:00 Wind speed: Constant as 0.6 m/s	Operate the weather condition until deviation of solar radiation, and relative humidity and wind speed $\leq$ 5%, and air temperature $\leq$ 0.3 °C Install the specimens with slots at 06:00 and start mass record by electronic balance (range 0–4 kg, accuracy 0.01 g; time step: 10 s) Hourly mass values m_i are calculated from the average values of all the instantaneous measured values within the hour The hourly adsorption and desorption rate $U_{ad,i}$ value is calculated from: $U_{ad,i} = (m_i - m_{i-1})/A$	HHCWT: Hot-humid climatic wind tunnel, including control systems for: Air temperature: 2 air conditioner (KFRd-60LW/VA-ZXF, 6 kW, and KFR-72LW/08FZBPC-a, 7.2 kW) and 2 electric heating fans (1.5 kW and 5 kW), range 20–40 °C, accuracy 0.3 °C Relative humidity: 2 electrode humidifiers (BFD-01-04, 0.4 kg/h) and 1 dehumidifier (DH-890C, 90 L/d), range 40–90%, accuracy 3% Solar radiation: 8 infrared lights (305–3000 nm), range 0–1030 W/m^2, accuracy 10 W/m^2 Wind: axial fan, range 0–5 m/s, accuracy 0.2 m/s	A_{BFB}–0.0822 m^2 (30 × 27.4 cm) A_{HW}–0.0810 m^2 (30 × 27 cm) Side and bottom surfaces are sealed with 1–2 mm Vaseline The gaps between specimens and the slots are also sealed with Vaseline

Note The times in the table are condition hour
Source Sort out from the documents provided by the State Key Laboratory of Subtropical Building Science

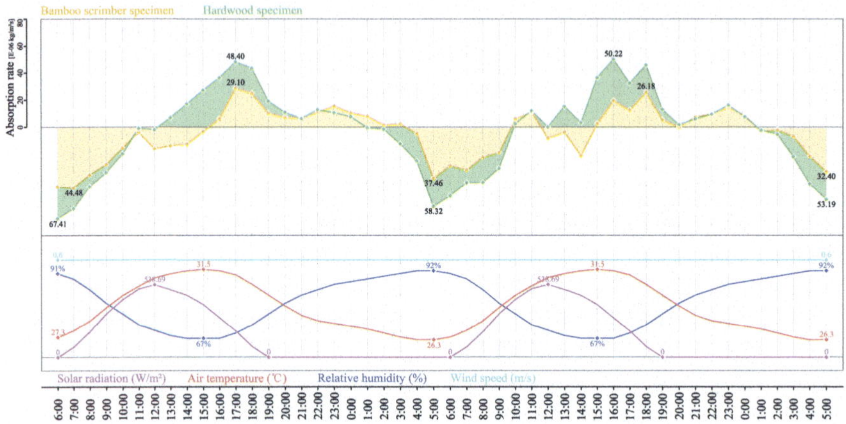

Fig. 4.5 Hourly mass change rate of bamboo scrimber and hardwood specimens of the wind tunnel test

50.22E^{-06} kg/(m^2 s) and 67.41E^{-06} kg/(m^2 s), much higher than the values 7.74E^{-07}–12.58E^{-07} kg/(m^2 s) resulted from the static drying test ($T = 23$ °C, RH = 50%). As a result of the comprehensive impacts of solar radiation, temperature, relative humidity, and wind, the wind tunnel test results show a much more mutative characteristic and contribute to describe the material in a way that is closer to the practical conditions (Fig. 4.5).

4.1.4 Discussion

Nine static test items for hygrothermal properties have been carried out for bamboo scrimber in Chap. 3. The test results by comparison with reference timbers show that bamboo scrimber has higher heat storage and heat transport properties, lower moisture storage, and moisture transport properties.

A dynamic test in wind tunnel is carried out to examine the moisture adsorption and desorption rate in practical outdoor weather condition. Results show that bamboo scrimber has lower moisture adsorption and desorption rate than the reference hardwood.

The significant magnitude difference between the static and dynamic test results shows the necessity of a comprehensive evaluation approach that can take more practical conditions, e.g., solar radiation, air temperature, relative humidity, and wind speed into consideration.

4.2 Performance Comparison Between Bamboo and Timber Building Envelope

The existing comparison studies on bamboo and timber generally concentrate on raw materials, focusing on certain microstructure characteristics rather than the hygrothermal properties necessary for the application in building envelope. Comparison on building component and enclosed space cannot be found; actual application conditions, such as climate, building function, construction type, and HVAC, have not been taken into consideration; to promote the strategy of 'substitute timber with bamboo', the discussion on substitutability of bamboo for timber in terms of forest production, prices, and mechanical performance are mostly stressed, but the physical performance of bamboo compared with timber is not sufficiently studied.

For clarifying the performance difference between the application of bamboo and timber in building envelope, timber and timber units are set as reference models, accordingly bamboo and bamboo units of the same construction and space size as evaluation models. Firstly, bamboo and timber exterior wall construction groups are designed according to the thermal requirement of typical climate regions; afterward, static performance evaluation and comparison are performed on the bamboo and timber component units. Secondly, bamboo and timber enclosed space units are constructed in WUFI Plus, in which the performance on levels of material, building component, and enclosed space are compared, with comprehensive consideration to typical conditions. Finally, the results comparison are carried out on different levels, where the reference timber units are set as benchmarks to provide performance assessment on the application of bamboo in building envelope.

4.2.1 Simulation Model Design

The bamboo properties parameters obtained in Chap. 3 and the timber parameters provided in WUFI Plus by Fraunhofer IBP are tested through the same methods, which is the basis for further parallel comparison between bamboo and timber units. Model groups for comparison are constructed in WUFI Plus, which are composed of external, internal, boundary, and HVAC conditions. Exterior walls with the same construction sizes are set as boundary condition in WUFI Plus to form the enclosed space units. Annual exterior walls hygrothermal performance, indoor hygrothermal environment, and HVAC demand are dynamically simulated (Fig. 4.6).

4.2.1.1 External Conditions

Given that climate data significantly affect the heat and moisture process of the building envelope, complete climate parameters are necessary to ensure the effectiveness of the simulation. The meteorological data in WUFI Plus are more complex than that

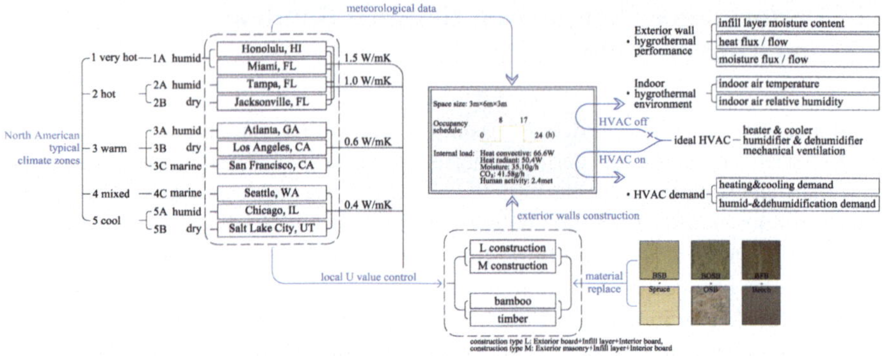

Fig. 4.6 WUFI Plus model design (external, internal, boundary and HVAC conditions)

in some other software for energy simulation, e.g. EnergyPlus, since meteorological parameter such as driving rain has been included. The currently available meteorological data in WUFI Plus are mainly from North America, Europe, and Japan. According to the Köppen climate classification, there are rich types of climate zones in North America, while in Europe and Japan, the climate conditions are relatively single and representative cities are not available in hot climate regions. Therefore, in the study, the meteorological data of the ten representative cities from nine typical climate zones of North America are selected as the exterior condition. Each climate data includes both the cool year and warm year and consists of annual hourly meteorological parameters covering air temperature, relative humidity, solar radiation, and driving rain that depends on the precipitation and wind. The ten cities are separated to different groups according to the local requirement for exterior wall heat transfer coefficient U_c (all the following values are calculated with material parameters corresponding to RH = 0%):

From Zone 5-cool to Zone 4-mixed, Chicago (Ch), Salt Lake City (Sl), Seattle (Se) are representative cities as the external conditions for U0.4 group [$U_c \leq 0.4$ W/(m^2 K)];
For Zone 3-warm, Atlanta (At), Los Angeles (Lo), San Francisco (Sa) are representative cities as the external conditions for U0.6 group [$U_c \leq 0.6$ W/(m^2 K)];
From Zone 2-hot to Zone 1-very hot, Tampa (Ta), Jacksonville (Ja), Honolulu (Ho), Miami (Mi) are representative cities as the external conditions for U1.0 group [$U_c \leq 1.0$ W/(m^2 K)]. In addition, in order to carry out the study in a wider construction range, the U1.5 group [$U_c \leq 1.5$ W/(m^2 K)] is set up and given the same representive cities from U1.0 group as the external conditions.

4.2.1.2 Internal Conditions

Enclosed space units of 3.0 m × 6.0 m × 3.0 m (width × depth × height) are constructed, of which the room occupancy period is set as 8:00–17:00. Standard office indoor heat and moisture load is given including convective heat: 66.6 W, radiant heat: 50.4 W, moisture: 35.10 g/h, CO_2: 41.58 g/h, and human activity: 2.4 met. The PMV/PPD related additional data are given including clothing: 0.7 clo and air velocity: 0.1 m/s.

4.2.1.3 Boundary Conditions

Boundary conditions. The exterior walls are selected as the study objects, while the floor and ceiling are set as partitions between the same interior conditions to avoid heat and moisture exchange. Two commonly used construction types are considered including type *L* (exterior board + infill + interior board) and type *M* (masonry + infill + interior board). The constructions are designed by controlling the heat transfer coefficient of approximately U_c = 0.4/0.6/1.0/1.5 W/(m K). (Note: the U_c values here are calculated with the material parameters in dry state and do not include the surface heat transfer resistance, i.e., the exterior surface heat transfer resistance R_e and the interior surface heat transfer resistance R_i. They are set as default values in WUFI as $R_e = 0.040$ m^2 K/W and $R_i = 0.130$ m^2 K/W) (Table 4.5).

For the partition boards, bamboo scrimber (BFB), bamboo laminated lumber (BSB) and bamboo oriented strand board (BOSB) are selected for the bamboo construction, while correspondingly Beech, Spruce, and OSB for timber construction, which forms three sets of alignment.

In order to ensure the comparability between bamboo and timber components, between full B/T construction and interior finishes construction types, and among different cities, the exterior walls are designed to reach the required U value by adjusting single factors:

From bamboo to timber components: replace the bamboo boards with timber boards, while maintain the construction layer dimensions;
From full B/T construction to interior finishes construction types: replace the 18 mm exterior board with 200-mm-brick wall, while maintain the remaining construction layer dimensions;
From one city to another city: adjust the thickness of the thermal insulation layer, or replace the insulation material with air, while maintain the remaining construction layer dimensions.

4.2.1.4 HVAC Conditions

Based on the external, internal, and boundary conditions arranged above, the HVAC is turned off first, and the indoor hygrothermal environment is simulated, involving the

Table 4.5 Boundary condition (Construction design of the exterior walls)

City	Type	1		2	3[a1]		4		5[a1]	6	7			Construction graph
		BSB	Spruce	PE	BF	air	BOSB	OSB	air	PE	BFB	C.beech	Brick	L type construction – Full bamboo/timber; M type construction – Bamboo/timber as internal finishes
U0.4	B-L	12.5		0.2	100		12.5		40		18			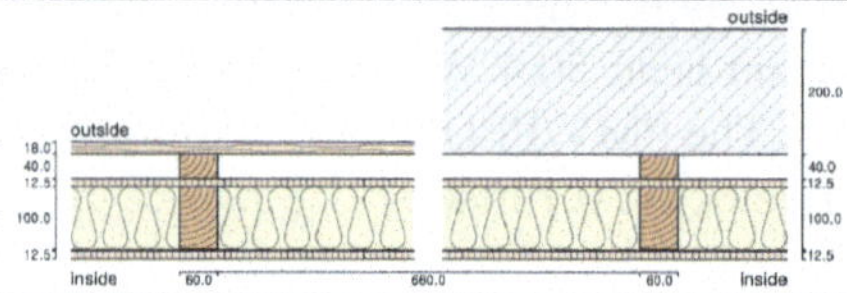
	T-L		12.5	0.2	100			12.5	40			18		
	B-M	12.5		0.2	100		12.5		40				200	
	T-M		12.5	0.2	100			12.5	40				200	
U0.6	B-L	12.5		0.2	60		12.5		40		18			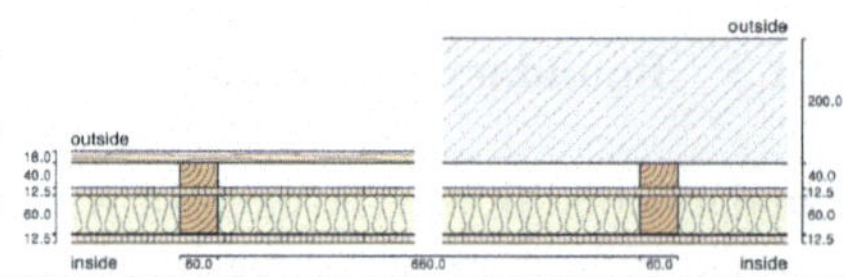
	T-L		12.5	0.2	60			12.5	40			18		
	B-M	12.5		0.2	60		12.5		40				200	
	T-M		12.5	0.2	60			12.5	40				200	
U1.0	B-L	12.5			30		12.5		40	0.2	18			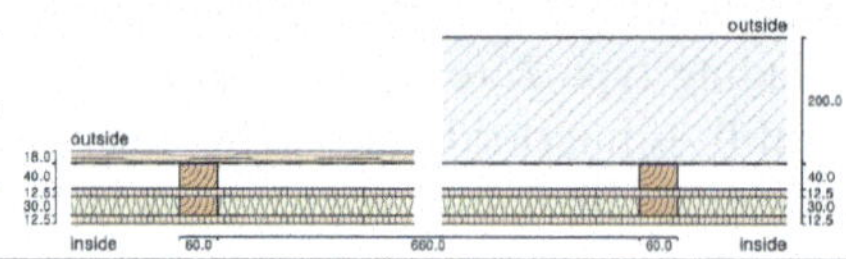
	T-L		12.5		30			12.5	40	0.2		18		
	B-M	12.5			30		12.5		40	0.2			200	
	T-M		12.5		30			12.5	40	0.2			200	
U1.5	B-L	12.5				30	12.5		40	0.2	18			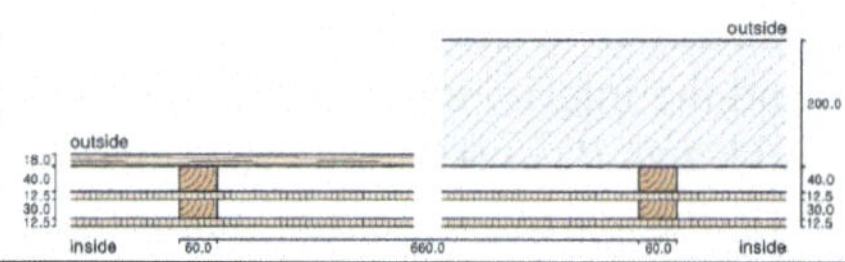
	T-L		12.5			30		12.5	40	0.2		18		
	B-M	12.5				30	12.5		40	0.2			200	
	T-M		12.5			30		12.5	40	0.2			200	

[a]1 Inner supports of internal distance $D = 720$ mm, width $d = 60$ mm, and the same thickness with the local layer. Material used: Bamboo construction—BFB, Timber construction—KVH Fichte
Construction type: *B*—bamboo, *T*—timber, *n*—full B/T construction, *r*—B/T as internal finishes
Material abbreviation (from left to right): *BSB*—bamboo laminated lumber; *PE*—PE foil; *BF*—natural bamboo fiber thermal insulation material; *air*—closed air layer; *BOSB*—bamboo oriented strand board; *OSB*—wood oriented strand board Platte (DIN EN ISO 10456); *BFB*—bamboo scrimber; *C.beech*—beech

inside surface temperature of the exterior walls, indoor air temperature, and relative humidity. Then the HVAC is turned on, maintaining the indoor air temperature within 20–26 °C, and the relative humidity 40–60%. The max. CO_2 concentration is set as 1550 ppmv. Natural ventilation is closed, the infiltration air change rate is given as 0, and the mechanical ventilation is set as 0.6 times per hour. The annual heating and cooling demand, and humidification and dehumidification demand are calculated for the evaluation of HVAC demand. For both HVAC on and HVAC off conditions, the annual moisture content of the exterior wall infill layer (represented by the south wall), heat and moisture flow through the exterior walls are indicators to describe the hydrothermal performance of the building component.

4.2.2 Results Analysis

4.2.2.1 Building Component Units

When the HAM model theory and its simulation results are used to evaluate the performance of building component, the heat and moisture transport performance would keep updating with the change of moisture content. Section 2.1 in Chap. 2 has carried out detailed calculation for the impact of the moisture content change caused by hygric environment on the building performance, so that there is no constant performance value at the level of building component. Therefore, the mean value and amplitude of annual simulation results are selected as the indicators for performance evaluation.

The annual moisture content of the south exterior wall infill layer mean value w-BF_{mean} and its amplitude w-BF_{amp} are used to characterize the impact of the hygric environment on the component, which shows the moisture resistance of bamboo and timber partition boards to the hygric environment. The heat flow through exterior walls H_{flow} and moisture flow M_{flow} are adopted to characterize the heat and moisture transport performance. In addition, as a hygroscopic building material, the adsorption and desorption performance is of special significance to the construction; therefore, the annual moisture flux through exterior surface $M_{e.flux}$ and interior surface $M_{i.flux}$ of south exterior wall are analyzed. There is latent heat effect accompanied with the process of $M_{e.flux}$ and $M_{i.flux}$, of which the former is getting attention in certain climate regions for its evaporative cooling effect, and the latter has a regulating function on the indoor hygrothermal environment.

(1) Hygric performance

For the M-type construction groups, the w-BF_{mean} difference between bamboo and timber units is not obvious. Except SI city group, the w-BF_{mean} ratio between bamboo and timber units is in the range 100.06–102.91%. In contrast, the difference among construction groups is highlighted in the L-type construction groups. For U0.4 groups, the w-BF_{mean} ratio between bamboo and timber units is in the range 108.24–125.50%, which shows a higher moisture content in the infill layer of bamboo construction. The reason behind is that, in this climate conditions, the moisture flow run mainly from indoor to outdoor and is easy to condensate when the air get in touch with the cold surface, so that the larger water vapor transfer resistance can on the one hand weaken the moisture flow, but on the other hand, cause condensated water between two construction layers and slow down its drying. For the U1.0 groups, the situation results are just the opposite that the w-BF_{mean} ratio between bamboo and timber units is in the range 94.33–97.72%. In this climate conditions, the moisture flow run mainly from outdoor to indoor and is influenced more by the driving rain. As an exterior board, BFB can efficiently block the outdoor moisture, especially the liquid water, from permeating into the inside of the construction, which results in lower moisture content of bamboo units than that of timber units.

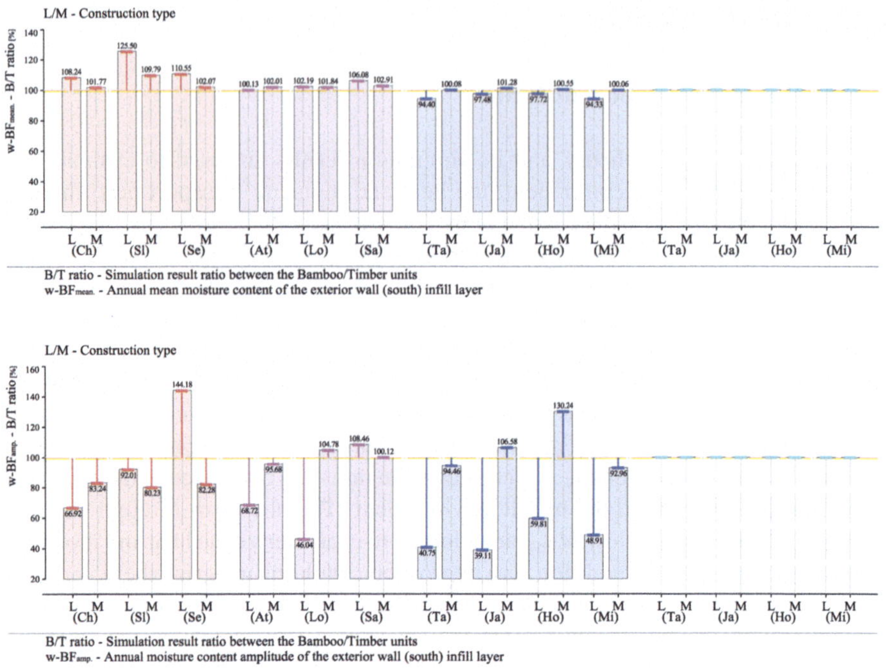

Fig. 4.7 Hygric performance comparison between bamboo and timber components (HVAC on): Annual mean moisture content of the south exterior wall infill layer $w\text{-BF}_{mean}$ (up) and the annual moisture content amplitude of the south exterior wall infill layer $w\text{-BF}_{amp}$ (down)

The comparison results of $w\text{-BF}_{amp}$ show a larger variability. For M-type construction, the $w\text{-BF}_{amp}$ ratio between bamboo and timber units is in the range 80.23–130.24%. For L-type construction, except in the city groups of Se and Sa, the $w\text{-BF}_{amp}$ of bamboo units is in the range 39.11–92.01% when compared with the timber units, which shows a better hygric stability in bamboo construction. Among this, the ratio for U1.0 group is 39.11–59.81%, which highlights a stronger barrier effect of BFB to the driving rain (Fig. 4.7).

The comparison results of M_{flow}, characterizing the moisture transport performance, show that bamboo units possess larger moisture resistance, which results from the lower moisture transport rate of bamboo. For the L-type construction, the M_{flow} ratio between bamboo and timber units is in the range 87.39–96.00%, while for the M-type construction, it is 90.48–96.07%. However, this is dialectical when judging from the perspective of enclosed space. In hot climate regions where there is large external heat and moisture load and the moisture flow through building envelope run mostly from outdoor to indoor, the large moisture resistance of the exterior walls is beneficial for weakening the influence of the outdoor heat and moisture load, especially the impact of liquid process on the indoor environment. On the contrary, in cold regions, driven by the temperature drop, the moisture flows mainly from indoor

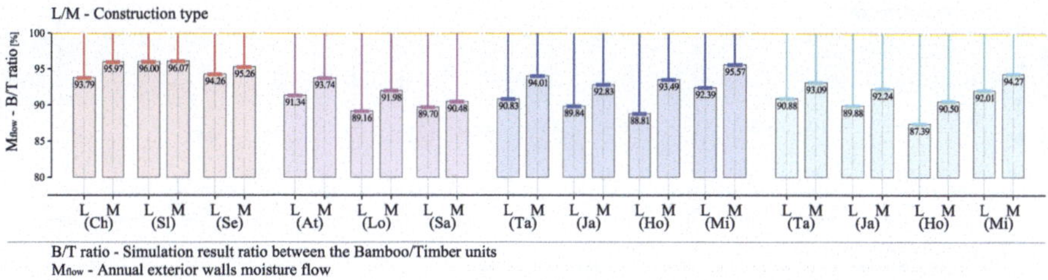

Fig. 4.8 Hygric performance comparison between bamboo and timber components (HVAC on): Annual moisture flow through exterior walls M_{flow}

to outdoor, so that for those conditions with heavy indoor moisture load, the large moisture resistance of the building envelope might obstruct the indoor moisture from exhausting and increase the indoor dehumidification demand (Fig. 4.8).

For the moisture exchange ability of the panels, simulation results show a significant smaller annual moisture flux through the surface of bamboo units than that of the timber units. For the exterior surface of L-type construction, setting the timber units with Beech as exterior board as 100%, the $M_{\text{e.flux}}$ of bamboo units with BFB as exterior board is in the range 37.73–53.59%. This results mainly from the smaller water absorption coefficient and water vapor transfer coefficient of bamboo, which reduces its moisture adsorption and desorption rate. On the one hand, this contributes a better moisture content stability to bamboo construction and can reduce the damage caused by hygric stress. On the other hand, it might weaken the evaporative cooling effect in certain conditions such as the Hot-Humid climate regions.

For the interior surface, the $M_{\text{i.flux}}$ of bamboo units with BSB as interior board is in the range 77.67–86.53%, while setting the timber units with Spruce as the interior board as 100%. Similarly, the bamboo units display a weaker moisture exchange ability, resulting from the smaller water vapor transfer coefficient on material level. This is not conducive to playing the moisture buffering effect of building envelope to regulate the indoor hygrothermal environment, but might help to block the water vapor from permeating into the inside of the construction in certain conditions such as the cold climate regions (Fig. 4.9).

(2) Thermal performance

The heat storage performance related indicator, annual inside surface temperature amplitude of the exterior walls $T_{\text{is.amp}}$, of the bamboo units is significantly smaller, which is due to the larger thermal capacity of bamboo. The $T_{\text{is.amp}}$ ratio between bamboo and timber units is in the range 96.21–99.10%, which is affected by the material of the inside surface and the total heat capacity of the construction. The difference magnitude between bamboo and timber units is similar in L-type and M-type construction (Fig. 4.10).

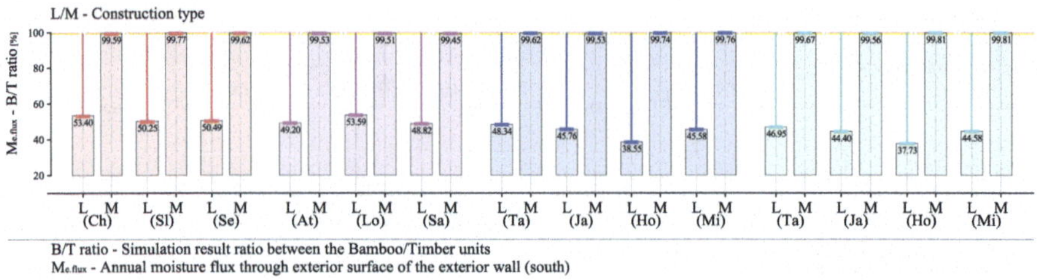

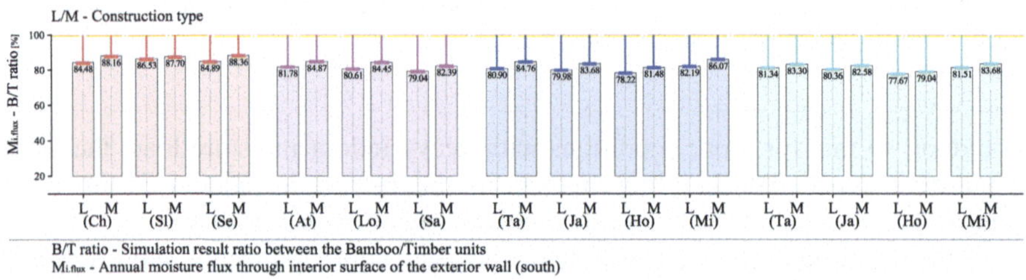

Fig. 4.9 Hygric performance comparison between bamboo and timber components (HVAC on): Annual moisture flux through exterior surface $M_{\text{e.flux}}$ (up) and interior surface $M_{\text{i.flux}}$ (down) of south exterior wall

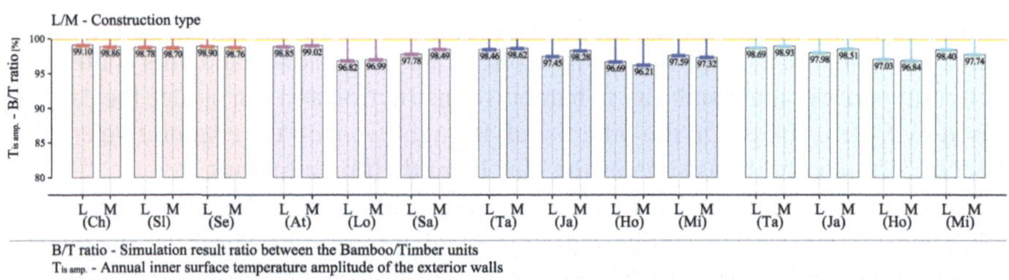

Fig. 4.10 Thermal performance comparison between bamboo and timber components (HVAC off): Annual inside surface temperature amplitude of the exterior walls $T_{\text{is.amp}}$

The heat transport performance related indicator, annual heat flow through exterior walls H_{flow}, has a significant impact on the heating and cooling demand. The H_{flow} of bamboo units is generally larger than that of the timber units. The H_{flow} ratio between bamboo and timber units is in the range 101.66–106.06% for U0.4 and U0.6 construction groups, but the difference is narrowed down in U1.0 and U1.5 groups. In few conditions such as the Ho city group, the H_{flow} of bamboo units is even slightly smaller than that of the timber units, of which the lowest reaches 98.74% (Fig. 4.11).

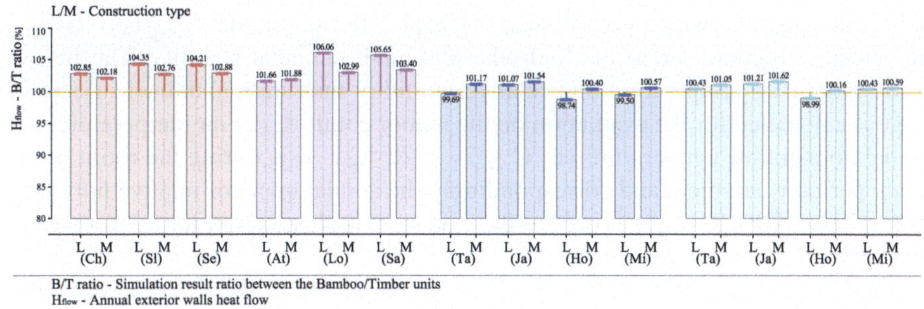

Fig. 4.11 Thermal performance comparison between bamboo and timber components (HVAC on): Annual heat flow through exterior walls H_{flow}

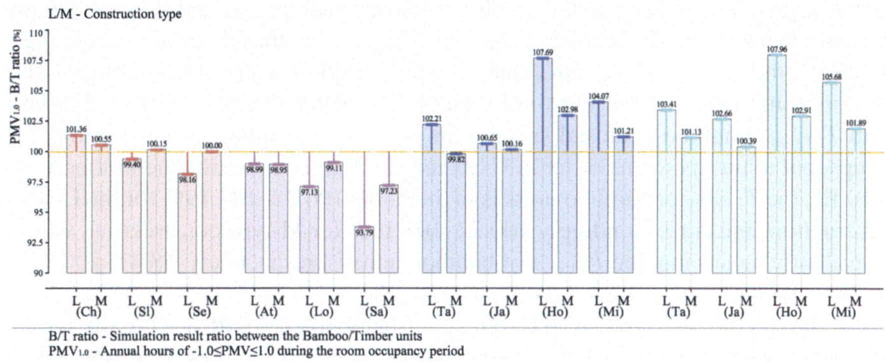

Fig. 4.12 Indoor hygrothermal environment comparison between bamboo and timber enclosed space units (HVAC off): Annual hours of $-1.0 \leq \text{PMV} \leq 1.0$ during the room occupancy period $\text{PMV}_{1.0}$

4.2.2.2 Enclosed Space Units

(1) HVAC off conditions

Taking the annual hours of $-1.0 \leq \text{PMV} \leq 1.0$ during the room occupancy period $\text{PMV}_{1.0}$ as indicator, bamboo units perform worse than timber units in U0.4 and U0.6 model groups, of which the $\text{PMV}_{1.0}$ ratio between is in the range 93.79–101.36% for L-type construction. The comparison results show just the opposite in U1.0 and U1.5 model groups that the $\text{PMV}_{1.0}$ ratio rises to 100.65–107.69%. For M-type construction, the difference between bamboo and timber units is narrowed down to a ratio range of 97.23–102.98% (Fig. 4.12).

The PMV is influenced by various factors including the indoor air temperature, relative humidity and inside surface temperature amplitude of the building envelope. The last one has been discussed in Sect. 4.2.2.1, and the former two factors would be analyzed below. building envelope Since the annual mean values of them are decided by the outdoor air temperature mean value, relative humidity mean value and the driving rain, which is explained in Chap. 2, the amplitude $T_{i.amp}$ and $RH_{i.amp}$ are selected to characterize the hydrothermal environmental stability. The period of the indoor air temperature and relative humidity that are maintained in certain range is calculated for the evaluation on the indoor comfort. For air temperature, the annual hours of air temperature 18 °C $\leq T_i \leq$ 28 °C $T_{i.18\text{-}28}$ during the room occupancy period is adopted, and correspondingly the annual hours of relative humidity 30% $\leq RH_i \leq$ 70% $RH_{i.30\text{–}70}$ is used as the indicators to characterize the hygrothermal environment of the enclosed space.[1]

Similar with the inside surface temperature amplitude of the exterior walls, the $T_{i.amp}$ of the bamboo enclosed space units is smaller than that of the timber units. The ratio between is in the range 96.91–99.25%, which shows better stability of the indoor air temperature in bamboo units. However, only the air temperature amplitude $T_{i.amp}$ is considered together with the mean value $T_{i.mean}$ and can the indoor thermal environment be defined. When the $T_{i.mean}$ is within a certain range, e.g., 20–26 °C, the smaller $T_{i.amp}$ can ensure longer period of temperature falling in the comfort zone; otherwise, it does not guarantee a better thermal comfort. Judging from $T_{i.18\text{–}28}$, bamboo enclosed space units show no advantage in U0.4 and U0.6 groups, but in turn perform better in U1.0 and U1.5 groups. Setting the timber units as 100%, the $T_{i.18\text{–}28}$ of bamboo units is in the range 100.22–103.19% for the L-type construction, and correspondingly 100.22–103.19% for M-type construction, which equivalently means that the $T_{i.18\text{–}28}$ of bamboo units is, respectively, 58.47–200.71 h and 7.2–104.79 h longer than the timber units (Fig. 4.13).

For the hygric environment, the $RH_{i.amp}$ of bamboo enclosed space units is significantly larger than that of the timber units. The ratio between is in the range 101.88–114.72% for L-type construction and 101.11–108.67% for M-type construction, which shows a weaker stability of hygric environment among the bamboo units. As for the $RH_{i.30\text{–}70}$, bamboo units also perform worse than the timber units. The ratio between is in the range 94.06–99.49% for L-type construction and 94.18–101.22% for M-type construction, which shows a shorter period of relative humidity maintained within certain comfort zone.

Combining this with the conclusion at the level of building component in Sect. 4.2.2.1, it can be seen that the large moisture transport resistance plays a positive role mainly at the level of building component where it can weaken the influence of the hygric environment on the inside of the construction. However, judging from the indoor air relative humidity of the enclosed space, bamboo enclosed space units show disadvantage compared with timber units in the HVAC off conditions. This is due to the large moisture resistance of bamboo units that blocks the indoor moisture from exhausting and reduces the moisture buffering effect of building envelope to regulate the indoor hygrothermal environment (Fig. 4.14).

---

[1]Generally, the indoor air temperature T_i in the range 20–26 °C and the relative humidity RH_i in the range 40–60% are regarded as the corresponding comfort range. However, in this study, the annual hours of T_i and RH_i in these ranges are too little, and the hours of RH_i between 40 and 60% are even 0 in certain cases, which make it unable to be compared by the ratio between the BC and NC model groups. Therefore, as the evaluation indicators, the range of RH_i and T_i are, respectively, enlarged to 30–70% and 18–28 °C.

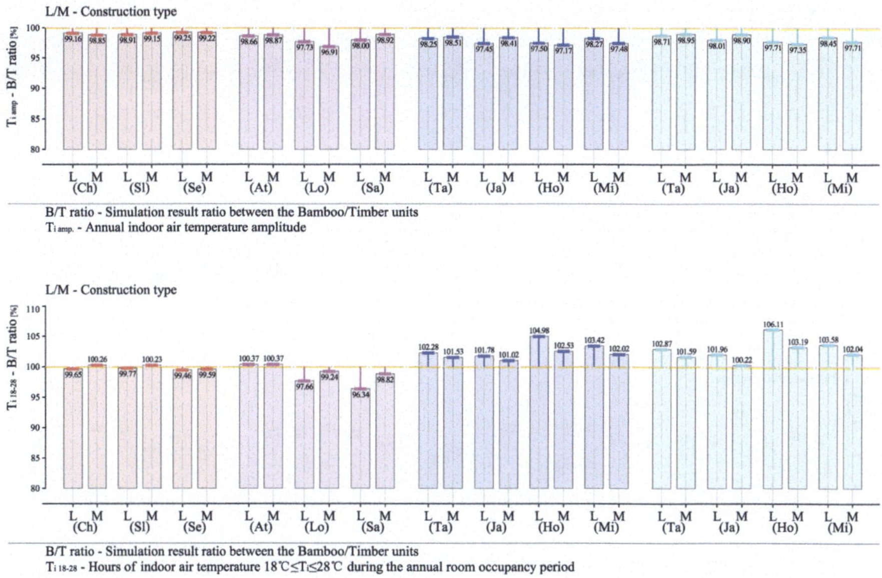

Fig. 4.13 Thermal environment performance comparison between bamboo and timber enclosed space units (HVAC off): Annual amplitude $T_{i.amp}$ (up) and hours of air temperature 18 °C $\leq T_i \leq$ 28 °C $T_{i.18-28}$ (down) during the room occupancy period

(2) HVAC on conditions

The *P* value ratio between bamboo and timber units is in the range 100.20–103.14% in U0.4 and U0.6 model groups, which show the weakness of bamboo units. On the contrary, the ratio is 98.35–99.73% in U1.0 and U1.5 groups, which means that bamboo overcome timber units in these conditions. This phenomenon is mainly due to the compensation of the heat storage advantage of bamboo for its heat transport disadvantage, but this advantage is weakened in M-type construction.

The comparison results of *H* value are different from the former conclusion of $RH_{i.30-70}$ in the HVAC off conditions that there is not obvious strengths and weakness between bamboo and timber units. Except certain cases, e.g., Sa city group, the *H* ratio between bamboo and timber units is in the range 96.80–102.94% and 98.35–104.23%, respectively, for L-type and M-type construction. The reason behind is that in the HVAC on conditions, the indoor relative humidity is maintained within a certain zone, so that the larger moisture transport resistance of bamboo construction is conducive to weakening the influence of the outdoor hygric environment, and therefore reducing the *H* demand to a certain extent (Fig. 4.15).

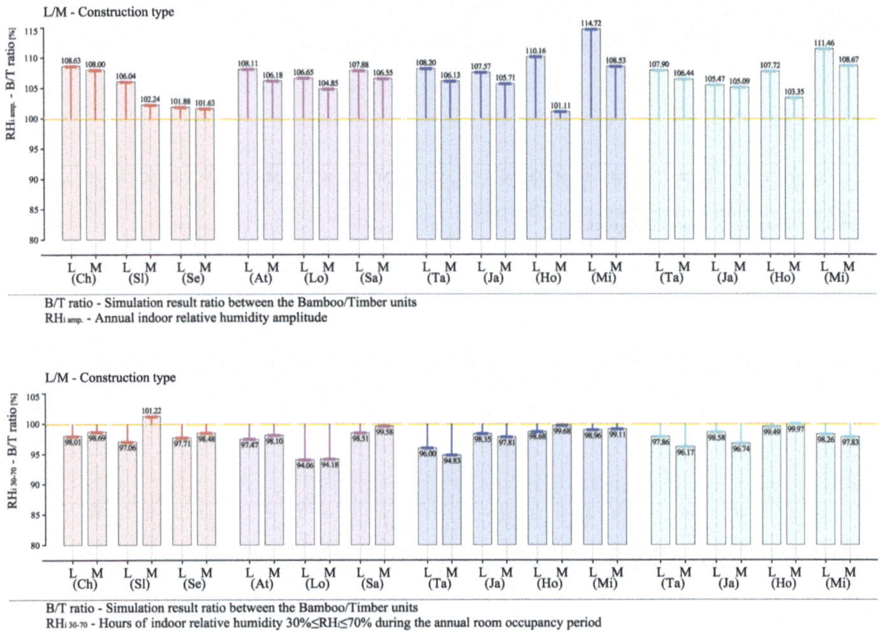

Fig. 4.14 Hygric environment performance comparison between bamboo and timber enclosed space units (HVAC off): Annual amplitude $RH_{i.amp}$ (up) and hours of relative humidity $30\% \le RH_i \le 70\%$ $RH_{i.30-70}$ (down) during the room occupancy period

4.2.3 Discussion

4.2.3.1 Computer Tool for Simulation

Before the coupled simulation of heat and moisture process for the building envelope in WUFI Plus, the bamboo and timber construction and enclosed space units with the same sizes are constructed and simulated in EnergyPlus. The annual heating and cooling demand P value in the HVAC on conditions, and annual hours of $-1.0 \le \text{PMV} \le 1.0$ during the room occupancy period $\text{PMV}_{1.0}$ in the HVAC off conditions are calculated. EnergyPlus is a comprehensive energy consumption simulation computer program developed by the Department of Energy (DOE) together with the Lawrence Berkeley National Laboratory (LBNL). It is one of the most widely used building energy consumption simulation programs in the world, which is based on the BLAST and DOE-2. Due to the completeness of function and reliability of calculation results, it is adopted in the energy efficiency assessment, e.g., LEED and used as a reference for judging the accuracy of other simulation software [1].[2]

[2]The Hygrothermic department of Fraunhofer IBP used to compare the simulation results of WUFI Plus in parallel with the comprehensive energy consumption simulation computer programs including EnergyPlus and TRANSYS, showing that the difference of all indicators is within 5%.

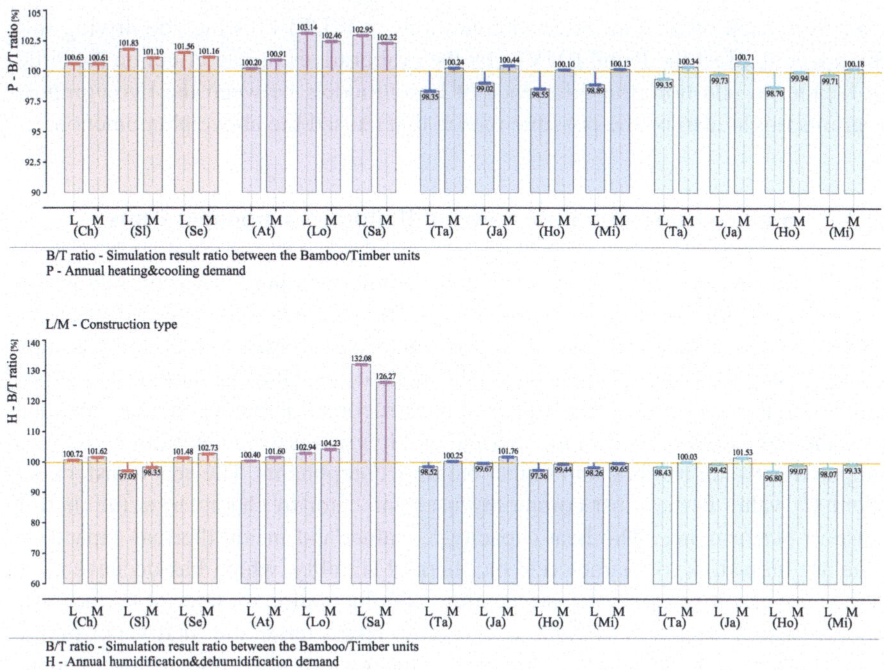

Fig. 4.15 HVAC demand comparison between bamboo and timber enclosed space units (HVAC on): Annual heating and cooling demand P (up) and humidification and dehumidification demand *H* (down)

On the one hand, for the *P* and $PMV_{1.0}$ values, the coupled heat and moisture process simulation in WUFI Plus and the thermal simulation in EnergyPlus show similar conclusion. Both simulations display a law that in cold climate regions, where thermal insulation is the main requirement for building envelope, bamboo units perform worse than timber units due to the weakness of heat transport performance. On the contrary, in hot climate regions where cooling is the dominated requirement, the heat storage performance of bamboo can make up for the shortage in heat transport performance, which enables bamboo units to overcome timber units in L-type construction. However, this strength is weakened in M-type construction, due to the already existing large thermal capacity of the construction.

On the other hand, the difference margin of bamboo and timber units is different between the simulation results of WUFI Plus and EnergyPlus, specifically the former is generally smaller than the latter. The simulation results in WUFI Plus have taken the influence of moisture content on the heat transport performance into consideration and been keeping updating the performance values during the simulation process. Therefore, even though the construction layer arrangement and material selection are the same in WUFI Plus and EnergyPlus, the instantaneous heat transfer coefficient and vapor diffusion resistance, etc., of the bamboo and timber constructions differ from each other. The external conditions in WUFI Plus include driving rain, which would affect the P and $PMV_{1.0}$ by the evaporative cooling effect on the building façade. In addition, the difference of basic models between the two computer programs might also be the possible reason that caused the above phenomenon.

4.2.3.2 Static Assessment at the Level of Building Component Units

As described in Sect. 4.2.3.1, in WUFI Plus simulation, there is no constant performance value at the building component level. Here the material state corresponding to RH = 0% is selected for the static performance assessment on building component units, and the comparison between the methods of static evaluation and coupled simulation. Except for the basic values of construction thickness and areal bulk density, nine indicators related to the thermal and hygric performances are chosen and compared between bamboo and timber building components. The heat transfer coefficient U_c value is the most commonly used indicator to characterize the thermal transport performance. The heat capacity S_c value, and inside thermal capacity S_i value are indicators to characterize the thermal stability, which has decisive effect on the temperature phase shift, attenuation, and fluctuation on interior surface. The vapor diffusion thickness s_d value is chosen to characterize the total vapor permeation resistance of the whole components; meanwhile, the condensation and the drying time can partly describe the hygric defects of the constructions. The drying time is calculated for the summer period according to the DIN 4108-3: 2014–11. (Table 4.6, Fig. 4.16).

(1) Heat transport properties. Calculated results show that with the same construction dimensions and layer arrangement, the U_c values of bamboo components are slightly higher than the timber components, which is caused by the higher thermal conductivity of bamboo. Setting timber construction as 100%, the U_c value of the bamboo construction groups is in the range 100.94–103.72%. From U0.4 to U1.5 groups, the Bamboo–Timber U value magnitudes are enlarged from 0.009 to 0.010 W/(m^{2}K) to 0.119–0.124 W/(m^2 K), which is enlarged as the contribution ratio of bamboo/timber boards to the whole performance increased.
(2) Heat storage properties. Calculated results show that the S_c values and S_i values of bamboo components is higher than the timber components, with magnitudes correspondingly of 15–24 kJ/(m^2 K) and 7.8–11 kJ/(m^2 K). Setting the timber units as 100%, the S_i value of bamboo units is in the range 144.23–164.63% for L-type construction, resulting from the significantly larger bulk density of bamboo compared with timber, and 106.24–131.43% for M-type construction, of which the difference range is narrowed due to the already large thermal capacity of the masonry layer. The temperature phase shifts of bamboo components are 0.7–1.2 h longer than timber components. The temperature attenuation shows the similar law with the S values, that the bamboo components are 0.19–1.31 larger than timber components in L-type construction, and 0.51–3.99 in M-

Table 4.6 Building components thermal and hygric performance calculation results (material properties corresponding to RH = 0%)

City	Type	Heat transfer coefficient	Heat capacity	Thermal capacity inside	Temperature phase shift	Temperature attenuation	Temperature fluctuation on interior surface	Vapor diffusion thickness	Condensation	Drying time
		[W/(m² K)]	[kJ/(m² K)]	[kJ/(m² K)]	[h]	[–]	[°C]	[m]	[kg/m²]	[d]
U0.4	B-L	0.334	86	27	7.5	4.18	4.5	34.19	0.072	62
	T-L	0.323	62	16.4	6.5	2.87	6.7	25.02	0.07	26
	B-M	0.324	455	46	13.3	12.05	1.2	46.45	0.073	93
	T-M	0.314	434	35	12.5	8.06	1.8	41.44	0.078	92
U0.6	B-L	0.478	80	24	6.5	2.80	6.8	33.81	0.014	11
	T-L	0.465	58	15.3	5.5	1.99	9.6	24.75	0	0
	B-M	0.457	449	51	12.3	7.87	1.8	45.99	0.032	6
	T-M	0.446	431	41	11.5	5.62	2.6	41.16	0.042	50
U1.0	B-L	0.715	74	23	5.5	1.89	10.0	33.46	0	0
	T-L	0.701	56	15	4.3	1.46	13.1	24.54	0	0
	B-M	0.668	444	60	11.3	5.29	2.7	45.58	0	0
	T-M	0.657	428	52	10.3	4.13	3.5	40.95	0	0
U1.5	B-L	1.196	74	25	4.0	1.40	13.3	33.31	0	0
	T-L	1.181	55	17	3.0	1.21	15.7	24.40	0	0
	B-M	1.072	443	83	9.7	3.97	3.6	45.43	0	0
	T-M	1.062	428	73	9.0	3.46	4.1	40.81	0	0

Construction type: *B*—bamboo, *T*—timber, *L*—L-type construction, *M*—M-type construction

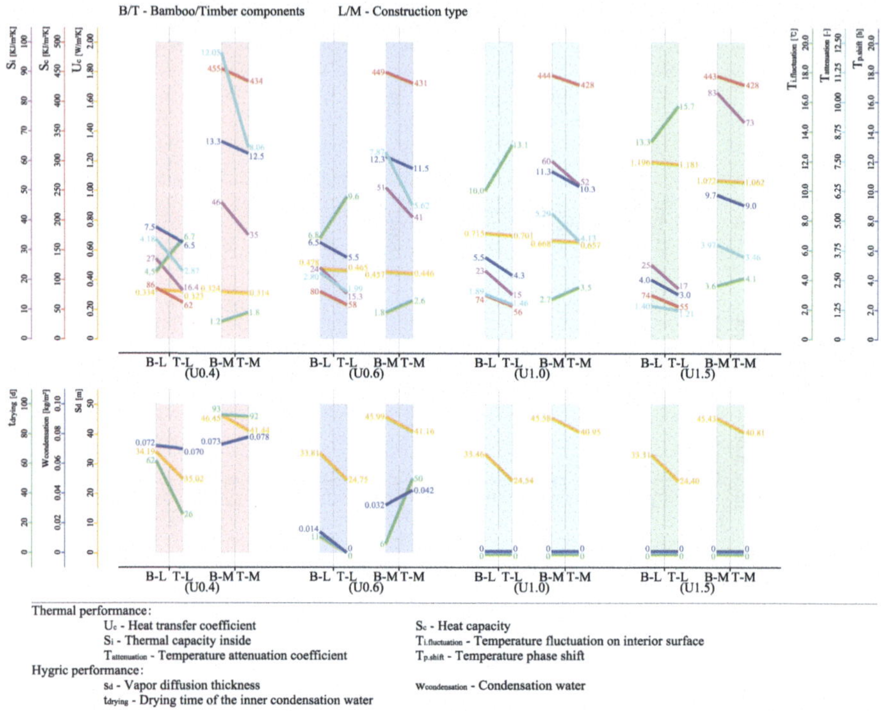

Fig. 4.16 Thermal and hygric performance comparison between bamboo and timber components (The bamboo values and timber values are marked on the boundaries of each column, when the connection lines go down, it means that the bamboo values is higher than the timber values, and vice versa)

type construction. The temperature fluctuations on interior surface of bamboo components are 2.2–3.1 °C and 0.5–0.8 °C smaller than timber components, respectively, for the L-type and M-type construction.

(3) Hygric performance. Calculated results show that the s_d values of bamboo components is larger than the timber components, with magnitudes of 4.62–9.17 m, which is caused by the higher water vapor permeation resistance factor—μ value of bamboo. It means that the bamboo components are of higher resistance to moisture transport. However, in U0.4 and U0.6 groups, where vapor transports mainly from indoor to outdoor, the condensation happens on boundaries between two layers of disparate μ values. The values of condensation and drying time show that the results from higher μ value can be dialectical that might help to weaken the vapor transport on one hand, while on the other hand might also cause water retardation on the boundaries and slow down the condensed water from drying out.

4.2.3.3 Difference Among the Assessment at the Levels of Material, Building Component and Enclosed Space

The performance of bamboo and timber application in building envelope is compared in aspects of material, building component and enclosed space.

(1) In terms of material. Raw bamboo and five typical bamboo-based panels are compared with the corresponding 18 reference timber products. The basic properties of raw bamboo are closer to hardwood and quite different from softwood that is normally the raw material for wood-based panels. The comparison between each bamboo-based panel with corresponding reference timber products shows that, except BPB, the other bamboo-based panels are strengthened in the distinction with timber and had higher bulk density (ρ) and lower porosity (Φ), lower moisture storage (u) and transport properties (δ), higher heat storage (S_{24h}) and transport properties (λ) than RT. For those main parameters that have a significant impact on building envelope, bamboo shows assignable difference from timber. Judging from the bulk density, open porosity, and microscopic fiber arrangement, it can be deduced that substituting bamboo with timber parameters would cause calculation errors to the performance evaluation of building envelope.

(2) In terms of building component. With the same construction dimensions, three groups of bamboos and timbers are chosen as the exterior boards, interior boards, and interlayer boards of the exterior wall constructions. Among the performance indicators, the U_c value is normally used to describe the heat transport performance of building component, while the S_c and D values are related with both heat transport and heat storage performance that can be characterized the thermal stability of the building envelope, which is highlighted in lightweight construction. The calculation results from the material parameters in the state corresponding to RH = 0% show that both the heat transport and the heat storage performances of bamboo components are higher than the timber components. In L-type construction, bamboo units show a significant advantage in S_c and D values, but the advantage is weakened in M-type construction.

(3) In terms of enclosed space unit. The impacts of different factors on the indoor environment and energy consumption performance of bamboo and timber space units are compared. The results show that the performance is comprehensively affected by the heat transfer U_c values and the heat storage S_c values of the building envelope, and the impact weight of the values varies in different operating conditions. The bamboo building components have both higher U_c and S_c values than timber components, and the bamboo units perform slightly worse than the timber units in cold climate zones where the weight of U_c value is higher. In the L-type construction in hot climate zones, the weight of S_c values rises, which highlights the advantage of bamboo space units. However in the M-type construction, the heat storage advantage of bamboo components is weakened due to the already existing large heat capacity of the walls.

(4) The comparison between the application of bamboo and timber in building envelope is carried out on three progressive levels of material, building component and enclosed space, which shows good correlation between the results obtained at the first two levels. However, affected by the climate condition, building function, construction type, and HVAC control, the indicators on material and building component levels cannot directly and accurately describe the performance of the enclosed space, which shows the insufficiency of describing the building envelope merely with the material and building component indicators. The physical performance of enclosed space is closely related to the building material, construction type, building function, and local climate, etc., which is comprehensively affected by various factors and could change dynamically. Traditionally, simplified and steady evaluation process is used in the design of building envelope to ensure eligible physical performance of the enclosed space. With the improvement of evaluation models and development of computer tools, it would be feasible to carry out dynamic and lifecycle physical performance' simulation on the building space. And this requires a great deal of fundamental research in detail on the climate and building material, etc.

4.3 Summary

With the meteorological data of the ten representative cities from nine typical climate zones of North America as the external conditions, timber and timber units are set as reference models, accordingly bamboo and bamboo units of the same construction and space size as evaluation models, by which the performance on levels of material, building component and space unit are compared, with comprehensive consideration to the representative conditions in terms of the external climate, building function, construction type, and HVAC. The reference timber units for comparison are set as benchmark to provide performance assessment on the application of bamboo in building envelope.

(1) Comparison between bamboo and the corresponding reference timber (RT) shows that the basic properties (bulk density and open porosity) of raw bamboo are closer to hardwood and far different from softwood which is normally the raw material for wood-based panels; except BPB, the other BBPs have higher bulk density and lower porosity, lower moisture storage and transport properties, and higher heat storage and transport properties than RT. The bamboo-based panel process strengthened the distinction between bamboo and timber.
(2) A dynamic test in wind tunnel with a climate data of Guangzhou summer typical meteorological day is performed to examine the moisture adsorption and desorption rate difference between bamboo scrimber (BFB) and corresponding antiseptic hardwood (HW). Results show that BFB has lower moisture adsorption & desorption rate than the reference HW; the significant magnitude difference between the steady and dynamic test results shows the necessity of a

comprehensive evaluation approach that can take more representative practical conditions into consideration.

(3) The annual simulation of thermal performance and coupled heat and moisture process for building component and enclosed space units, respectively, in WUFI Plus shows that, bamboo units have heat storage advantage and heat transport disadvantage that can vary with the operating conditions; judging from the annual energy consumption and PMV indicators, they perform worse in cold and severe cold regions and surpass timber units in L-type construction (lightweight) in hot and temperate regions, but the advantages are weakened in M-type construction (heavyweight). 'Substitute timber with bamboo' in tropical and subtropical regions, where bamboo forest widely distributed, is conducive to playing the strengths of heat storage properties of bamboo, which is beneficial for both utilizing the local forest resources and improving the building physical performance.

(4) The larger moisture transport resistance of bamboo on material level shows dialectical features when it comes to the building component and enclosed space levels. On the one hand, it can help to weaken the water vapor transport and the influence of external hygric environment on the inside of the construction; on the other hand, it might also cause water retardation on the boundaries, slow down the condensed water and indoor moisture from exhausting outwards, and weaken the moisture buffering effect of the building envelope.

Reference

1. 潘毅群. 实用建筑能耗模拟手册 [M]. 北京: 中国建筑工业出版社, 2013/Pan YQ (2013) Building energy simulation handbook. China Architecture & Building Press, Beijing

Chapter 5
Hygrothermal Performance Optimization on Bamboo Building Envelope in Hot-Humid Climate Region

The comparison between bamboo and timber units in Chap. 4 shows that the heat transfer coefficient requirement in the hot climate zone is relatively low, and the contribution of the partition boards to the whole performance of the component is increased. The effect of the heat capacity is highlighted, which is beneficial to play the heat storage advantage of bamboo. In comparison, the construction in the cold climate zone is relatively thick, so the performance is mainly guaranteed by the infill material, which is not conducive to presenting the difference caused by the partition boards. In the heat and moisture processes of the building envelope in the Hot-Humid climate zone, due to the high external temperature and relative humidity, strong solar radiation and heavy rainfall, the external heat and moisture load is large, so the direction of heat and moisture flow is relatively simple under the ideal HVAC condition, which is suitable to establish the basic model for the research of heat and moisture process in building envelope. Therefore, this chapter selects the Hot-Humid climate region as an external condition for the optimization studies of bamboo construction. Firstly, clarify the research variables following an analysis of the layered construction system; secondly, set comparison model groups and carry out a large sample of the coupled heat and moisture process simulation; finally, provide suggestions for the optimization of material and construction parameters of the bamboo building envelope, as well as the guidance for climate adaptive construction design in the Hot-Humid climate zone, after a single-factor impact analysis of the material and construction parameters on the hygrothermal performance bamboo layered construction (Fig. 5.1).

Z. Huang, *Application of Bamboo in Building Envelope*,
Green Energy and Technology, https://doi.org/10.1007/978-3-030-12032-0_5

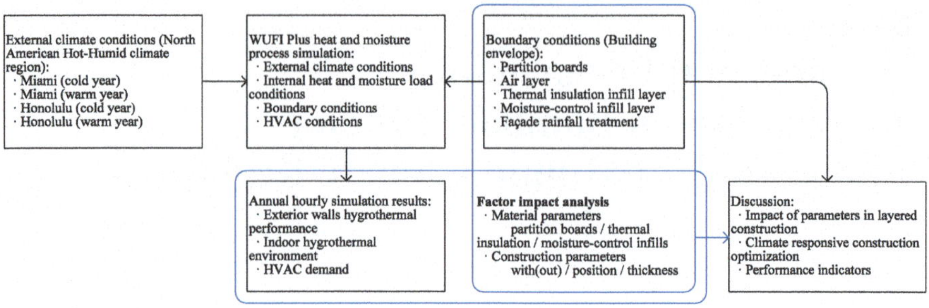

Fig. 5.1 Framework of Chap. 5

5.1 Layered Construction System

5.1.1 Timber Layered Construction System

The masonry construction and the log construction systems normally use a single material to undertake all functions such as the load bearing, thermal insulation, air-tightness, and sound insulation. Layered construction is a construction method that emerges with wood-based panel technologies, as well as the timber construction systems, such as the timber frame construction, skeleton construction, and timber panel construction. The layered construction assigns different functions to the appropriate material layers and achieves better overall performance through the control of the construction layer order and connection method.

Due to the potential of saving energy consumption and carbon footprint in the life cycle, timber layered construction systems have received extensive attention in Europe and the USA, such as the Chair of Timber Structures and Building Construction at the Technical University of Munich, Germany, which hosted the project 'TES EnergyFacade—prefabricated timber based building system for improving the energy efficiency of the building envelope' in 2008–2009, studying the application of timber to design high-level prefabricated building components and their industrial production. In conjunction with the universities and institutions in Germany, Finland, and Norway, the prefabricated timber panel exterior wall/facade is developed using a layered construction method. The physical performance of the building envelope is improved by construction design that makes good use of the material properties. The exterior cladding layer has a wide range of materials for selection and can integrate well with the solar systems [1].

The timber layered construction is constantly evolving as the building technologies, the requirements of energy efficiency standards, and indoor comfort increase, in which timber also changes both its form and function. Two sets of timber construction design atlas in Germany, the '*Holzbau Altas*' [2] and 'Holzrahmenbau: Bewährtes Hausbau-System' (referred to as '*Holzrahmenbau*') [3] are selected for quantitative analysis. The U-Wert-Rechner is used as the thermal performance calculation tools, analyzing the contribution ratio of each material layer in the timber layered construction to the overall thermal resistance *R*, the thermal inertia index *D*, and the water vapor diffusion resistance equivalent air layer thickness s_d. These selected three indicators are respectively used to characterize the heat transport, heat storage, and moisture transport performance of the construction.[1] (Table 5.1; Fig. 5.2).

Calculation results in the table show that in the selected 18 sets of exterior wall layered constructions, the contribution ratio of timber to the overall thermal resistance *R*, thermal inertia index *D*, water vapor diffusion resistance equivalent air layer thickness s_d value is 1.56–18.38%, 9.95–49.25%, and 43.50–66.73%, respectively. It can be seen that the contribution rate of the timber in the layered construction is significantly reduced, and especially, the contribution rate to the thermal resistance is greatly reduced. The reduction magnitude is inversely correlated with the overall thermal resistance of the construction, which is due to the other functional materials that contribute a share of the overall performance.

On the other hand, most of the added functional materials in layered construction, such as thermal insulation infills, vapor barriers, have a small mass, while timber still accounts for a large proportion of the overall mass, reaching 44.81–80.98%, which makes the timber still bear the main thermal storage performance.

Bamboo has higher heat storage and heat transport properties, as well as lower moisture transport properties, so it can be derived that the contribution of bamboo, in terms of the heat storage and moisture transport (moisture resistance) performance, in the construction of the same size would further increase compared to the timber units, while the contribution to the heat transport (thermal resistance) performance would decline.

From the perspective of the thermal and hygric performance, the functional weight of each construction layer of the timber exterior walls is quantitatively analyzed above. In layered construction, the bamboo- or wood-based panel largely replaces the solid log as the main material of the construction and plays the role as the construction framework and the partition boards. The overall construction is changed from the use of a single material to the combination of different functional materials. The function of bamboo and timber boards retains heat storage and moisture transport (moisture resistance), but their contribution to heat transport (thermal resistance) decreases. Therefore, the research on the application of bamboo and timber to layered construction should take not only the heat transport performance, but also the heat storage and moisture transport performance into account. The heat storage performance affects the thermal stability of the construction itself and the enclosed space, while the moisture transport performance has an impact on the moisture protection of the infill material inside the construction and the hygric environment quality of the enclosed space.

[1] s_d, Water vapor diffusion thickness (equivalent air layer thickness), *m*. It characterizes the resistance against water vapor diffusion of the construction layer and is numerically equal to the thickness of the air layer with the same resistance. There is $s_d = \mu \times s$, where μ is the vapor diffusion resistance factor of the material, and *s* is the thickness of the construction layer.

Table 5.1 Pperformance contribution of timber boards to the whole components in layered exterior walls' construction cases

Construction group		Thermal resistance R (m^2 K/W)	Performance contribution of timber boards to the whole components [%]			
			Thermal resistance R	Thermal inertia index D	Vapor diffusion resistance s_d	Areal density *Weight*
'*Holzrahmenbau*'—ventilated	1	4.6437	2.07	12.82	44.07	29.41
	2	4.7399	4.06	22.73	61.18	45.45
	3	4.6587	2.06	12.54	44.01	26.85
	4	4.7399	4.06	22.73	61.18	45.45
	5	6.1587	1.56	9.95	43.82	23.92
	6	6.2399	3.08	18.38	61.00	41.18
'*Holzbau Altas*'—ventilated	1	2.1360	18.07	49.25	64.75	80.98
	2	2.1985	9.97	41.32	64.47	54.67
	3	2.2216	10.91	43.77	66.73	57.13
	4	2.7485	7.98	36.30	64.41	52.55
'*Holzrahmenbau*'—unventilated	1	3.6062	14.80	41.81	44.02	30.71
	2	5.1062	10.45	34.36	43.84	29.34
	3	5.1062	10.45	34.36	43.84	29.34
	4	6.6062	8.08	29.17	43.66	28.08
'*Holzbau Altas*'—unventilated	1	2.7924	10.47	47.22	59.71	80.57
	2	2.2703	9.15	38.82	43.50	53.89
	3	2.9041	11.76	46.27	65.20	63.54
	4	2.7462	18.38	48.34	60.58	33.55
Mean		3.9791	8.74	32.79	54.44	44.81
Max		6.6062	18.38	49.25	66.73	80.98
Min		2.1360	1.56	9.95	43.50	23.92

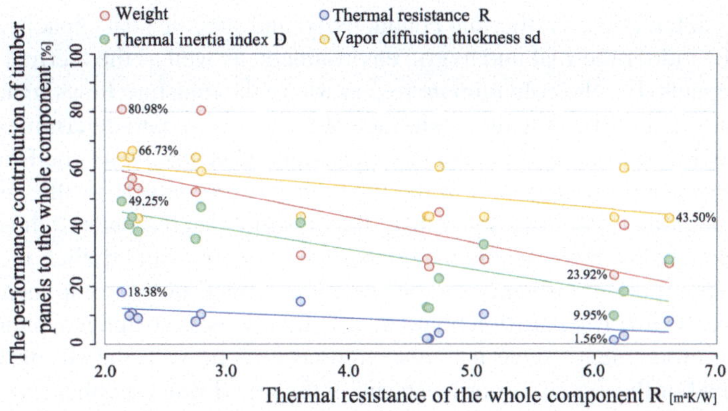

Fig. 5.2 Contribution ratio of timber boards to the whole hygrothermal performance in exterior walls layered construction cases (the cases are taken from the reference collection of '*Holzbau Altas*' and '*Holzrahmenbau: Bewährtes Hausbau-System*')

It is worth mentioning that the existing experience of the timber layered construction is mainly from the cold climate zone. For example, the s_d values are arranged from the inside to the outside in descending order when the material of each construction layer is selected [4]. The changes in indoor and outdoor climatic conditions can lead to variable characteristics of the heat and moisture processes in building envelope and affect the rational arrangement of construction layers. Therefore, the climate adaptive design method of layered construction is particularly worthy of investigation.

5.1.2 Related Parameters of Layered Construction

(1) Partition board

Exterior board. In addition to the solar radiation, the exterior board is also in contact with the rainfall, which acts as a barrier to the outdoor liquid water. Generally, it is necessary to use materials of low liquid water transport properties, such as the hardwood and bamboo scrimber, with exterior surface treated by asphalt, paint, etc., to achieve the desired effect as an outer moisture barrier. When the lightweight exterior board is replaced by a heavyweight material such as masonry, it is converted from an L-type construction (exterior board + infill layer + interior board) to an M-type construction (masonry layer + infill layer + interior board). Normally, the latter has a significantly larger overall heat capacity.

Interior board. Contrary to the exterior board, the interior board interchanges heat with the indoor air on the one hand and performs heat radiation exchange with indoor occupants on the other hand. The surface thermal properties of the interior board, as well as the difference in heat capacity of the inner side of the component due to the differences in bulk density, thermal conductivity, and specific heat capacity, have an effect on the indoor thermal and hygric environment, as well as the thermal comfort of the occupants. For the cold climate region where the moisture flows mainly from inside to the outside, the moisture resistance of the interior board determines the rate of moisture flow into the component to some extent. Conversely, for the Hot-Humid climate zone where the moisture flow direction is from outside to the inside, the interior board affects the discharge rate of the moisture that has entered the inside of the construction. For hygroscopic materials such as bamboo and timber, the interior board has an important influence on the 'breathing effect' of the components.

Interlayer board. The partition board mainly functions as chamber separation and structure reinforcement. Since it is not required to cope with the strong heat and moisture load of the exterior board, especially the liquid water, neither the aesthetic requirements of the interior finish, there are many optional materials, including the type of plate material that is composed of small-sized raw materials, such as the particleboard, oriented strand board (OSB), fiberboard, wood wool board, and so on.

(2) Frame

The frame, as a load-bearing member of the layered construction, is usually built into the cavity between the partition boards together with the insulation infill. Due to the requirement of load bearing, materials with higher mechanical strength, such as hardwood, bamboo scrimber, and metal, are mostly used. These materials have a larger bulk density and a higher thermal conductivity, which on the one hand contribute to the heat capacity improvement of the component, and on the other hand, result in problems such as the 'heat bridge' or 'cold bridge.'

General experience to deal with the 'cold/heat bridges' involves selecting a frame material with a lower thermal conductivity and appropriately reducing the contact area of the frame with the partition boards, for example, using the narrower face of the frame for fixing the boards, the type of point node, or grading the frame by erecting the vertical bar that bears the main force on the inner side, and arranging the small size purlin laterally outside the vertical bar, thereby the contact area among the frame bars, and between them and their connection to the partition boards is reduced.

(3) Construction infill

The infill layer generally refers to the thermal insulation material that is placed within a construction cavity to reduce the heat transfer coefficient of the component. There are generally EPS, XPS, and plant fibers as the insulation material. EPS, XPS, and other rigid foam board materials do not match the expansion rate of the timber partition boards, which can be effectively relieved by adding some thermal insulation fiber material to the foam [4]. In addition, the loose fiber thermal insulation material requires the ambient environment to be of airtightness and therefore is particularly suitable for the layered construction in which the cavity is enclosed by the partition boards. Compared with the non-hygroscopic insulation materials, plant fiber infills interact with the environment, especially the hygric environment. This feature of plant fibers helps to play their 'breathing effects' for heat and moisture regulation on the building components and their enclosed space on the one hand [5], but on the other hand, it may adversely affect the thermal performance due to the changes in its moisture content.

In addition to the thermal insulation infill, as described in Chap. 1, the moisture-control materials such as bamboo charcoal are also used in the cavity as an infill for controlling the moisture content of the indoor environment and within the construction, such as the 'wall construction method' in Japanese timber construction system. In comparison, thermal insulation infills have been paid more attention to the improvement of heat transport performance, while the effects of moisture-control infills on the improvement of the moisture transport performance and the moisture protection of the thermal insulation materials are worthy of investigation.

(4) Air layer

Compared to the single-layer wall (also known as a single leaf), the double-layer wall (also known as a double leaf) or even the multi-layer wall has the air layer that separates the wall into two or more leaves, and each of these leaves takes on different functions. In the cold climate zone, the outer leaves undertake the weatherproof and ventilation functions, while the inner leaves play their effects as the thermal insulation and water vapor barrier [4].

The closed air layer is placed inside the layered construction to help improve the thermal performance, mainly manifested by an increase in the thermal resistance. The Fraunhofer IBP quantifies the thermal parameters of a closed air layer with a thickness of 5–150 mm, bulk density of 1.3 kg/m^{3,} and specific heat capacity of 1000 J/(kg K), to become a basic parameter that can be used for calculation.

When the reckoning of these parameters is carried out, the parameter-fitting results show that the thermal conductivity of the air layer increases linearly with the increase of thickness. The fitting result is approximately $\lambda = 6.1\text{d} - 0.0065$ ($R^2 = 0.9964$), where d [m] is the thickness of the construction layer, and λ [W/(m K)] is the thermal conductivity. In addition, the thermal resistance of the closed air layer of different thickness is calculated, and the obtained R value is between 0.1064 and 0.1786 m^2 K/W, of which the maximum value appears at $d = 50$ mm. In general, the multiple closed air layers having the thickness of less than 50 mm will improve the overall thermal performance of the construction to a larger extent than a single air layer of the same overall thickness (Table 5.2).

The closed air layer is almost negligible in terms of its moisture storage and moisture transport resistance and thus cannot improve the overall hygric performance of the building component. Conversely, a properly arranged ventilated air layer is able to remove moisture by the flowing air, which helps to exhaust the condensed water and slow down the accumulation of moisture at the boundary between two construction layers. On the other hand, the pressure balance caused by the air passage helps to block the infiltration of exterior rainwater.

From the perspective of thermal performance, under certain conditions, the ventilated air layer can take away the heat inside the construction and improve the overall thermal performance. However, this is affected by many factors such as the airflow rate and temperature within the construction cavity, so that it is difficult to be mea-

Table 5.2 Parameters of closed air with gradient thickness and the thermal and hygric transport performance calculation

Thickness d [mm]	Thermal conductivity λ [W/(m K)]	Water vapor permeation resistance factor μ [–]	Thermal resistance R [m^2 K/W]	Water vapor diffusion thickness s_d [m]
5	0.047	0.79	0.1064	0.004
10	0.071	0.73	0.1408	0.0073
20	0.13	0.56	0.1538	0.0112
25	0.155	0.51	0.1613	0.0128
30	0.18	0.46	0.1667	0.0138
40	0.23	0.38	0.1739	0.0152
50	0.28	0.32	0.1786	0.016
60	0.337	0.27	0.178	0.0162
70	0.4	0.23	0.175	0.0161
80	0.46	0.2	0.1739	0.016
90	0.523	0.17	0.1721	0.0153
100	0.59	0.15	0.1695	0.015
110	0.655	0.13	0.1679	0.0143
120	0.723	0.11	0.166	0.0132
130	0.79	0.1	0.1646	0.013
140	0.864	0.09	0.162	0.0126
150	0.94	0.07	0.1596	0.0105

The source of the thermal conductivity and water vapor permeation resistance factor: the material database in WUFI Plus of the Fraunhofer IBP

sured like a closed air layer, and clarified as definite parameters for calculation, e.g., the thermal conductivity.

(5) Functional enhancement (thin) layer

The thin layer referred to herein, due to its small thickness, is generally on the order of millimeters and has ignorable effect on the overall heat and moisture storage performance. Normally, it uses materials with specific functions to block the heat and moisture transport process. For example, in the cold climate regions, if a construction is provided with a water vapor barrier layer on the indoor side of the thermal insulation layer, the moisture penetration can be effectively prevented to reduce the influence on the thermal performance. When the aluminum foil is attached to both sides and one side within the air chamber, it can effectively reduce the radiant heat exchange between the partition boards on both sides of the air layer, thereby reducing the effective heat transfer coefficient of the building envelope.

In addition, the other material and construction factors can also have an impact on the overall performance. Material modification, such as carbonization, bleaching, dyeing, coating treatment, and changes in surface conditions caused by contaminants, can cause changes in the surface heat transfer and internal heat and moisture transport properties of the material. The air passage formed by the boards splicing gap may reduce the heat and moisture transport performance of the construction layer, and the degree of reduction is affected by the size and number of the gaps.

This study is mainly aimed at the material selection, thickness, and arrangement of the construction layers and is based on the one-dimensional heat and moisture process. The accurate description on the heat and moisture process of the frame in the construction requires a two-dimensional model. Therefore, the discussion of the frame in this study is only evaluation at the construction level in a steady state and does not conduct by dynamic simulation at the enclosed space level. The material parameters are obtained from Chap. 3, and the effects of carbonization, bleaching, coating, filming, and contaminants on the surface and internal properties of the materials are not considered. In addition, due to the functional limitations of the computer simulation tools, it is not able to simulate the effects of a ventilated air layer on the thermal and hygric performance of the construction.

5.2 Building Envelope Coupled Heat and Moisture Process Simulation

5.2.1 Simulation Method

The bamboo properties parameters obtained from Chap. 3 are input into the construction layer in boundary condition of the model groups in WUFI Plus. Different conditions for comparison are set up for full-year simulation, which provides annual hourly data including exterior walls' hygrothermal performance, indoor hygrothermal environment, and HVAC demand (Fig. 5.3):

(1) Exterior walls' hygrothermal performance. Both thermal performance and hygric performance are considered, of which the former includes the heat flow through the exterior walls, and the latter includes the moisture flow through the exterior walls and the moisture content of the infill layers.
(2) Indoor hygrothermal environment. Related indicators involve indoor air temperature, relative humidity, exterior wall inside surface temperature, and the predicted mean vote (PMV) of the enclosed space units.
(3) HVAC demand. Related indicators include annual HVAC period, heating and cooling demand, and humidification and dehumidification demand. The peak value of HVAC demand is important to the design of HVAC devices; however, the related discussion is shown in Chap. 2 and would not be carried out in this chapter.

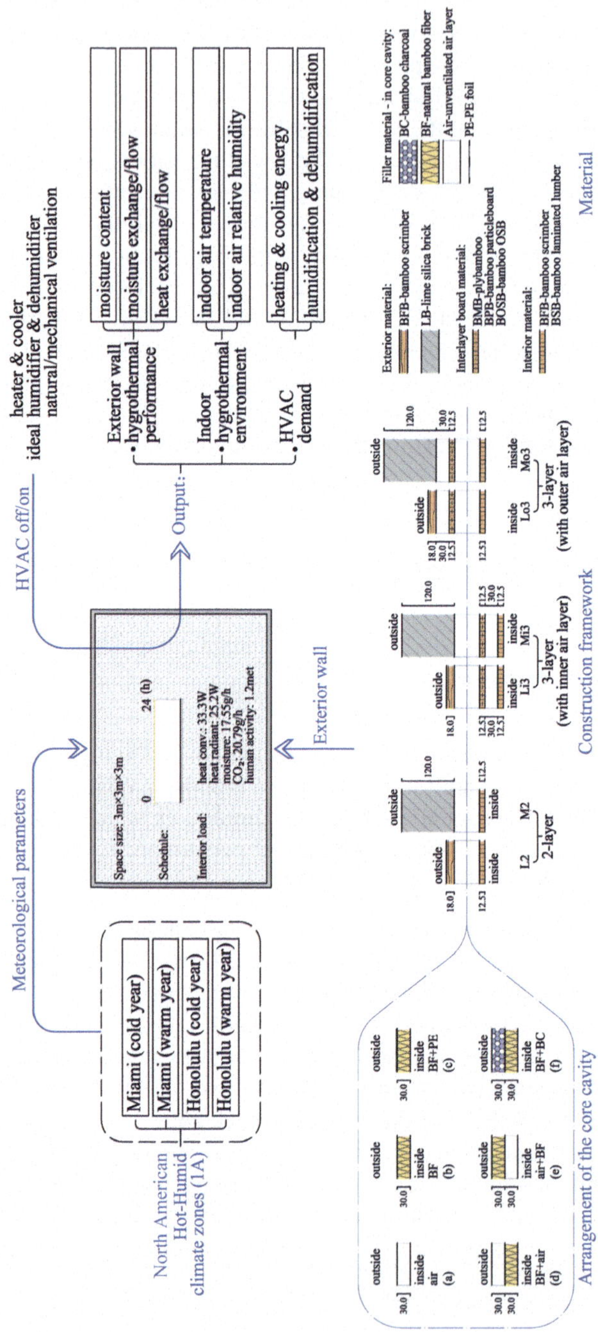

Fig. 5.3 WUFI Plus model design (external, internal, boundary, and HVAC conditions)

5.2.2 *Simulation Model Design*

Model groups for comparison are constructed in the coupled heat and moisture process simulation program WUFI Plus, which is developed by the Fraunhofer IBP. The models are composed of the external condition, internal condition, boundary condition, and HVAC condition. The annual hourly simulation is carried out, involving the building component hygrothermal performance, the indoor hygrothermal environment, and the HVAC demand.

5.2.2.1 External Conditions

Since the climate condition significantly affects the heat and moisture process of the building envelope, the complete meteorological parameters are necessary to ensure the effectiveness of the simulation results. However, among these databases, the driving rain is not available in many regions. In this study, the two cities, Miami and Honolulu, from the North American 1A (very hot and humid) climate region, where complete meteorological parameters are available, are selected as the exterior condition. Each city includes both cool year and warm year and consists of annual hourly meteorological parameters covering the solar radiation, air temperature, relative humidity, and the driving rain coefficient that depends on the precipitation and wind conditions.

The annual average outdoor air temperatures of Miami and Honolulu are, respectively, 23.43 and 24.25 °C in cool year, while they are correspondingly 25.08 and 25.21 °C in warm year. As a reference, the annual average outdoor air temperature of Guangzhou, a typical city in Hot-Humid climate regions of China, is 22.23 °C (TMY—typical meteorological year data from EnergyPlus 'Guangzhou.592,870'), which means both cool year and warm year of Miami and Honolulu can reach that value. However, when the climate regions are expanded to other areas, for example, northward to 2A (hot and humid region, including Tampa and Houston), 2B (hot and dry region, including San Antonio, New Orleans, and Jacksonville) and 3A (warm and humid region, including Charleston and Wilmington), the annual average air temperature of the seven cities decreases to the range 16.12–20.88 °C in cool year and 18.44–22.85 °C in warm year. Related simulation results show that, compared with the 2 cities from 1A, the heating demands rise in these seven cities, which even exceed the corresponding cooling demands, and the direction of heat and moisture flow through the exterior walls becomes uncertain. Therefore, the simulation results and the following analyses can have appropriate reference value only in Hot-Humid climate regions.

5.2.2.2 Internal Conditions

Enclosed space units of 3.0 m × 3.0 m × 3.0 m (width × depth × height) are constructed in the program, of which 24 h is set as the operation schedule, and standard office profile is given as the indoor heat and moisture load, including convective heat:

33.3 W, radiant heat: 25.2 W, moisture: 17.55 g/h, CO_2: 20.79 g/h, and human activity: 1.2 met. The PMV/PPD-related additional values are given, including clothing: 0.7 clo and air velocity: 0.1 m/s.

5.2.2.3 Boundary Conditions

Construction arrangement. The exterior walls (each 9 m^2) are selected as the study objects, while the floor and ceiling are set as partitions between the same interior conditions to avoid heat and moisture exchange. Two commonly used construction types are considered, including L-type (exterior board + infill layer + interior board) and M-type (masonry layer + infill layer + interior board). Both type L and M included two-layer construction (two partition boards with one cavity) and three-layer construction (three partition boards with two cavities). The constructions are designed by controlling the heat transfer coefficient (U_c value) as 0.9745–1.1375 W/(m^2 K) for the two-layer groups, and 0.7654–0.9270 W/(m^2 K) for the three-layer groups (Note: the U_c values are calculated with the material parameters in dry state and do not include the surface heat transfer resistance, i.e., the exterior surface heat transfer resistance R_e and the interior surface heat transfer resistance R_i, which are given the default values in WUFI Plus as R_e = 0.040 m^2 K/W and R_i = 0.130 m^2 K/W) (Table 5.3).

In order to ensure the comparability between corresponding construction groups, the boundary condition (exterior walls) is designed by adjusting single factors:

From L-type construction to M-type construction: replace the 18 mm thick BFB with the 120 mm thick LB, which forms comparative groups for exterior boards;

From 2-layer construction to 3-layer construction: add '12 mm thick interlayer board + 30 mm thick air layer' to the inside of the cavity to form the Li3/Mi3 groups. Correspondingly, add the interlayer board and air layer to the outside of the cavity to form the Lo3/Mo3 groups. The comparative groups for the position of air layer/thermal insulation layer are formed between the 2-layer and 3-layer construction, and between the Li3/Mi3 and Lo3/Mo3 construction;

The replacement among different partition boards: the replacement of the 12 mm thick interior BFB/BSB forms 2 comparative groups for the interior boards, while the replacement of the 12 mm thick interlayer BMB/BPB/BOSB forms 3 comparative groups for the interlayer boards;

The arrangement within the core cavity: 4 conditions within the core cavities are set, including a (30 mm air layer), b (30 mm BF thermal insulation layer), c (inside 30 mm BF thermal insulation layer + outside 0.2 mm PE foil), d (inside 30 mm BF thermal insulation layer + outside 30 mm air layer), e (inside 30 mm air layer + outside 30 mm BF thermal insulation layer) and f (inside 30 mm BF thermal insulation layer + outside 30 mm BC moisture-control layer). Among them, the comparative groups for vapor barrier layer, air layer and moisture-control layer are formed respectively between group c and b, d/e and b, f and b; Comparative groups for the position of thermal insulation layer within single cavity is formed between the group e and d; Group a is set as a benchmark for the comparison among groups with air layer of different thicknesses.

Table 5.3 Design of boundary conditions—exterior walls' construction design

Group	Number of partition boards	Exterior board	Interior board	Interlayer board	Core cavity (inside to outside)	
					Inside	Inside
L2	2	BFB (18 mm)	BFB (12 mm) BSB (12 mm)	-	Air (30 mm) BF (30 mm) BF+PE (30.2 mm) BF+Air (60 mm) Air+BF (60 mm) BF+BC (60 mm)	
M2	2	LB (120 mm)	BFB (12 mm) BSB (12 mm)	-	Air (30 mm) BF (30 mm) BF+PE (30.2 mm) BF+Air (60 mm) Air+BF (60 mm) BF+BC (60 mm)	
Li3	3	BFB (18 mm)	BFB (12 mm) BSB (12 mm)	BMB (12 mm) BPB (12 mm) BOSP (12 mm)	Air (30 mm)	Air (30 mm) BF (30 mm) BF+PE (30.2 mm) BF+Air (60 mm) Air+BF (60 mm) BF+BC (60 mm)
Mi3	3	LB (120 mm)	BFB (12 mm) BSB (12 mm)	BMB (12 mm) BPB (12 mm) BOSP (12 mm)	Air (30 mm)	Air (30 mm) BF (30 mm) BF+PE (30.2 mm) BF+Air (60 mm) Air+BF (60 mm) BF+BC (60 mm)
Lo3	3	BFB (18 mm)	BFB (12 mm) BSB (12 mm)	BMB (12 mm) BPB (12 mm) BOSP (12 mm)	Air (30 mm) BF (30 mm) BF+PE (30.2 mm) BF+Air (60 mm) Air+BF (60 mm) BF+BC (60 mm)	Air (30 mm)

(continued)

Table 5.3 (continued)

Group	Number of partition boards	Exterior board	Interior board	Interlayer board	Core cavity (inside to outside)	
					Inside	Inside
					BF+Air (60 mm) Air+BF (60 mm)	BC (30 mm)
Mo3	3	LB (120 mm)	BFB (12 mm) BSB (12 mm)	BMB (12 mm) BPB (12 mm) BOSP (12 mm)	Air (30 mm) BF (30 mm) BF+PE (30.2 mm) BF+Air (60 mm) Air+BF (60 mm) BF+BC (60 mm)	Air (30 mm)
					BF+Air (60 mm) Air+BF (60 mm)	BC (30 mm)

Abbreviation
BSB Bamboo laminated lumber
BMB Plybamboo
BFB Bamboo scrimber
BFB Bamboo particleboard
BOSB Bamboo oriented strand board
BF Natural bamboo fiber
BC Bamboo charcoal
LB Lime silica brick
For the two-layer construction (two partition boards with one cavity), the air layer thickness in the cavity is set as 10/20/30/40/50/60/70 mm and 80/100/120/140 mm; while for the three-layer construction (three partition boards with two cavities), the air layer thickness in both cavities is set as 10/20/30/40/50/60/70 mm. All the conditions included groups with BF in the cavity as insulation layer (+BF) and the groups without BF (–BF)

5.2.2.4 HVAC Conditions

Based on the models that consist of the above conditions, the HVAC is firstly turned off to evaluate the indoor hygrothermal environment. Subsequently, the HVAC is turned on, and the indoor air temperature is maintained within 20–26 °C. The relative humidity is maintained within 40–60%. The maximum CO_2 concentration is set as 1550 ppmv. Natural ventilation is closed. The infiltration air change rate is set as 0, while the mechanical ventilation is set as 0.6 times per hour. The operation data including the annual heating and cooling, humidification and dehumidification demand are calculated. The operation period is set from 2015-01-01 to 2017-01-01, in which the hourly data for the second complete year are collected for further analysis.

5.3 Layered Construction Parameters Impact Analysis

The annual BF moisture content mean value $w\text{-BF}_{mean}$ and amplitude $w\text{-BF}_{amp}$, the annual moisture flow M_{flow} and heat flow H_{flow} through the exterior walls, the annual moisture exchange between the inner surface of the interior board and the indoor air $M_{i.flux}$, and respectively the annual moisture exchange between the outer surface of the exterior board and the outdoor air $M_{e.flux}$ are selected as indicators to characterize the hygrothermal performance of the exterior walls. When the HVAC is off, indicators to characterize the indoor hygrothermal environment include the annual hours of $-1.0 \leq \text{PMV} \leq 1.0$ during the room occupancy period $\text{PMV}_{1.0}$, the annual indoor air temperature mean value $T_{i.mean}$ and amplitude $T_{i.amp}$, exterior walls inside surface temperature amplitude $T_{is.amp}$, annual relative humidity mean value $\text{RH}_{i.mean}$ and amplitude $\text{RH}_{i.amp}$, as well as the annual hours of the indoor air temperature and relative humidity that are in certain comfort range. For the air temperature, the $T_{i.18\text{-}28}$ value counts in the annual hours of air temperature between 18 and 28 °C. Correspondingly, for the relative humidity, the $\text{RH}_{i.30\text{–}70}$ value calculates into the annual hours of relative humidity between 30 and 70%. When the HVAC on, the annual heating and cooling demand P value and humidification and dehumidification H value are chosen as indicators to characterize the HVAC demand of the enclosed space.[2]

A comparative analysis on the simulation results of the model groups would be carried out in the following sections, giving single-factor effect studies on the partition boards, air layer, thermal insulation layer, moisture-control layer, and the rainfall on façade.

5.3.1 Partition Boards

5.3.1.1 Exterior Board

L-type construction is arranged with bamboo scrimber (BFB) as exterior board, while the M-type construction with lime silica brick (LB). The hygrothermal properties difference between BFB and LB result in the hygrothermal performance difference

[2]Generally, the indoor air temperature T_i in the range 20–26 °C and the relative humidity RH_i in the range 40–60% are regarded as the corresponding comfort range. However, in this study, the annual hours of T_i and RH_i in these ranges are too little, and the hours of RH_i between 40 and 60% are even 0 in certain cases, which makes it unable to be compared by the ratio between the BC and NC model groups. Therefore, as the evaluation indicators, the range of RH_i and T_i are, respectively, enlarged to 30–70% and 18–28 °C.

between L- and M-type constructions. With a larger moisture resistance, BFB is conductive to blocking the influence of the external hygric environment, especially the driving rain, on the inside of the construction. In comparison, LB belongs to heavyweight material, which can substantially enlarge the thermal capacity of the building component. Moreover, due to the rapid moisture adsorption and desorption rate, it is potential to improve the indoor hygrothermal environment and reduce the HVAC demand by taking advantage of its evaporative cooling effect in certain conditions.

(1) Exterior wall hygrothermal performance

The south wall is considered as an example to analyze the difference between LB and BFB as exterior board to the annual moisture content of the BF layer. In HVAC on conditions, the w-BF_{mean} ratio between LB and BFB model groups is in the range 106.30–352.92% and 103.03–275.12%, respectively, for BFB and BSB as interior board. Correspondingly, the w-BF_{amp} ratio is 63.03–3126.80% and 78.81–1590.98%. It means that, compared with BFB, LB as exterior board is unfavorable to the control of the inner moisture content, which is caused by the lower moisture resistance of LB that allows the exterior moisture, especially the liquid water to enter the construction more easily.

Further analysis finds that among the model groups of LB as exterior board, those groups of BSB as interior board, compared with BFB, perform better in the w-BF_{mean} and w-BF_{amp} values, which is due to the lower moisture resistance of the BSB that allows the inner moisture to exhaust.

For the model groups with 0.2 mm PE foil arranged on the outer side of the BF layer, the w-BF_{mean} ratio of LB and BFB model groups is significantly narrowed to the range 106.30–113.20% and 103.03–110.23%, respectively, for BFB and BSB as interior board, which shows that the setting of vapor barrier on the upstream of the moisture flow can effectively improve the hygric performance of the construction with LB as exterior board (Fig. 5.4).

Due to the smaller overall moisture resistance, there is larger M_{flow} through the walls with LB as exterior board. In HVAC on conditions, the M_{flow} ratio between the model groups with LB and BFB as exterior board is in the range 73.36–260.98%, in which the ratio of the groups with BFB as interior board is 80.89–188.33% that is overall smaller than the 73.36–260.98% of the groups with BSB as interior board. The simulation result of the H_{flow} is just the opposite, that the H_{flow} ratio between the model groups with LB and BFB as exterior board is in the range 80.35–106.55%, in which the ratio of the groups with BFB as interior board is 81.56–106.55% that is overall smaller than the 80.35–100.62% of the groups with BSB as interior board.

The BFB as interior board makes it harder for the inner moisture to exhaust, which results in higher moisture content of the BF layer and finally the larger annual moisture flow through the walls. Conversely, the combination of BSB and LB can achieve better performance (Fig. 5.5).

(2) Indoor hygrothermal environment

In HVAC off conditions, taking $T_{i.18\text{–}28}$, $RH_{i.30\text{–}70}$ and $T_{is.amp}$ as indicators for comparison, simulation results show that the $T_{i.18\text{–}28}$ of the model groups with LB as exterior board is 99.49–117.62% while setting the groups with BFB as exterior

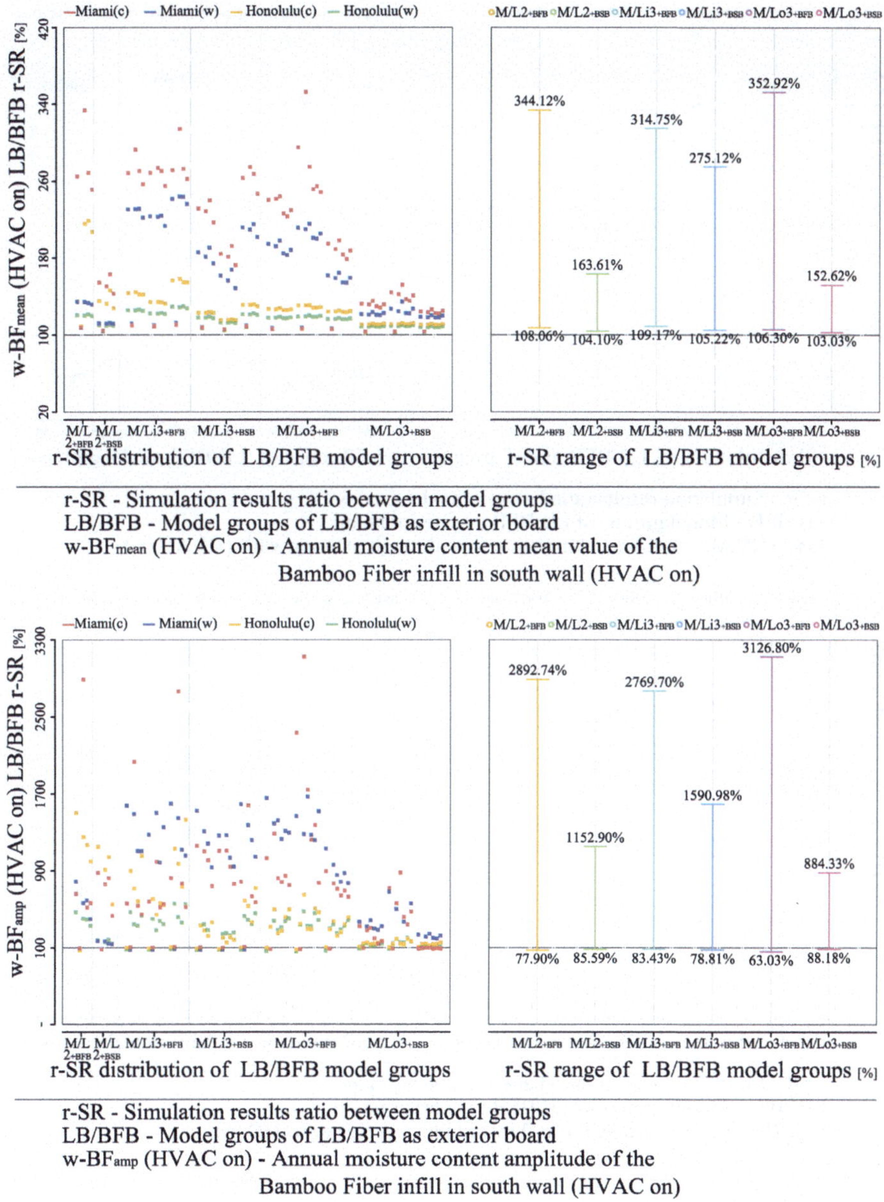

Fig. 5.4 Simulation results' comparison between model groups of LB/BFB as exterior board: annual BF infill moisture content mean value w-BF$_{mean}$ (up) and amplitude w-BF$_{amp}$ (down)

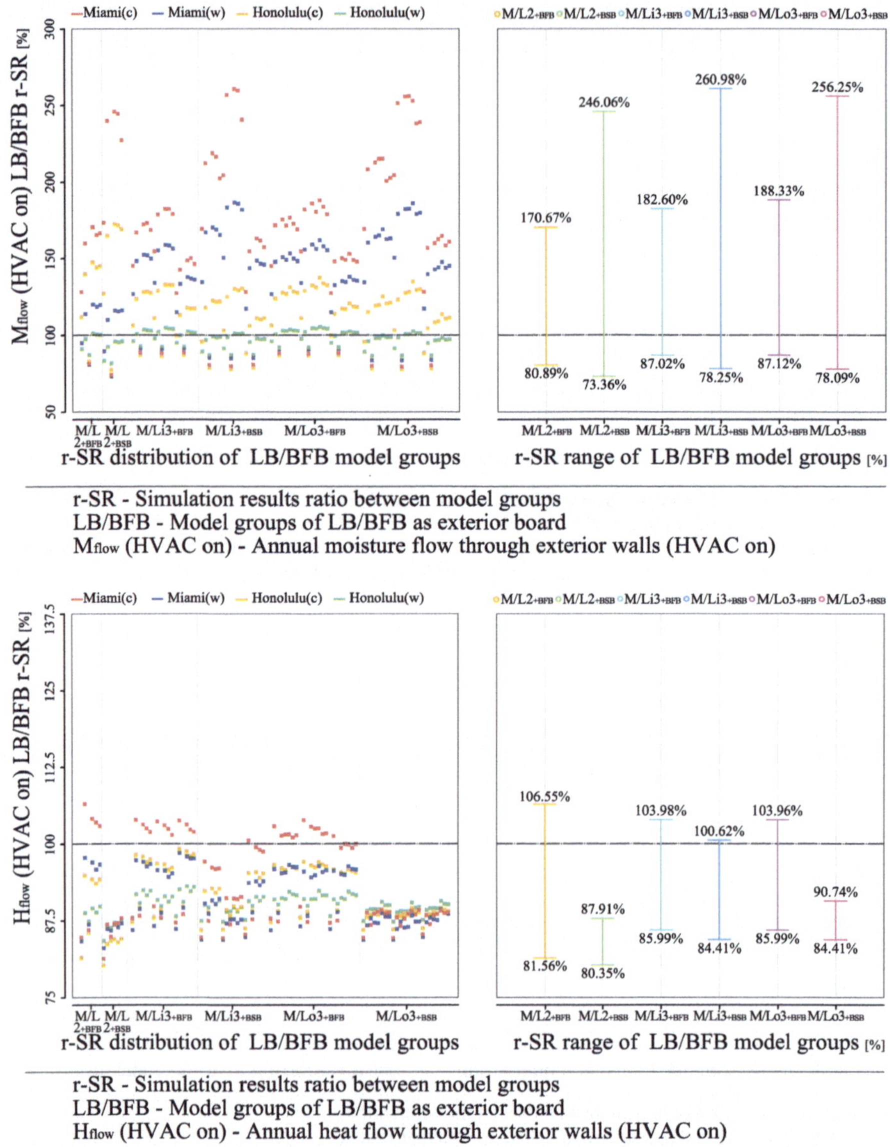

Fig. 5.5 Simulation results' comparison between model groups of LB/BFB as exterior board: annual moisture flow M_{flow} (up) and heat flow H_{flow} (down) through exterior walls

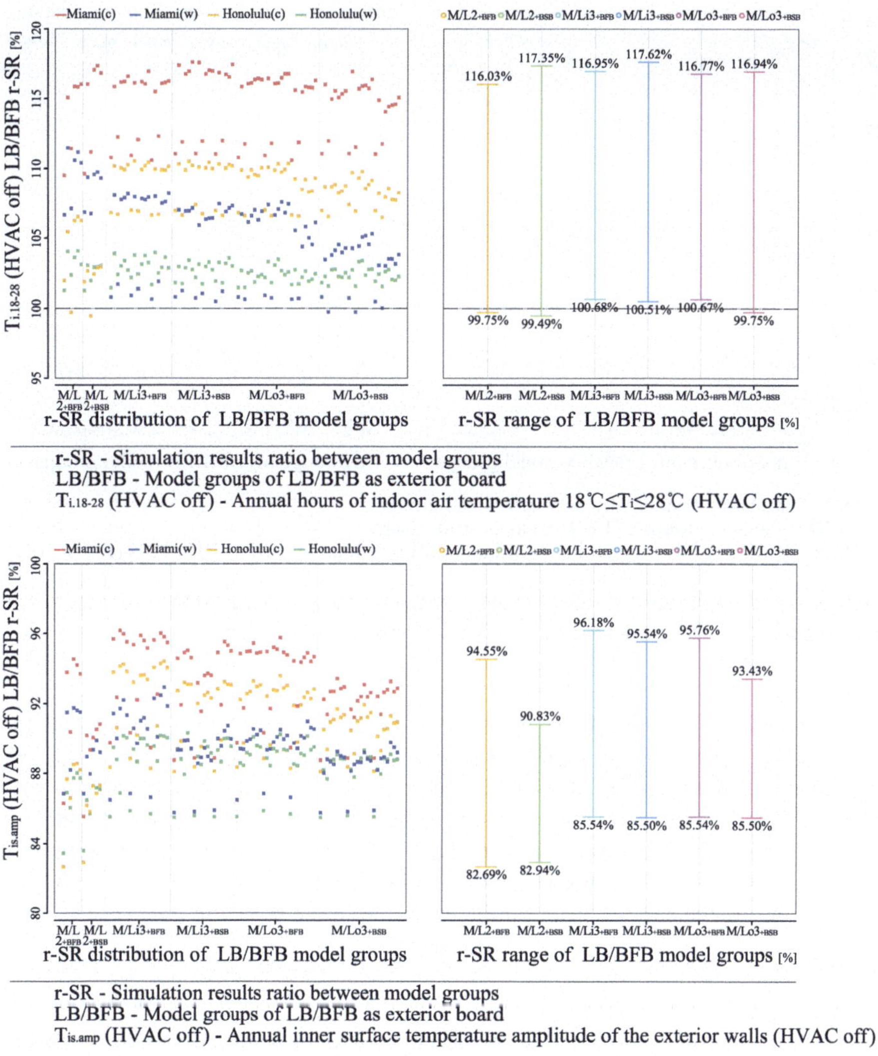

Fig. 5.6 Simulation results' comparison between model groups of LB/BFB as exterior board: annual hours of air temperature 18 °C ≤ T_i ≤ 28 °C $T_{i.18\text{–}28}$ (up) and inner surface temperature amplitude of the exterior walls $T_{is.amp}$ (down)

board as 100%, which shows better indoor thermal environment of the LB groups. The $T_{is.amp}$ ratio between the model groups with LB and BFB as exterior board is in the range 99.49–117.62%, which shows better thermal stability of the LB groups. On the one hand, these result in the larger thermal capacity of the LB groups. On the other hand, the evaporative cooling effect of the porous masonry layer contributes improvement to the indoor thermal environment (Fig. 5.6).

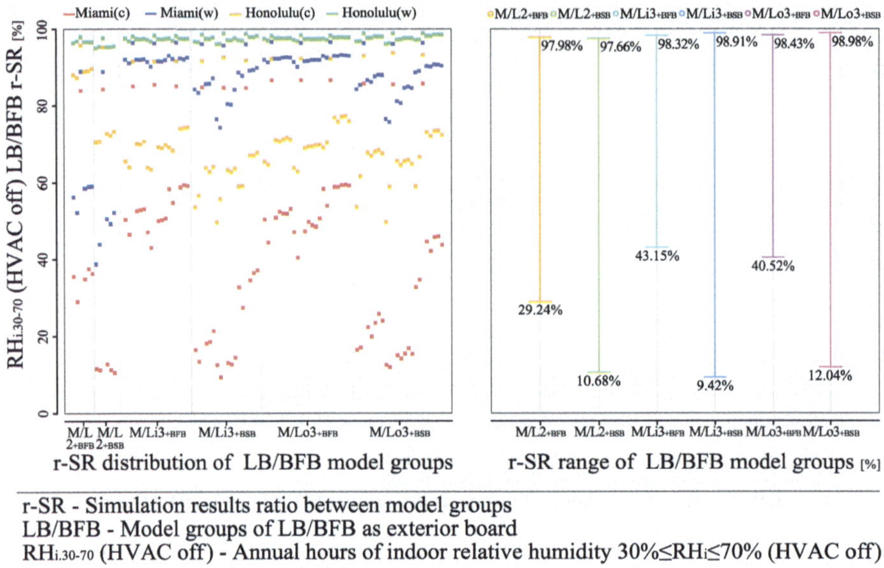

Fig. 5.7 Simulation results' comparison between model groups of LB/BFB as exterior board: annual hours of relative humidity 30% $\leq$ RH_i $\leq$ 70% $RH_{i.30\text{–}70}$

The simulation result of the hygric environment is just the opposite that the $RH_{i.30\text{–}70}$ ratio between the model groups with LB and BFB as exterior board is, respectively, in the range 29.24–98.43% and 9.42–98.98%. For the model groups with 0.2 mm PE foil arranged on the outer side of the BF layer, the ratio is 84.06–98.43% for the groups with BFB as interior board and 84.42–98.98% for the groups with BSB as interior board, which shows a significant improvement effect to the indoor hygric environment by the vapor barrier on the upstream of the moisture flow (Fig. 5.7).

(3) HVAC demand

Similar to the H_{flow} simulation results, due to the larger thermal capacity and evaporative cooling effect, the P value of the model groups with LB as exterior board performs better than that of the BFB groups, of which the ratio is in the range 80.63–102.84%. Except for that two-layer construction with BFB as interior board, the P values of the model groups with LB as exterior board are lower than that of the BFB groups. The H-value comparison results show similar with the M_{flow} value that the ratio between the model groups with LB and BFB as exterior board is in the range 102.17–500.39%.

For the model groups with 0.2 mm PE arranged on the outer side of the BF layer, the H ratio between model groups with LB and BFB as exterior board is in the range 104.00–129.07% and 102.17–133.33%, respectively, for the groups with BFB and BSB as interior board, which shows a significant improvement to the humidification and dehumidification demand by the vapor barrier on the upstream of the moisture flow (Fig. 5.8).

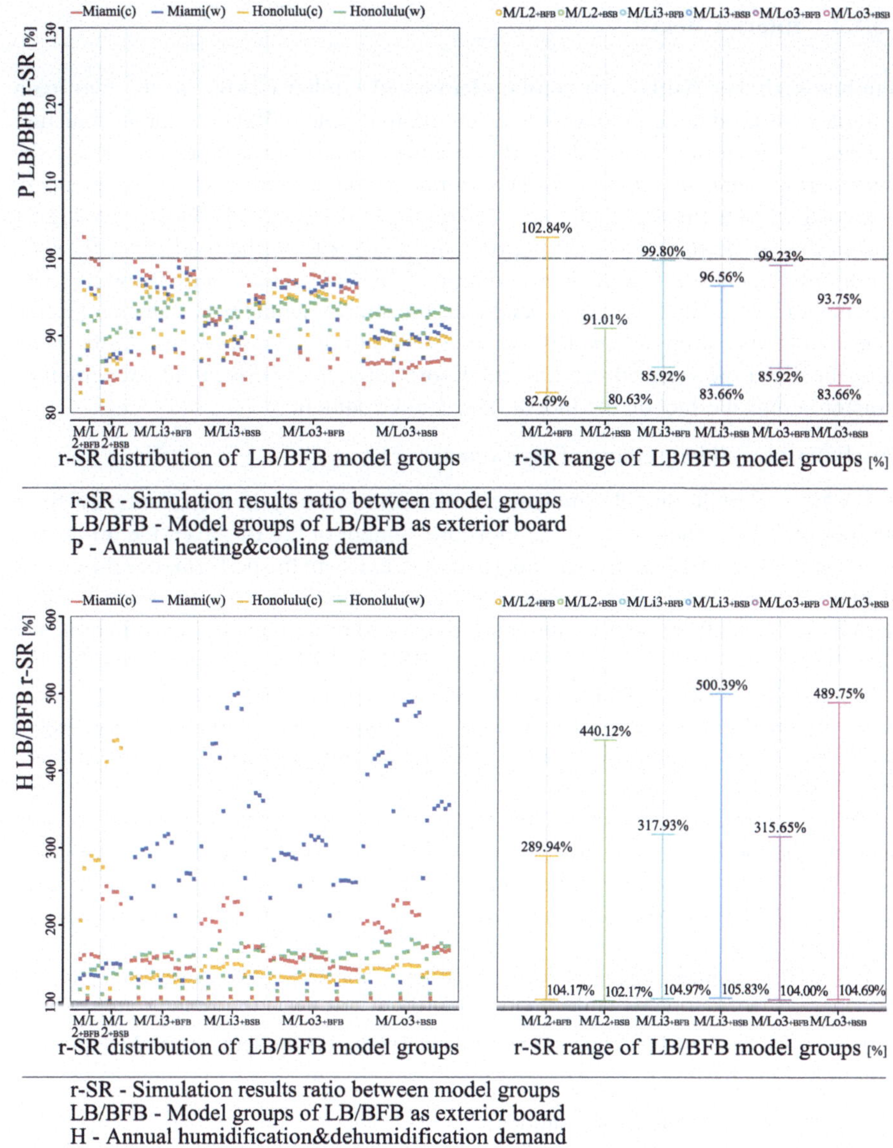

Fig. 5.8 Simulation results' comparison between model groups of LB/BFB as exterior board: annual heating and cooling demand *P* (up) and humidification and dehumidification demand *H* (down)

5.3.1.2 Interior Board

Bamboo scrimber (BFB) and bamboo laminated lumber (BSB) are the two main materials for producing indoor bamboo decorative panels. BFB has larger moisture resistance, which might slow down the exhaust of moisture that permeates into the construction when it is applied in Hot-Humid climate regions where the moisture flows mainly from outdoor to indoor. With higher bulk density, BFB possesses higher heat storage property, which can improve the inside surface thermal capacity when it is applied as interior board. On the contrary, BSB has smaller thermal conductivity and moisture resistance, which is beneficial for improving the heat transport performance and exhausting the inside moisture of the building components. At the same time, the faster moisture adsorption and desorption rate of BSB might strengthen its moisture-control effect on the indoor hygric environment.

(1) Exterior wall hygrothermal performance

In HVAC on conditions, the moisture flow occurs mostly in the orientation from outdoor to indoor; therefore for the moisture content of the BF layer, the difference between BSB and BFB as interior board mainly exists in the performance difference of blocking and exhausting the moisture. For L-type construction, since there is already very small amount of moisture that penetrates into the construction, the difference between the model groups with BSB and BFB is insignificant that the w-$\mathrm{BF}_{\mathrm{mean}}$ ratio between BSB and BFB model groups is in the range 96.41–99.61%. However, the difference becomes obvious for M-type construction that it is enlarged to 39.15–99.40%, which shows the significant advantage of BSB as interior board (Fig. 5.9).

The similar law is shown in the H_{flow} simulation results that the H_{flow} ratio between the model groups with BSB and BFB as interior board is in the ignorable range 99.62–101.10% for L-type construction. However, the corresponding ratio is 81.43–99.69% for M-type construction, which shows the advantage of BSB as interior board.

Due to the smaller moisture resistance of BSB, the M_{flow} ratio between the model groups with BSB and BFB as interior board is 114.76–212.00%, of which it is 120.10–151.48% and 114.76–212.00%, respectively, for the L-type and M-type construction (Fig. 5.10).

(2) Indoor hygrothermal environment

Due to the larger bulk density of BFB ($\rho_{\mathrm{d}} = 1108.77$ kg/m^3), compared with BSB ($\rho_{\mathrm{d}} = 563.81$ kg/m^3), the BFB has a significant advantage of thermal capacity that contributes larger inside thermal capacity of the construction. In HVAC off conditions, the $T_{\mathrm{is.amp}}$ ratio between the model groups with BSB and BFB as interior board is 101.10–103.63% and 97.50–103.06%, respectively, for the L-type and M-type construction, in which the difference between BSB and BFB as interior board is narrowed due to the already existing larger thermal capacity of the M-type construction. The $T_{\mathrm{i.amp}}$ simulation results show almost the same law with the $T_{\mathrm{is.amp}}$ values (Fig. 5.11).

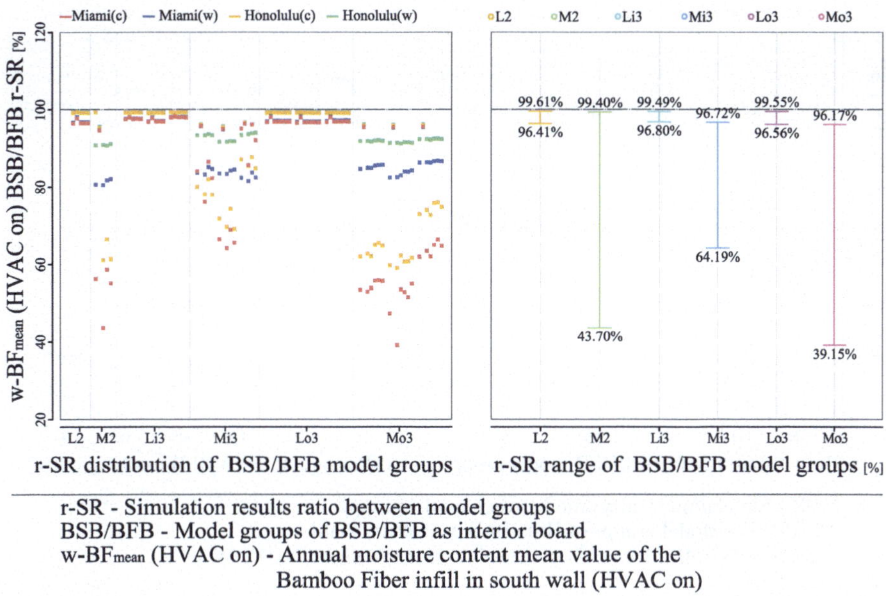

Fig. 5.9 Simulation results' comparison between model groups of BSB/BFB as interior board: annual BF infill moisture content mean value $w\text{-BF}_{mean}$

The larger water vapor permeability coefficient of BSB [$\delta_{RH=50\%} = 27.99E^{-13}$ kg/(m s Pa)], compared with BFB [$\delta_{RH=50\%} = 4.37E^{-13}$ kg/(m s Pa)], contributes better moisture-control ability as interior board, which results in the reduction of the $RH_{i.amp}$. For L-type construction, the $RH_{i.amp}$ ratio between model groups with BSB and BFB as interior board is in the range 75.07–87.08%, and for M-type construction, it is 63.93–109.45%. Generally, the $RH_{i.amp}$ of the BSB groups is smaller than that of the BFB groups, except in several cases of the M-type construction, where the very low overall moisture resistance of the construction with BSB as interior board has weak resistance to the exterior moisture load (Fig. 5.12).

(3) HVAC demand

Similar to the results in Chap. 4, for L-type construction, the heat storage capacity contributes more important proportion to the overall performance, which can, to a certain extent, make up the weakness of the heat transfer resistance. In this study, the *P*-value ratio between the model groups with BSB and BFB as interior board is in the range 99.89–101.10%, where BFB shows a slight advantage, and for the M-type construction, this is 81.82–100.29% that shows the significant advantage of the BSB.

For M-type construction, due to the smaller moisture resistance of the overall construction, the M_{flow} of the model groups with BSB as interior board is larger than that of the BFB groups, which results in larger *H* values of the BSB groups, and the *H*-value ratio between the BSB and BFB groups is in the range 87.61–155.40% (Fig. 5.13).

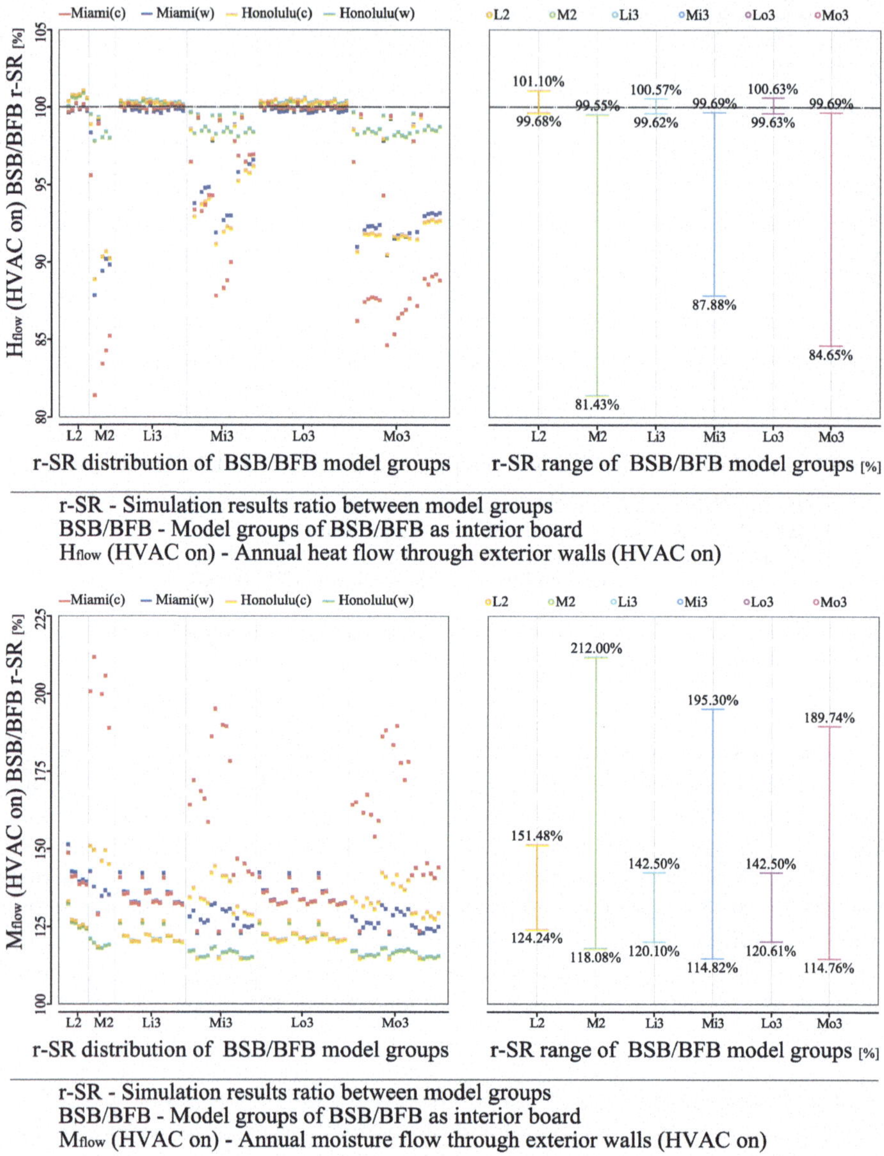

Fig. 5.10 Simulation results' comparison between model groups of BSB/BFB as interior board: annual heat flow H_{flow} (up) and moisture flow M_{flow} (down) through exterior walls

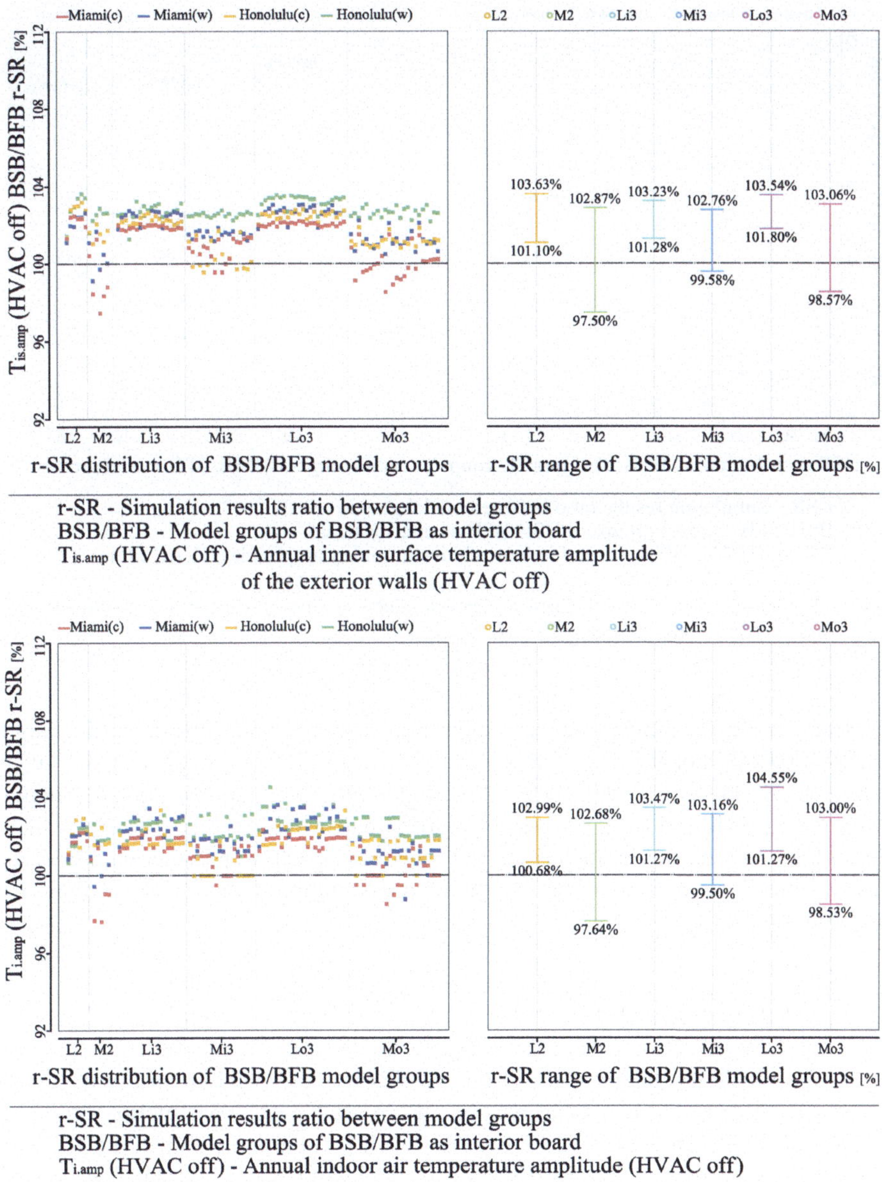

Fig. 5.11 Simulation results' comparison between model groups of BSB/BFB as interior board: annual exterior walls' inner surface temperature amplitude $T_{\text{is.amp}}$ (up) and indoor air temperature amplitude $T_{\text{i.amp}}$ (down)

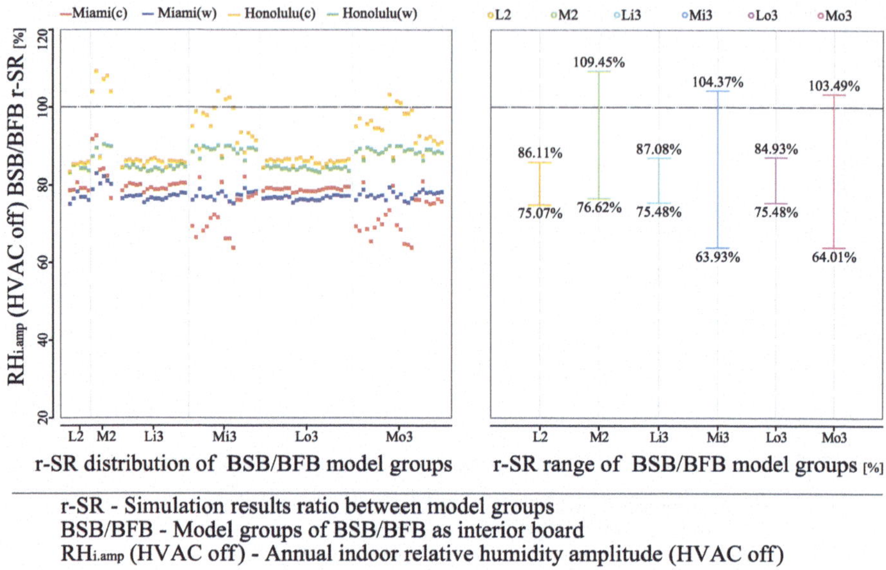

Fig. 5.12 Simulation results' comparison between model groups of BSB/BFB as interior board: annual indoor relative humidity amplitude $RH_{i.amp}$

However, for the three-layered L-type construction, this ratio is in the range 84.19–100.00%. For the model groups with 0.2 mm PE foil arranged on the outer side of the BF layer of the M-type construction, the *H*-value ratio between the model groups with BSB and BFB as interior board is narrowed to 87.61–99.17%, which shows the vapor barrier on the upstream of the moisture flow, together with BSB as interior board, can help to lower the annual indoor humidification and dehumidification demand.

5.3.1.3 Interlayer Board

Plybamboo (BMB), bamboo particleboard (BPB) and bamboo oriented strand board (BOSB) are used as interlayer boards in multi-layer construction (number of layer $\geq$ 3). When compared in material level, for thermal conductivity there is BMB > BOSB > BPB, where BPB shows advantage of thermal resistance; however, for bulk density, there is BOSB > BMB > BPB, which makes the thermal capacity of BMB and BOSB to be better than BPB; for water vapor transfer resistance, there is BOSB > BMB > BPB. Considering that when HVAC off, the effect of interlayer boards on the indoor hygrothermal environment is weaker than the interior boards, meanwhile their impact on moisture content of the infill layer in the construction is smaller than the exterior boards, the annual heat and moisture flow through exte-

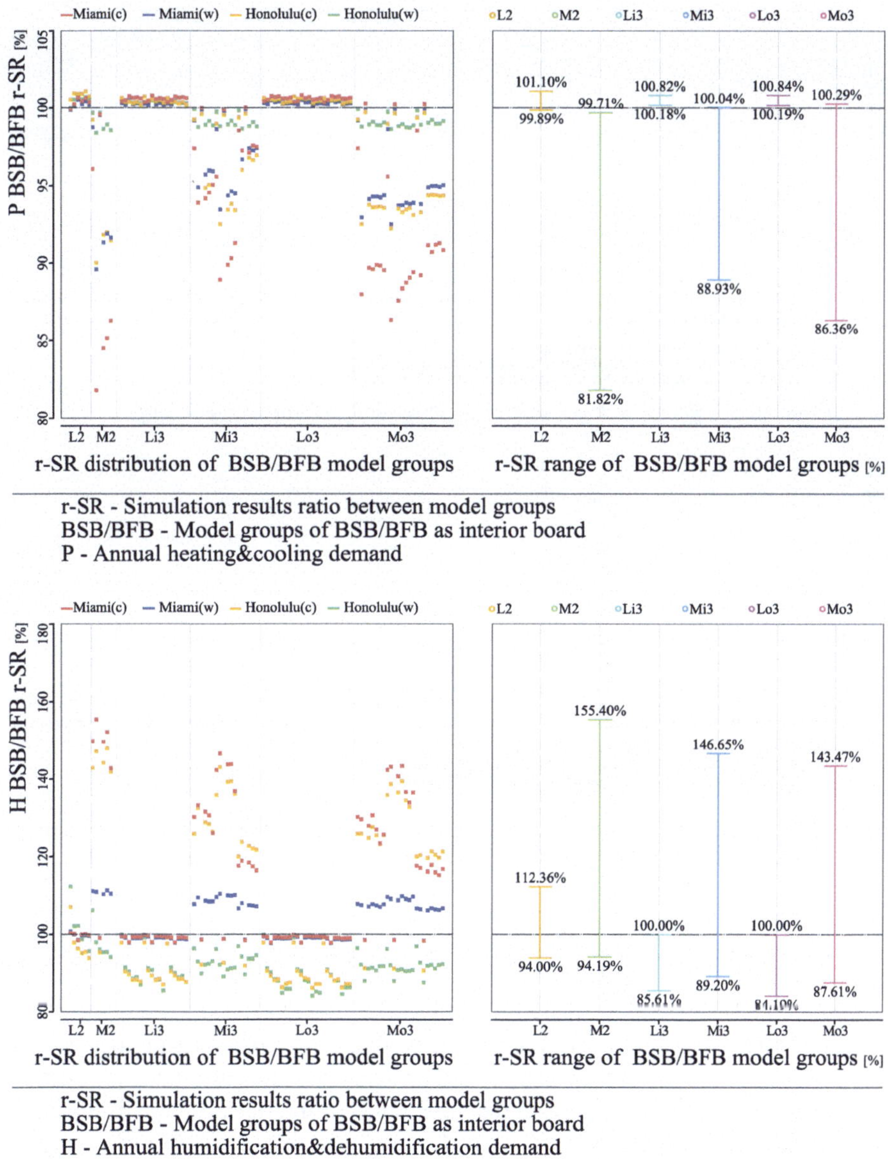

Fig. 5.13 Simulation results' comparison between model groups of BSB/BFB as interior board: annual heating and cooling demand *P* (up) and humidification and dehumidification demand *H* (down)

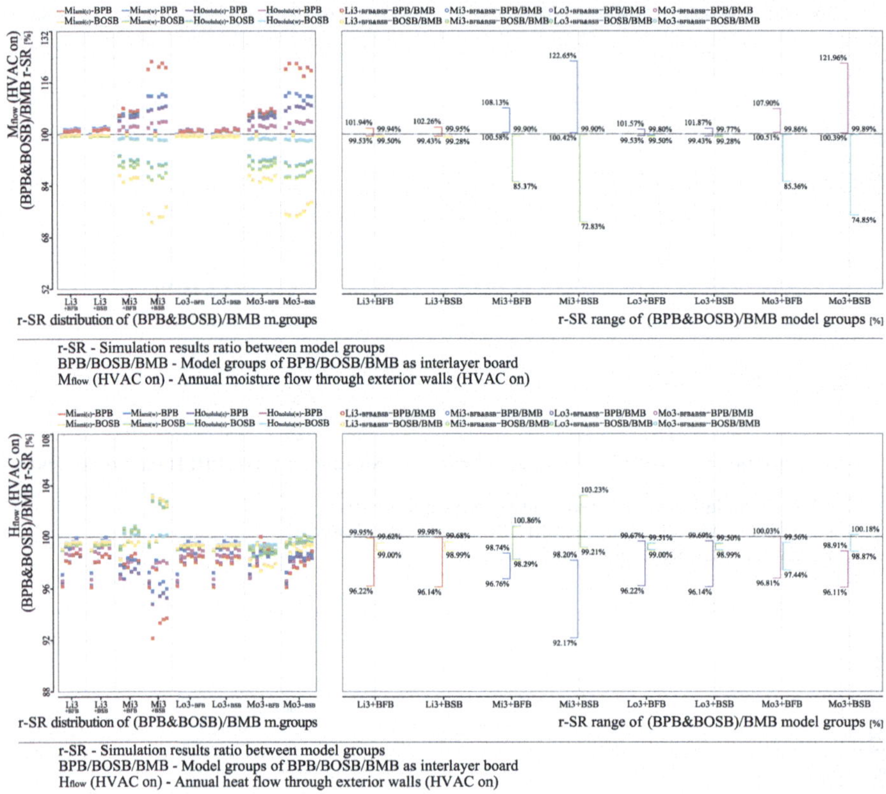

Fig. 5.14 Simulation results' comparison between model groups of BMB/BPB/BOSB as interlayer board: annual moisture flow M_{flow} (up) and heat flow H_{flow} (down) through exterior walls

rior walls when HVAC on and the HVAC demand are chosen for comparison in the following sections.

Simulation results show changeable advantage and disadvantage among BMB, BPB, and BOSB as interlayer board for L-type and M-type construction. For the L-type construction, due to the already large overall moisture resistance, the difference among the model groups with the three kinds of interlayer boards is insignificant. Setting BMB groups as 100%, the M_{flow} of BPB groups and BOSB groups is, respectively, 99.43–102.26% and 99.28–99.95%. Correspondingly, the H_{flow} ratio is 96.14–99.98% and 98.99–99.68%. For the M-type construction, the difference among the three boards is highlighted that the M_{flow} ratio between the model groups with BPB and BMB, and between BOSB and BMB as interlayer board is, respectively, 100.39–122.65% and 72.83–99.90%. Correspondingly, the H_{flow} ratio is 92.17–100.03% and 97.44–103.23% (Fig. 5.14).

The annual HVAC demand shows the similar law with the heat and moisture flow through the exterior walls. For the L-type construction, setting BMB groups

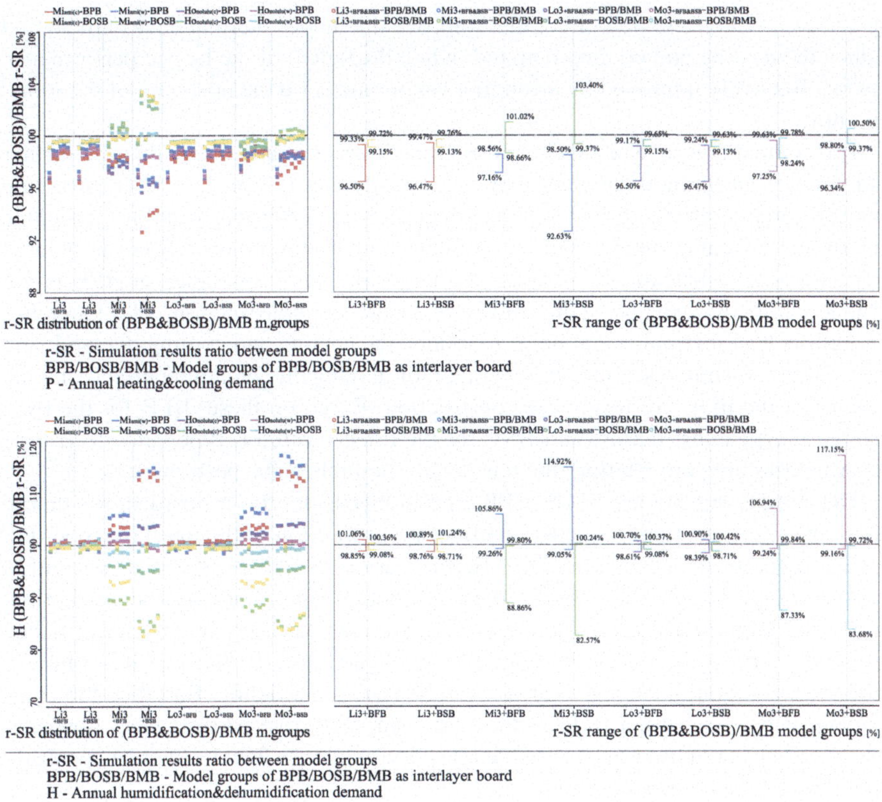

Fig. 5.15 Simulation results' comparison between model groups of BMB/BPB/BOSB as interlayer board: Annual heating and cooling demand *P* (up) and humidification and dehumidification demand *H* (down)

as 100%, the *P* value of BPB and BOSB groups is, respectively, in the range 96.47–99.47% and 99.13–99.76%. Correspondingly, the *H* ratio is 98.39–101.06% and 98.71-101.24%. For the M-type construction, the *P* ratio between the model groups with BPB and BMB, and between BOSB and BMB as interlayer board is, respectively, 92.63–99.63% and 98.24–103.40%. Correspondingly, the *H* ratio is 99.05–117.15% and 82.57–100.24% (Fig. 5.15).

5.3.1.4 Discussion

The comparison between BFB (L-type construction) and LB (M-type construction) as exterior board shows that the former has an advantage in moisture resistance that is beneficial to blocking the influence of exterior moisture on the inside of the construction and the indoor hygric environment, and reducing the humidification and dehumidification demand. The latter has a significant advantage in thermal capacity and evaporative cooling effect that improve the indoor thermal environment and reduce the heating and cooling demand, while the defect of the hygric performance can be effectively improved by arranging vapor barrier on the upstream of the moisture flow.

The comparison between BFB and BSB as interior board shows that the former had strength in thermal capacity that plays prominent role in L-type construction that can compensate for the weakness of heat transfer performance and form a stable indoor thermal environment, which is weakened in M-type construction where large thermal capacity has already existed; the larger moisture resistance of BFB is not conductive to the exhaust of the moisture inside the construction. The latter has strength in heat resistance that helps to reduce the heating and cooling demand, and the smaller moisture resistance is beneficial for playing the 'breathing effect' and the exhaust of the moisture inside the construction. Based on these, BFB for the two-layered L-type construction, and BSB for the three-layered L-type, and all M-type constructions are combination that has better comprehensive performance.

The comparison among BMB, BPB, and BOSB as interlayer board shows that in three-layer construction, since the contribution proportion of the interlayer board to the overall performance is litter, the difference among the three boards shows within a narrow range, which is affected by the construction types. For L-type construction, due to the large moisture resistance of the exterior board, the hygric performance difference among the interlayer boards is weakened and the thermal performance difference is highlighted, and vice versa in M-type construction. Based on the comparison, BPB is regarded to have the best comprehensive performance; however, vapor barrier is required on the upstream of the moisture flow in M-type construction to avoid the hygric weakness.

5.3.2 Air Layer

Simulation results show that, compared with the model groups without air layer (NA groups), the HVAC demand indicators of the groups with closed air layer (CA groups) are improved. Since the moisture capacity and moisture resistance of the closed air layer are very small, it has almost no impact on the *w*-BF values, but as an independent cavity, it contributes to the vapor pressure balance so that can reduce the moisture flow through the construction. In addition, as described in Sect. 5.1.2, the thermal conductivity of the closed air layer with thickness of 10–50 mm is in the range 0.071–0.28 W/(m K), which can improve the heat transport performance of the construction.

5.3.2.1 Independent Closed Air Layer

The simulation results comparison between the CA and NA model groups show the impact of independent closed air layer on the hygrothermal performance of the

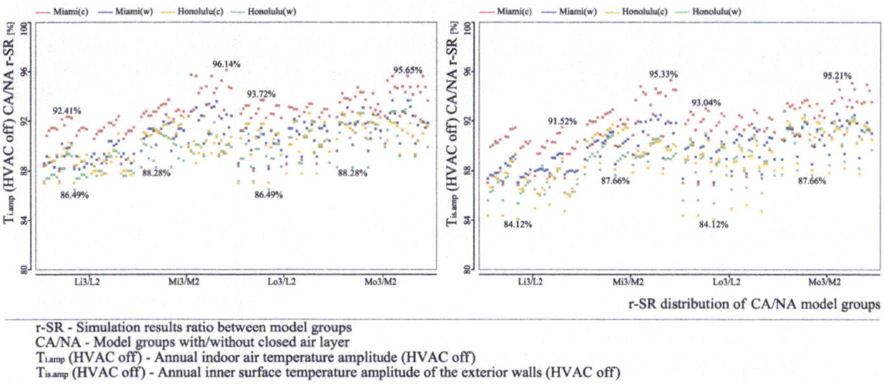

Fig. 5.16 Simulation results' comparison between model groups with independent closed air layer (CA) and without air layer (NA): annual indoor air temperature amplitude $T_{\mathrm{i.amp}}$ (left) and inner surface temperature amplitude of the exterior walls $T_{\mathrm{is.amp}}$ (right)

construction. Among them, the comparison of Li3 to L2 and Mi3 to M2 model groups characterizes the impact of air layer arranged on the inner side; correspondingly, the comparison of Lo3 to L2 and Mo3 to M2 groups shows the impact of air layer arranged on the outer side.

(1) Construction type

In the HVAC off conditions, the $T_{\mathrm{i.amp}}$ and $T_{\mathrm{is.amp}}$ value simulation results' ratio between CA and NA groups are 86.49–93.72% and 84.12–93.04% for L-type construction, correspondingly 88.28–96.14% and 87.66–95.33% for M-type construction. This is caused by the additional air layer that reduces the total heat transfer coefficient, and to some extent improves the heat storage performance, which results in the improvement of stability of the indoor thermal environment (Fig. 5.16).

In the HVAC on conditions, the H_{flow} and P values of CA model groups are in the range 66.42–83.40% and 68.28–87.70% for L-type construction while setting NA groups as 100%, and correspondingly, they are 69.35–92.99% and 71.17–94.95% for M-type construction, which shows an significant improvement of the thermal performance (Fig. 5.17).

Since the additional arrangement of the 30 mm independent air layer requires an interlayer board in the meantime, the improvement range shown in the above comparison results of L3/L2 and M3/M2 model groups includes the contribution of the additional interlayer board. In these cases, besides the heat transport performance, the heat storage performance (thermal capacity) of the L-type construction has also been improved significantly. On the other hand, due to the already existing larger thermal capacity, the improvement of M-type construction is smaller when compared with the L-type construction.

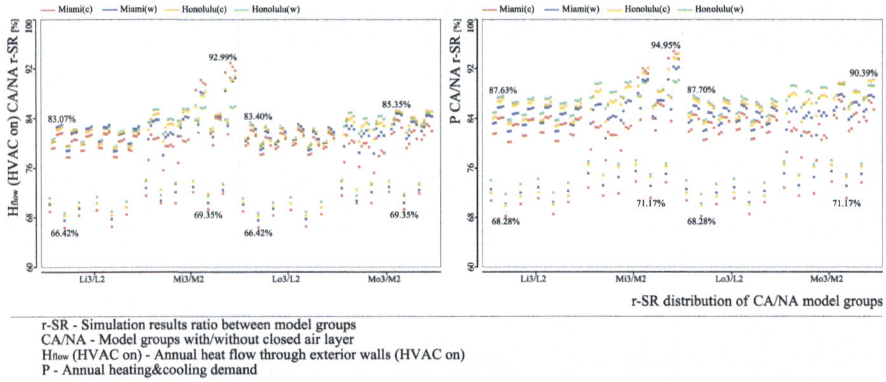

Fig. 5.17 Simulation results' comparison between model groups with independent closed air layer (CA) and without air layer (NA): annual heat flow through exterior walls H_{flow} (left) and heating and cooling demand P (right)

(2) Position of the air layer

For three-layer construction (three partition boards with two cavities), the arrangement of independent closed air layer on the outer side or inner side of the construction is essentially equivalent to the arrangement of thermal insulation layer on the inner cavity or the outer cavity correspondingly, which would be discussed in Sect. 5.3.3.2. The model groups of thermal insulation layer in the inner cavity show better hygrothermal performance than that in the outer; therefore, the arrangement of independent closed air layer on the outer side of the construction is considered to have a better effect.

5.3.2.2 Closed Air Layer Within Insulation Cavity

In the HVAC off conditions, the additional arrangement of the 30 mm closed air layer in the core cavity improves the stability of indoor thermal environment to a certain extent. The $T_{\text{i.amp}}$ value of CA model groups is in the range 95.77–98.29% and 96.04–99.12%, respectively, for the L-type and M-type construction while setting the NA groups as 100%. The $T_{\text{is.amp}}$ value shows the similar law with $T_{\text{i.amp}}$ that it is 95.42–97.86% and 95.87–98.64%, respectively, for the L-type and M-type construction. These two values show similar law with the results displayed in Sect. 5.3.2.1, but the extent of improvement is smaller, since there is no additional interlayer board to increase the thermal capacity of the construction (Fig. 5.18).

In the HVAC on conditions, the improvement effect of the closed air layer is highlighted. For L2 and M2 model groups, the H_{flow} and P values' simulation results of CA groups are in the range 84.99–90.12% and 85.96–91.82% while setting NA groups as 100%. Correspondingly, they are 88.55–92.79% and 89.86–94.72% for L3 and M3 groups (Fig. 5.19).

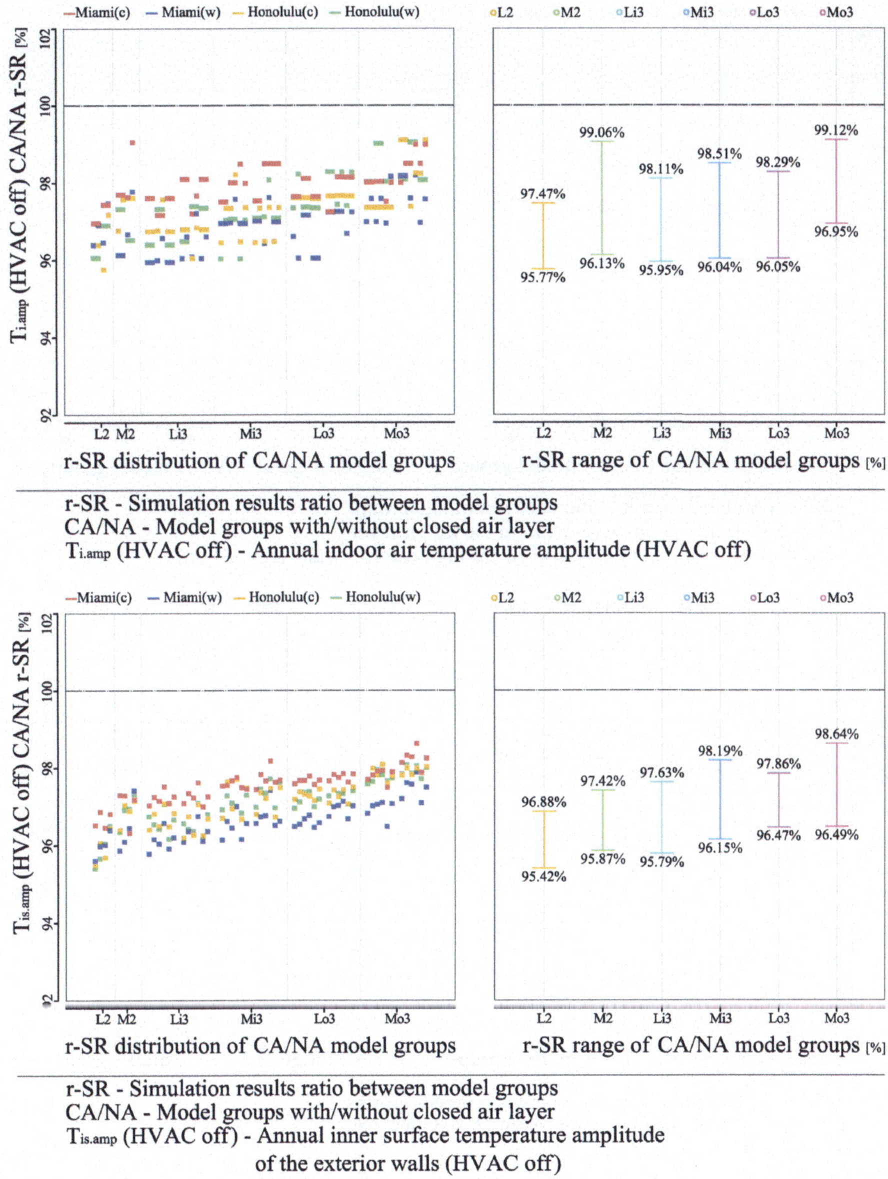

Fig. 5.18 Simulation results' comparison between model groups with closed air layer in insulation infill cavity (CA) and without air layer (NA): annual indoor air temperature amplitude $T_{i.amp}$ (up) and inner surface temperature amplitude of the exterior walls $T_{is.amp}$ (down)

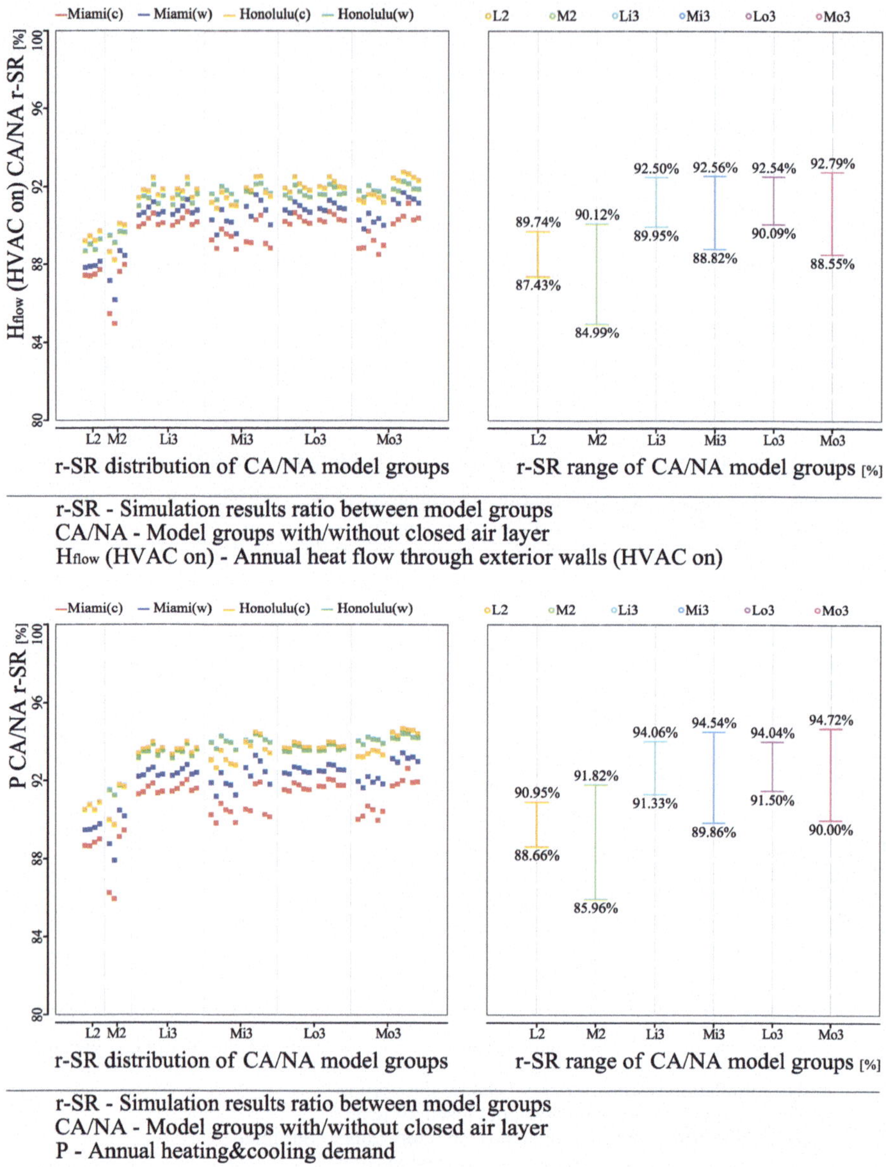

Fig. 5.19 Simulation results' comparison between model groups with closed air layer in insulation infill cavity (CA) and without air layer (NA): annual heat flow through exterior walls H_{flow} (up) and heating and cooling demand P (down)

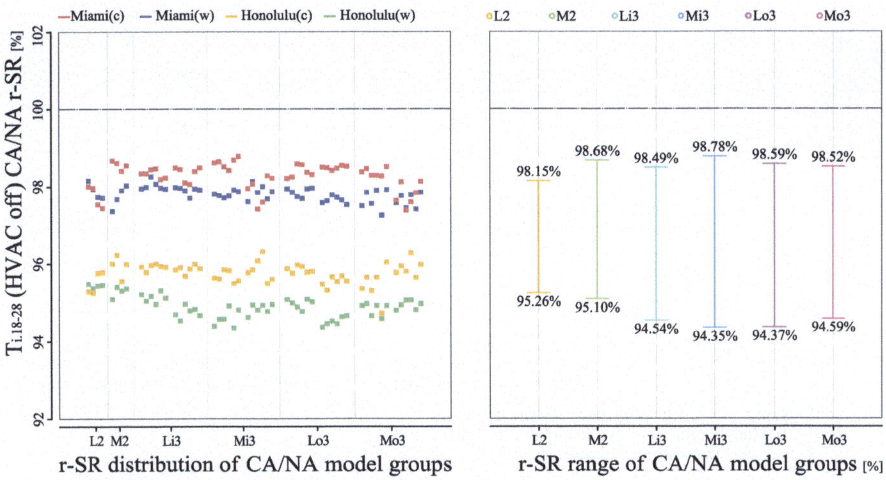

r-SR - Simulation results ratio between model groups
CA/NA - Model groups with/without closed air layer
Ti.18-28 (HVAC off) - Annual hours of indoor air temperature 18℃≤Ti≤28℃ (HVAC off)

Fig. 5.20 Simulation results' comparison between model groups with closed air layer in insulation infill cavity (CA) and without air layer (NA): annual hours of air temperature 18 °C $\leq T_{\mathrm{i}} \leq$ 28 °C $T_{\mathrm{i.18–28}}$

But from another perspective, this improvement of the heat transport performance on building component level does not bring about the improvement of indoor thermal environment in the HVAC off conditions. Taking the annual hours of air temperature 18 °C $\leq T_{\mathrm{i}} \leq$ 28 °C $T_{\mathrm{i.18–28}}$ value as indicator, the $T_{\mathrm{i.18–28}}$ ratio between CA and NA model groups is in the range 95.10–98.68% and 94.35–98.78%, respectively, for the two-layer and three-layer constructions, which means that the period of indoor air temperature maintained in certain comfort zone is instead shortened as the improvement of the heat transport performance of the exterior walls. From this point of view, it can be seen that adding closed air layer is not appropriate for those building envelope projects that are mainly in HVAC off conditions (Fig. 5.20).

5.3.2.3 Air Layer Thickness

For the two-layer construction, an air layer with thickness of 10/20/30/40/50/60/70 mm and 80/100/120/140 mm is added to the core cavity, while for the three-layer construction, two air layers with the same thickness of 10/20/30/40/50/60/70 mm are added to each of the two cavities. For these constructions, the conditions with BF layer (+BF) and without BF layer (−BF) in the core cavity are both taken into consideration.

In the HVAC on conditions, the annual heating and cooling demand P value is chosen as the indicator to characterize the thermal performance of the exterior walls.

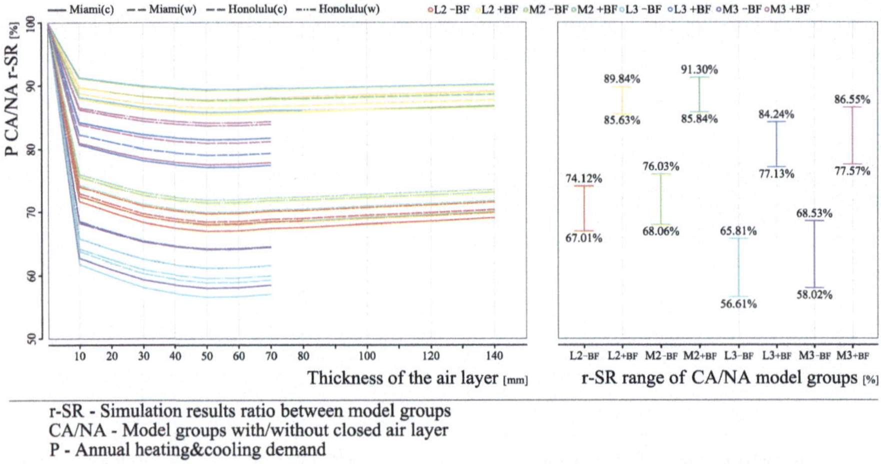

Fig. 5.21 Impact of air layer thickness on the thermal performance: annual heating and cooling demand *P*

With the increase of thickness of the air layer, the *P* ratio between CA and NA model groups shows a process of first decreasing and then increasing and reaches the lowest value approximately when the thickness $d = 40\text{–}50$ mm.

The comparison between the three-layer constructions with air layer of 10/20/30/40/50/60/70 mm in the double cavities and the two-layer constructions with 20/40/60/80/100/120/140 mm in the single cavity shows that the former had larger improvement range than the latter, which indicates that multi-layer air layers contribute better effect than single air layer of the same total thickness. This law is consistent with the conclusion generated from the discussion on building components (Fig. 5.21).

The difference is, similar to the discussion in Sect. 5.3.2.2, that in the HVAC off conditions, with the increase of thickness of the air layer, the $T_{i.18\text{–}28}$ ratio between CA and NA model groups shows a process of first decreasing and then increasing and reaches the lowest value approximately when the thickness $d = 40\text{–}50$ mm. The reason behind is that the annual indoor air temperature mean values $T_{i.mean}$ are, respectively, 25.79, 27.54, 26.81, and 27.82 °C, while the corresponding annual outdoor air temperature mean values $T_{e.mean}$ are 23.43, 25.08, 24.25, and 25.21 °C. In these cases, the heat flow runs mainly from indoor to outdoor; therefore, the improvement of the thermal transport performance caused by the additional air layer, on the contrary, blocks the indoor extra heat from exhausting (Fig. 5.22).

5.3.2.4 Discussion

In the HVAC on conditions, both independent closed air layer and the closed air layer within insulation cavity can improve the thermal performance effectively, and the

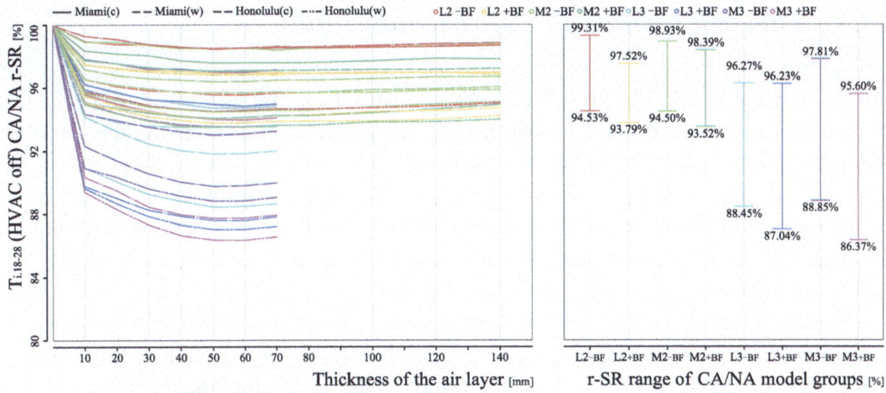

r-SR - Simulation results ratio between model groups
CA/NA - Model groups with/without closed air layer
$T_{i.18-28}$ (HVAC off) - Annual hours of indoor air temperature 18 ℃≤T_i≤28 ℃ (HVAC off)

Fig. 5.22 Impact of air layer thickness on the thermal performance: annual hours of air temperature 18 °C ≤ T_i ≤ 28 °C $T_{i.18–28}$

latter that requires lower construction costs is recommended. Multi-layer air layers contribute better effect than single air layer of the same total thickness.

In particular, for the HVAC off conditions, the performance improvement on the level of building component does not necessarily guarantee the improvement of all indicators on the level of enclosed space. As discussed in Sects. 5.3.2.2 and 5.3.2.3, in the HVAC off conditions, with the additional air layer, the annual hours of air temperature 18 °C ≤ T_i ≤ 28 °C $T_{i.18–28}$ value are instead shortened, so that adding closed air layer is not appropriate for those building envelope projects that are mainly in HVAC off conditions.

Since the heat storage, moisture storage, and moisture transport resistance of the air can almost be ignored, the main contribution of closed air layer is to improve the heat transport performance of the construction, which is the focus of discussion in this section. But in fact, the additional air layer can avoid the direct moisture transfer between two layers that contact side by side, which can contribute indirect effect on the improvement of the hygric performance of the construction. And this is worth further exploration.

In addition, the proper arrangement of ventilated air layer can carry away the inner moisture with the airflowing, which is conducive to exhausting the condensed water and slowing down the accumulation of moisture at the boundary between two layers. Under certain circumstances, ventilated air layer can bring away the extra heat inside the construction, so that improve the thermal performance. However, this effect is affected by the airflowing rate, temperature, and many other factors; therefore, the parameters such as thermal conductivity is difficult to be defined like a closed air layer that can be confirmed by actual measurement. Therefore, the independent ventilated air layer has not been discussed in this section and needed to be supplemented in further work.

5.3.3 *Thermal Insulation Construction Infill*

For the thermal insulation layer in layered construction, the heat storage, moisture storage and moisture transport performance contribution ratios are much smaller when compared with the partition boards, but it is decisive on the heat transport performance. The thermal insulation layer at a different position would be affected by the hygric environment to different degrees, which influences its moisture content, and results in the change of the heat and moisture flow through the exterior walls, as well as the heating and cooling demand of the enclosed space.

Simulation results show that the difference caused by the position of thermal insulation layer mainly displays in the w-BF_{mean}, H_{flow}, and P values. The action laws would be discussed on these three values in the following analyses.

5.3.3.1 Insulation Position Within Single Cavity

For L2 and M2 construction groups, the impact caused by position exchange of the 30 mm BF and the 30 mm air layer within the core cavity is analyzed. For L2 groups, the large moisture transport resistance of BFB exterior board preferably separates the cavity from the external hygric environment; therefore, the impact of position exchange on the hygrothermal performance of the construction is ignorable. When compared with IBF groups, which arrange the BF layer on interior side of the cavity, the w-BF_{mean}, H_{flow}, and P values of EBF groups, which arrange the BF layer on exterior side of the cavity, are respectively in the range 98.23–103.56% (mean value 100.77%), 99.97–101.47% (mean value 100.45%) and 99.98–101.39% (mean value 100.41%) (Fig. 5.23).

On the contrary, the position exchange of thermal insulation and air layers within the core cavity would cause obvious difference to the M2 groups. The w-BF_{mean}, H_{flow}, and P ratio between EBF and IBF model groups are, respectively, in the range 97.52–134.04% (mean value 108.43%), 98.87–112.35% (mean value 104.27%), and 99.06–111.82% (mean value 103.95%). This indicates that the arrangement of thermal insulation layer on the inner side within the cavity, which is in the downstream direction of the moisture flow, can improve the hygrothermal performance of the construction.

5.3.3.2 Insulation Position Within Double Cavity

(1) Comparison among the positions within interior or exterior cavity

For three-layer construction, the difference caused by position exchange of the thermal insulation and air layers within the interior or exterior cavity is relatively small. For the Li3, Lo3, Mi3, and Mo3 construction groups, the w-BF_{mean}, H_{flow}, and P ratio between EBF and IBF model groups are, respectively, in the range 96.40–107.42%, 99.15–100.86%, and 99.24–100.67% (Fig. 5.24).

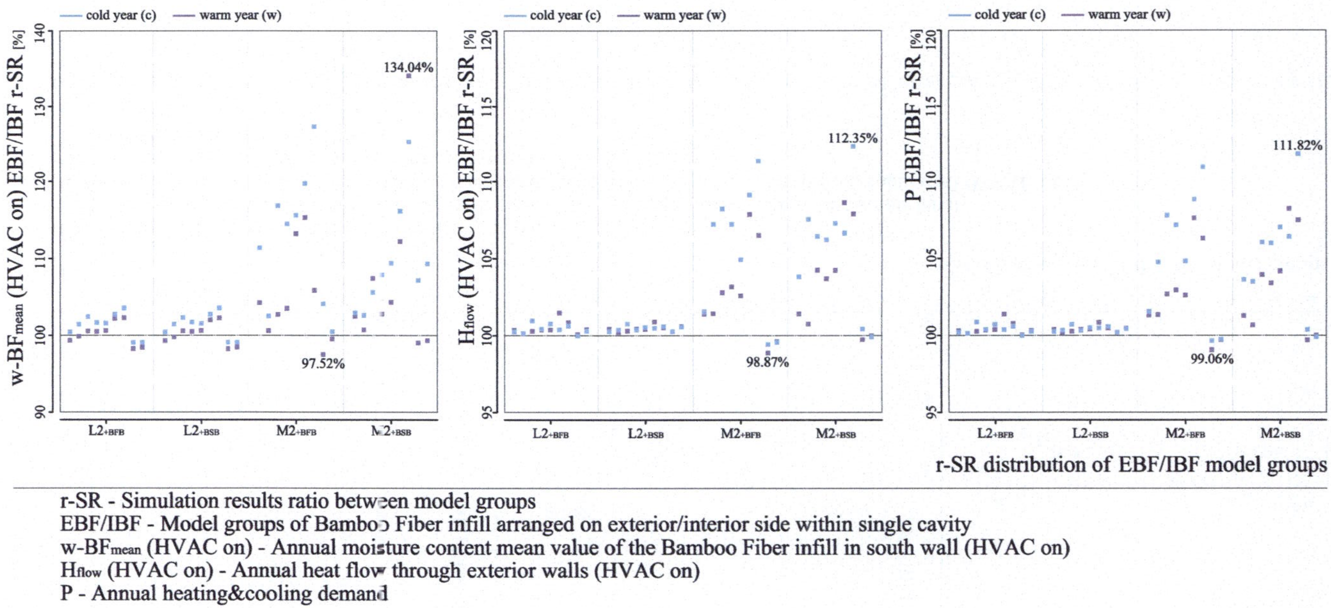

Fig. 5.23 Construction types L2 and M2: simulation results' comparison between model groups of BF arranged on exterior (EBF) and interior side (IBF) in the cavity

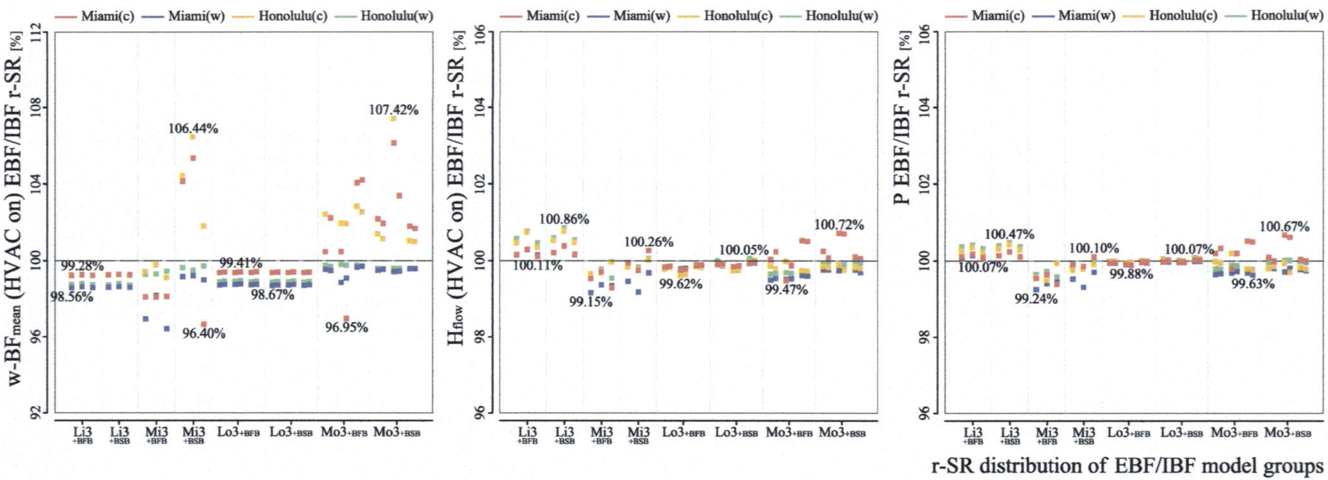

r-SR - Simulation results ratio between model groups
EBF/IBF - Model groups of Bamboo Fiber infill arranged on exterior/interior side within single cavity
w-BF_{mean} (HVAC on) - Annual moisture content mean value of the Bamboo Fiber infill in south wall (HVAC on)
H_{flow} (HVAC on) - Annual heat flow through exterior walls (HVAC on)
P - Annual heating&cooling demand

Fig. 5.24 Construction types L3 and M3: simulation results' comparison between model groups of BF arranged on exterior (EBF) and interior side (IBF) in the cavity

(2) Comparison between arrangements in the interior and in the exterior cavities

The w-BF$_{mean}$, $H_{flow,}$ and P simulation results for model groups with thermal insulation layer in the interior cavity (EBF, including Lo3 and Mo3 construction groups) and the groups with thermal insulation layer in the exterior cavity (IBF, including Li3 and Mi3 construction groups) are analyzed. Results show that for L-type construction, the ratios between EBF and IBF model groups are, respectively, in the range 97.59–102.23% (mean value 99.50%), 97.83–101.25% (mean value 100.11%), 99.06–100.81% (mean value 100.10%), of which the difference is ignorable.

For M-type construction, a similar law is found as shown in Sect. 5.3.3.1 that the arrangement of thermal insulation layer in the interior cavity is conducive to avoiding the influence of the external hygric environment. When compared with IBF groups, the w-BF$_{mean}$, $H_{flow,}$ and P values of EBF groups are, respectively, in the range 44.03–100.24% (mean value 85.95%), 88.16–100.71% (mean value 98.38%) and 89.88–100.61% (mean value 98.83%), which shows that the arrangement of thermal insulation layer on the interior cavity can improve the hygrothermal performance of the construction (Figure 5.25).

5.3.3.3 Discussion

In cold climate regions, where sufficient thermal insulation is required in the design of building envelope, the thermal insulation layer is normally thickened and had large contribution ratio to the thermal resistance of the whole component; therefore, the position would have more obvious impact on its heat transport performance of the construction. The moisture flow through the building envelope is mainly the gaseous moisture that runs from indoor to outdoor; correspondingly, the experience under such climate conditions is to put the thermal insulation layer to the outer side to achieve better performance.

In contrast, the moisture flows mainly from outdoor to indoor through the building envelope in Hot-Humid climate regions that are targeted in this chapter. Besides this, there is liquid water occurred in the outer part of the construction due to the heavy rain load in these regions. The conclusion from this study shows that the arrangement of the thermal insulation layer in inner side of the construction is beneficial for weakening the external hygric environmental impact, reducing heat flow through exterior walls and the heating and cooling demand.

For hygroscopic thermal insulation material, e.g., BF, the related discussion should include the treatment on the material itself. The calculation of the impact of moisture resistance treatment on the thermal performance in Chap. 3 shows that the reducing of the adsorption and desorption rate can contribute significant improvement to the thermal performance, as shown in Sect. 5.3.4.1 of Chap. 3. Since this is the discussion on the material level rather than the construction design level, it would not be extended here.

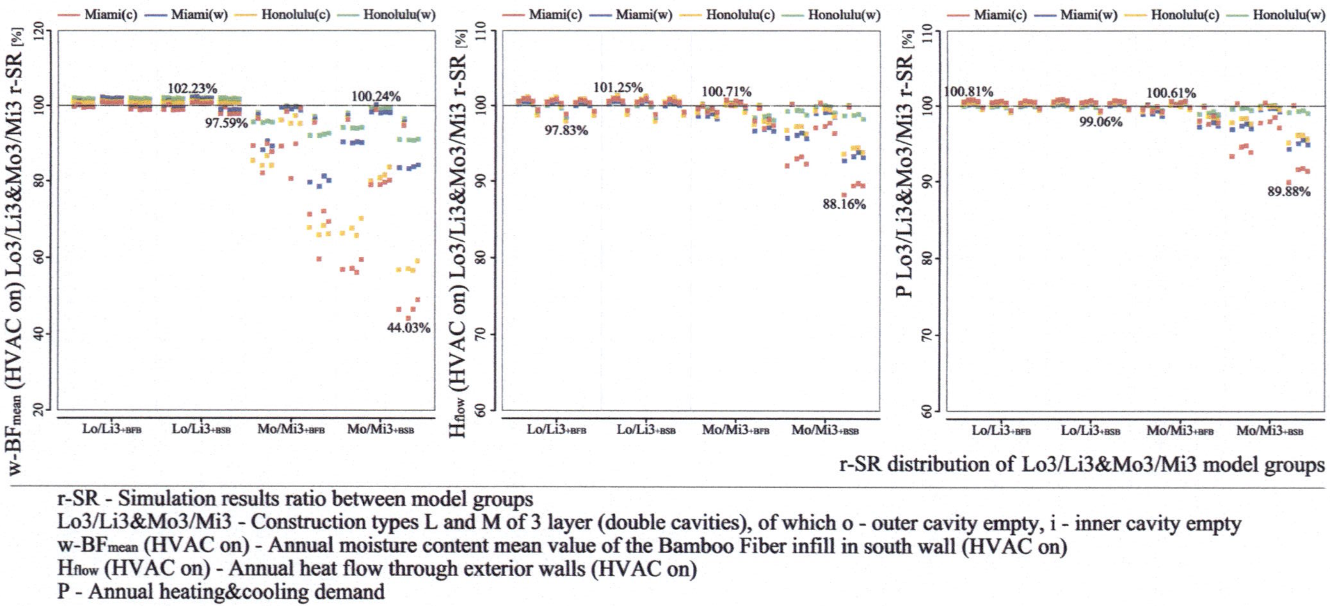

Fig. 5.25 Construction types L3 and M3: simulation results' comparison between model groups of BF arranged in interior cavity (Lo3, Mo3) and in exterior cavity (Li3, Mi3)

5.3.4 *Moisture-Control Construction Infill*

The simulation results' comparison between the model groups with BC layer (BC groups) and the model groups without BC layer (NC groups) shows the impact of BC layer on the hygrothermal performance of the building envelope.

5.3.4.1 Exterior Walls' Hygrothermal Performance

The south wall is considered as an example for analysis. Setting NC groups as 100%, the w-BF_{mean} of BC groups is in the range 99.95–100.66% for the L-type construction, and 80.51–100.42% for the M-type construction in the HVAC on conditions, which indicates that BC reduces the annual moisture content of the BF layer for the M-type construction (Fig. 5.26).

The adsorption and desorption process of the BC layer buffers the moisture flow through the exterior walls, which forms the 'moisture-buffering effect'. When compared with NC groups, the M_{flow} of BC groups is in the range 88.13–96.19% and 85.77–97.97%, respectively, for the L-type and M-type construction, which means that the moisture flow is weakened also for the L-type construction though its BF layer moisture content is not reduced. Correspondingly, the M_{flow} of BC groups is in the range 86.95–91.20% and 84.03–91.72% while setting NC groups as 100% (Fig. 5.27).

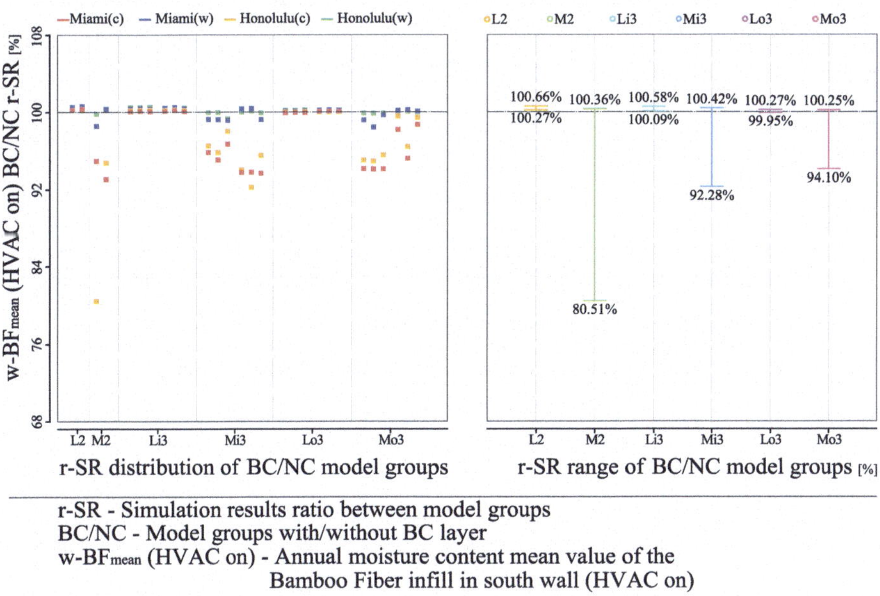

Fig. 5.26 Simulation results' comparison between model groups with (BC)/without (NC) BC layer arranged on the outer side of BF layer: annual BF infill moisture content mean value w-BF_{mean}

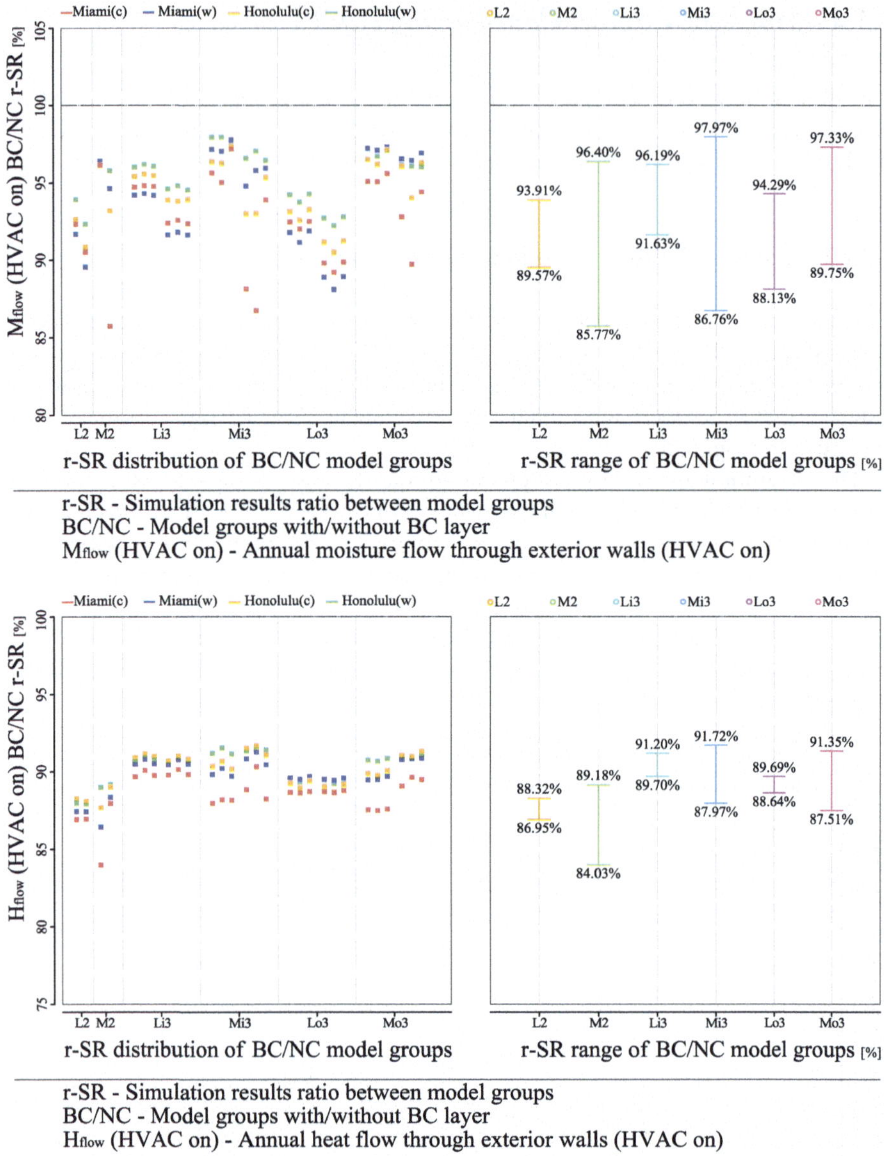

Fig. 5.27 Simulation results' comparison between model groups with (BC)/without (NC) BC layer arranged on the outer side of BF layer: annual moisture flow M_{flow} (up) and heat flow H_{flow} (down) through exterior walls

5.3.4.2 Indoor Hygrothermal Environment

Simulation results show that the BC layer does not significantly affect the annual mean value of indoor air temperature $T_{i.mean}$ and interior surface temperature $T_{is.mean}$, but narrow their amplitudes. In HVAC off conditions, the $T_{i.amp}$ of BC groups is lowered to 92.86–96.70% and 93.86–97.17%, respectively, for the L-type and M-type construction compared with the NC groups. Similarly, the $T_{is.amp}$ of BC groups is in the range 92.43–96.02% and 93.57–96.46% while setting NC groups as 100%. These mainly result from the setting of BC layer that improves the overall thermal capacity of the construction (Fig. 5.28).

The impact of BC layer on the indoor hygrothermal environment is worth discussing. Setting NC groups as 100%, the $M_{i.exchange}$ of BC groups is in the range 82.02–91.13% and 86.00–96.60%, respectively, for the L-type and M-type construction, which means that the BC layer does not strengthen the moisture exchange between the indoor air and the inside surface of the exterior walls; on the contrary, it weakens that effect to a certain extent. However, the $RH_{i.30–70}$ of BC groups increases to 100.29–103.95% and 97.67–164.58%, respectively, for the L-type and M-type construction compared with the NC groups, which shows a significant improvement of the BC layer to the indoor hygrothermal environment of the M-type construction.

The above contrasting phenomenon of $M_{i.exchange}$ and $RH_{i.30–70}$ shows that, unlike those moisture-control material applied to the inner surface of the building envelope that strengthens directly the moisture exchange between the building envelope and the indoor air to realize the moisture-control effect, the action mechanism of BC layer as construction infill is to improve the moisture capacity and moisture adsorption and desorption rate of the building envelope, which changes the moisture exchange process between the indoor and outdoor air, thereby affecting the indoor hygric environment indirectly (Fig. 5.29).

5.3.4.3 HVAC Demand

In HVAC on conditions, the P value of BC model groups is in the range 88.18–93.37% and 85.12–94.34%, respectively, for the L-type and M-type construction while setting the NC groups as 100%, which shows a significant effect of BC layer to reduce the energy consumption. On the one hand, this is due to the thermal capacity improvement of the building envelope; on the other hand, the indoor heat load is cut down, which results from the reduce of w-BF that weakens the heat flow through the exterior walls. Correspondingly, the H-value simulation results' ratio between BC and NC groups is 83.47–98.91% and 87.47–98.13%, respectively, for the L-type and M-type construction (Fig. 5.30).

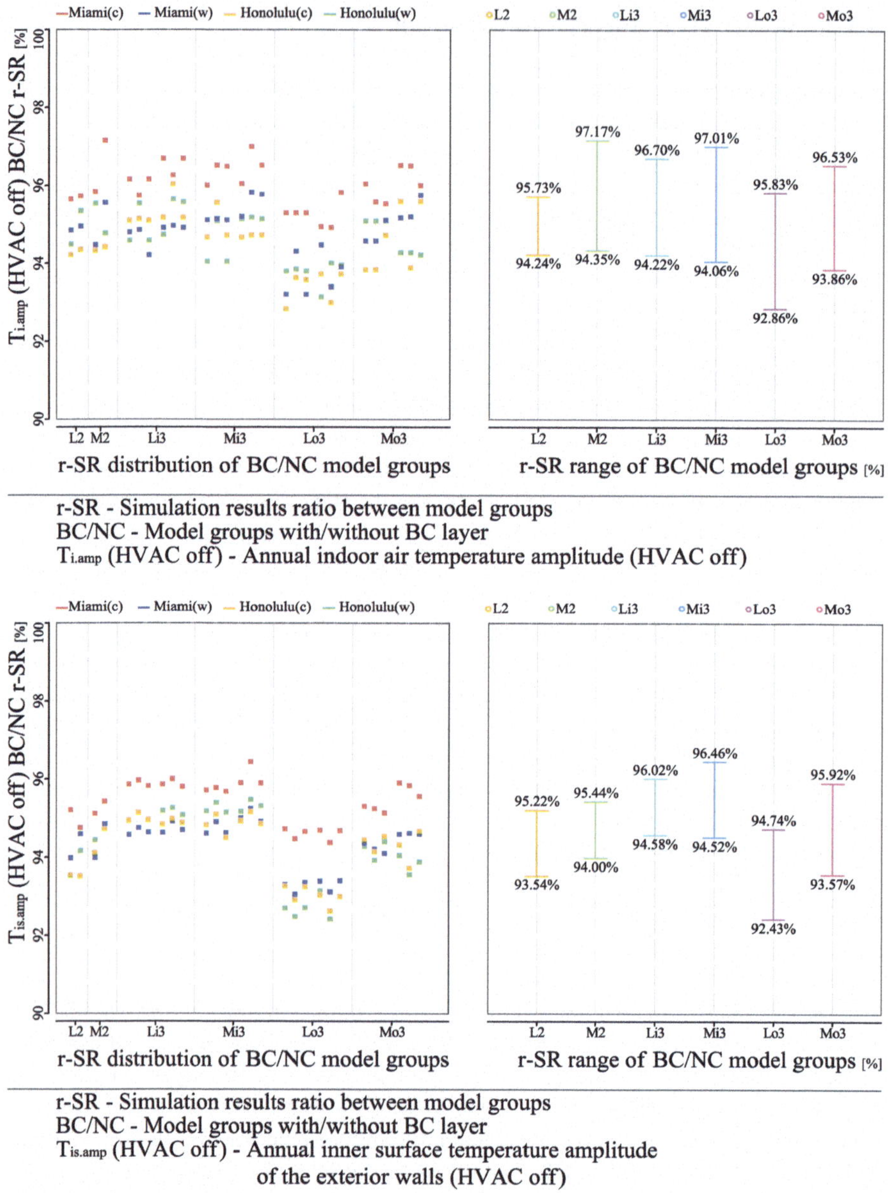

Fig. 5.28 Simulation results' comparison between model groups with (BC)/without (NC) BC layer arranged on the outer side of BF layer: annual indoor air temperature amplitude $T_{\text{i.amp}}$ (up) and inner surface temperature amplitude of the exterior walls $T_{\text{is.amp}}$ (down)

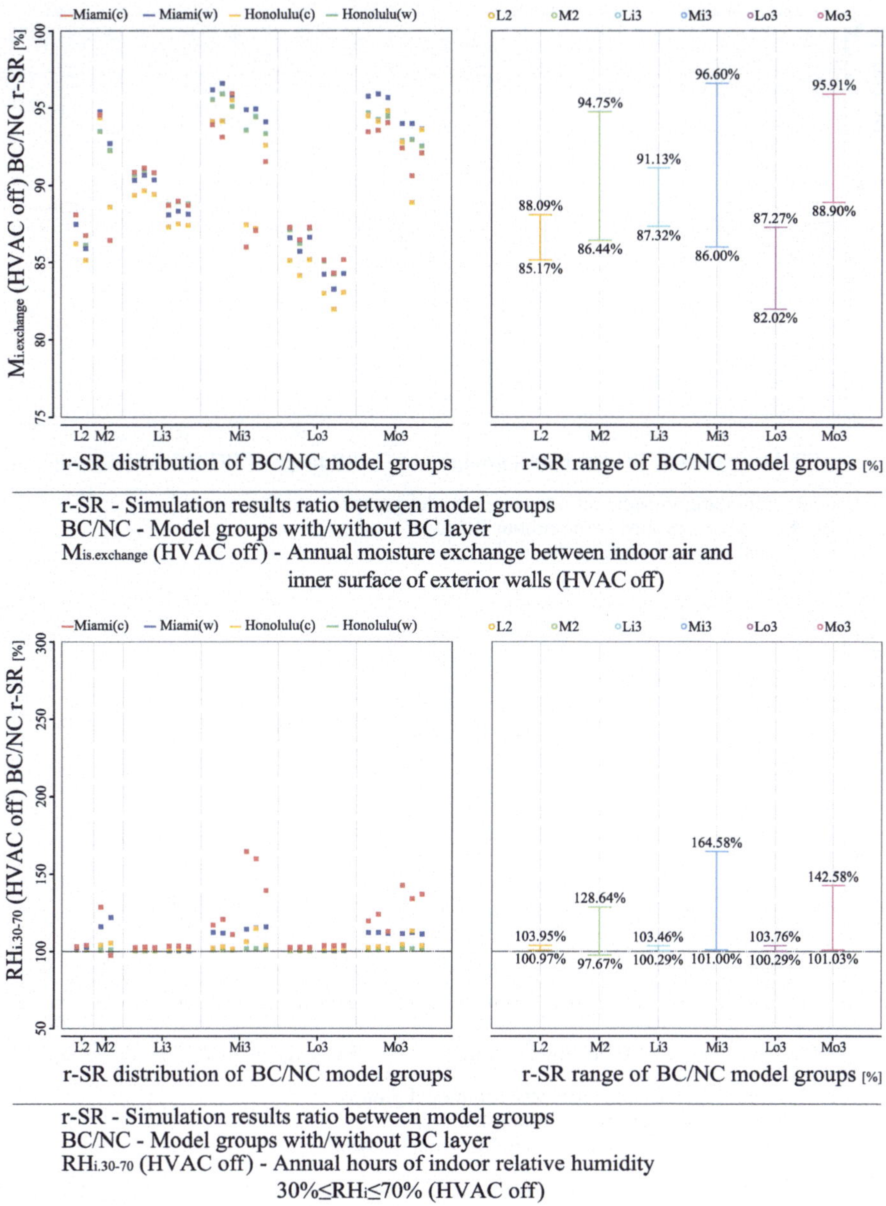

Fig. 5.29 Simulation results' comparison between model groups with (BC)/without (NC) BC layer arranged on the outer side of BF layer: annual moisture exchange between indoor air and the inner surface of the exterior walls (up), and hours of relative humidity $30\% \leq RH_i \leq 70\%$ (down)

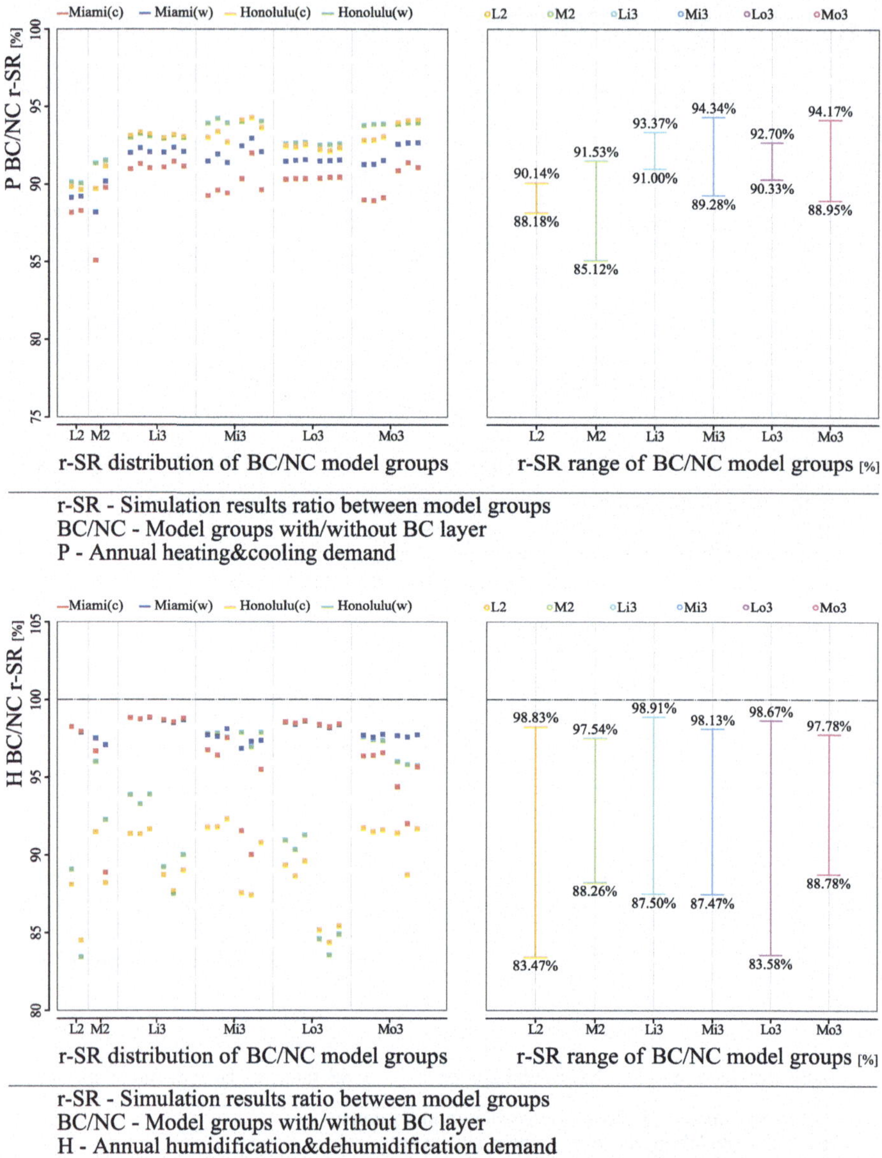

Fig. 5.30 Simulation results' comparison between model groups with (BC)/without (NC) BC layer arranged on the outer side of BF layer: annual heating and cooling demand P (up) and humidification and dehumidification demand H (down)

5.3.4.4 Discussion

Taking the south wall as an example, correlation analyses for BF moisture content *w*-BF are performed with the hygric environment and the hygric performance based on the simulation results. All indicators referring to the hygric performance of building component, indoor hygric environment, and the humidification and dehumidification demand display a significant correlation with the annual mean value and amplitude of *w*-BF at 0.01 level (2-tailed), and this proves the significance of controlling *w*-BF. All correlation coefficient values (r) are higher while the corresponding statistical significance values (sig) are lower when the correlation analyses are performed with w-BF mean value than with the w-BF amplitude, and this indicates the more decisive role of the former value.

The *w*-BF mean value exhibits a significant positive correlation with the outdoor annual relative humidity mean value $RH_{e.mean}$ and the driving rain DR_s at 0.01 level (2-tailed). Correspondingly, the *w*-BF amplitude exhibits a significant positive correlation with $RH_{e.mean}$ and DR_s at 0.05 level (2-tailed) (Fig. 5.31).

The setting of BC on the outer side of BF layer reduces significantly the moisture content of BF layer in M-type construction, heat, and moisture flow through the exterior walls, improves the indoor hygrothermal environment, and reduces the HVAC demand.

When the moisture-control materials are applied as interior finishes, they strengthen the direct moisture exchange between the inner surface of the building envelope and the indoor air. Different from this, the action mechanism of BC layer as construction infill is to improve the heat and moisture process of the building envelope, buffering the moisture exchange between the indoor and outdoor air, thereby affecting the indoor hygric environment indirectly.

BF and BC can be produced from the widely distributed bamboo forest that is not sufficiently utilized. The combination of BF and BC as construction infill shows better climate adaptive potential in this study. The application of BF and BC as construction infill in Hot-Humid climate regions can benefit both bamboo resources' utilization and improvement of the local building physical performance.

5.3.5 Treatment Against Rainfall on Facade

The simulation results are compared between the model groups with consideration to the driving rain on exterior board (DR) and the groups without consideration to the driving rain (NR). The simulation results' ratio between the NR and DR model groups shows the impact of the rain avoidance on the hygrothermal performance of the building envelope.

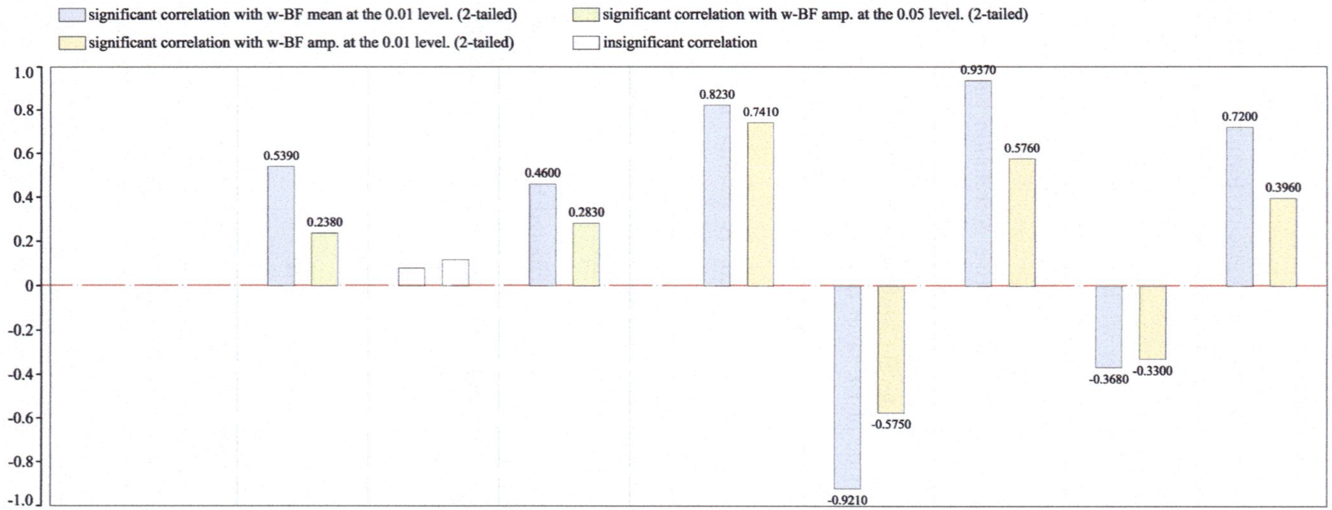

	Correlation indicator	RHe mean	RHe amp	DRs	Mflow-s	RHi 30-70	RHi mean	RHi amp	H
w-BF mean	Correlation coefficient-r:	0.5390	0.0802	0.4600	0.8230	-0.9210	0.9370	-0.3680	0.7200
	Statistical significance-sig:	1.0406E-06	0.502895	0.000049	7.7778E-19	2.1811E-30	9.5377E-34	0.001462	1.0625E-12
	Number-N:	72	72	72	72	72	72	72	72
w-BF amp.	Correlation coefficient-r:	0.2380	0.1176	0.2830	0.7410	-0.5750	0.5760	-0.3300	0.3960
	Statistical significance-sig:	0.043852	0.325027	0.015880	1.0279E-13	1.2652E-07	1.1639E-07	0.004651	0.000582
	Number-N:	72	72	72	72	72	72	72	72

Fig. 5.31 Correlation analyses for moisture content of BF layer, mean value $w\text{-BF}_{mean}$, and amplitude $w\text{-BF}_{amp}$, with the hygric environment and the hygric performance

5.3.5.1 Exterior Walls' Hygrothermal Performance

The driving rain factor has an impact on the heat and moisture flow by affecting the moisture content of the BF layer. Since the transport resistance to the liquid water of BFB as exterior board is much greater than that of LB, the impact of driving rain on L-type construction can be ignored. In contrast, this is worth investigating for the M-type construction, which would be discussed in the following sections.

In the HVAC on conditions, the w-BF_{mean} ratio between NR and DR model groups are in the range 12.27–95.83%, 30.03–94.79%, and 27.09–96.76%, respectively, for the M2, Mi3 and Mo3 construction groups, which indicates that the avoidance of rainfall on the façade has a significant effect on the control of the moisture content inside the construction.

For the model groups with 0.2 mm PE arranged on the outer side of the BF layer, the w-BF_{mean} ratio between NR and DR model groups is correspondingly reduced to the range 90.12–95.83%, 86.94–94.79% and 91.78–96.76%, which shows a significantly weakening influence of driving rain on the inside of the construction by the vapor barrier on the inside of the LB board (Fig. 5.32).

Generally, the avoidance of driving rain can reduce the annual moisture flow through the exterior walls. In the HVAC on conditions, the M_{flow} ratio between NR and DR model groups is affected by the interior board, which is in the range 50.70–105.28% and 33.08–105.40%, respectively, for the groups with BFB and BSB as interior board. Correspondingly, the H_{flow} ratio is 85.72–107.80% and 93.52–108.96%. In comparison with the moisture flow, the change of heat flow shows greater uncertainty.

For the model groups with 0.2 mm PE arranged on the outer side of the BF layer, the M_{flow} ratio between NR and DR model groups is in the range 100.85–105.28% and 99.80–105.40%, respectively, for the groups with BFB and BSB as interior board. Correspondingly, the H_{flow} ratio is 101.59–107.80% and 102.06–108.38%, which indicates that the contact of LB with external rainfall can reduce the annual heat and moisture flow through the exterior walls to a certain extent. This is a result of comprehensive effects, including the change of driving potential for the heat and moisture transport due to the change in moisture content, and the evaporative cooling effect. The arrangement of vapor barrier on the inside of the LB board can effectively block the moisture transport caused by the external rainfall, without hindering the improvement of evaporative cooling effect on and the heat transport performance of the building component (Fig. 5.33).

5.3.5.2 Indoor Hygrothermal Environment

In the HVAC off conditions, the annual hourly indoor air temperature T_i and relative humidity RH_i are analyzed. Setting DR model groups as 100%, the $T_{i.18–28}$ value of NR groups is in the range 84.45–97.43% and 83.79–97.05%, respectively, for the groups with BFB and BSB as interior board, in which NR model groups show obviously worse performance than the DR groups.

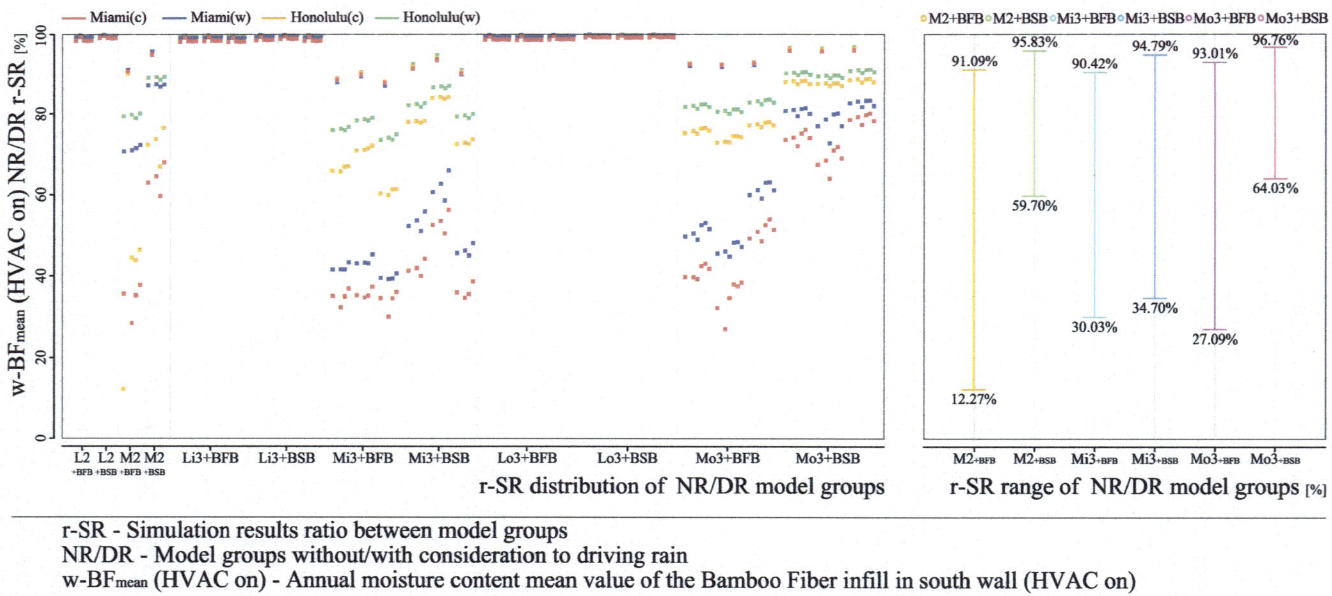

Fig. 5.32 Simulation results' comparison between model groups without (NR)/with (DR) consideration to the driving rain on exterior board: annual BF infill moisture content mean value $w\text{-BF}_{mean}$

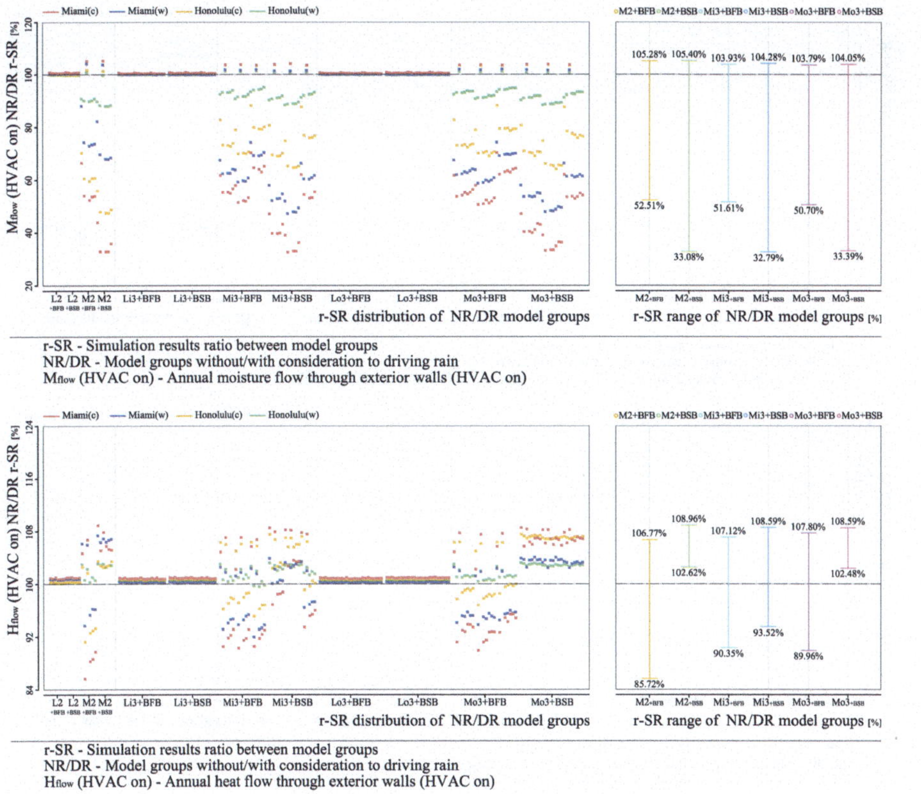

Fig. 5.33 Simulation results' comparison between model groups without (NR)/with (DR) consideration to the driving rain on exterior board: annual moisture flow M_{flow} (up) and heat flow H_{flow} (down) through exterior walls

On the contrary, the $RH_{i.30\text{–}70}$ ratio between the NR and DR model groups is in the range 101.89–354.54% and 101.08–1116.74%, respectively, for the groups with BFB and BSB as interior board, in which NR model groups show obviously better performance than the DR groups (Fig. 5.34).

The avoidance of rainfall can substantially improve the indoor hygric environment in the HVAC off conditions, but meanwhile worsen the indoor thermal comfort to a certain extent. For the model groups with 0.2 mm PE arranged on the inner side of the LB layer, the $RH_{i.30\text{–}70}$ ratio between NR and DR model groups is narrowed down to the range 101.89–121.02% and 101.08–120.99%, respectively, for the groups with BFB and BSB as interior board, which shows that the arrangement of vapor barrier on the inside of the LB board can improve the indoor hygric environment while effectively ensure the thermal comfort.

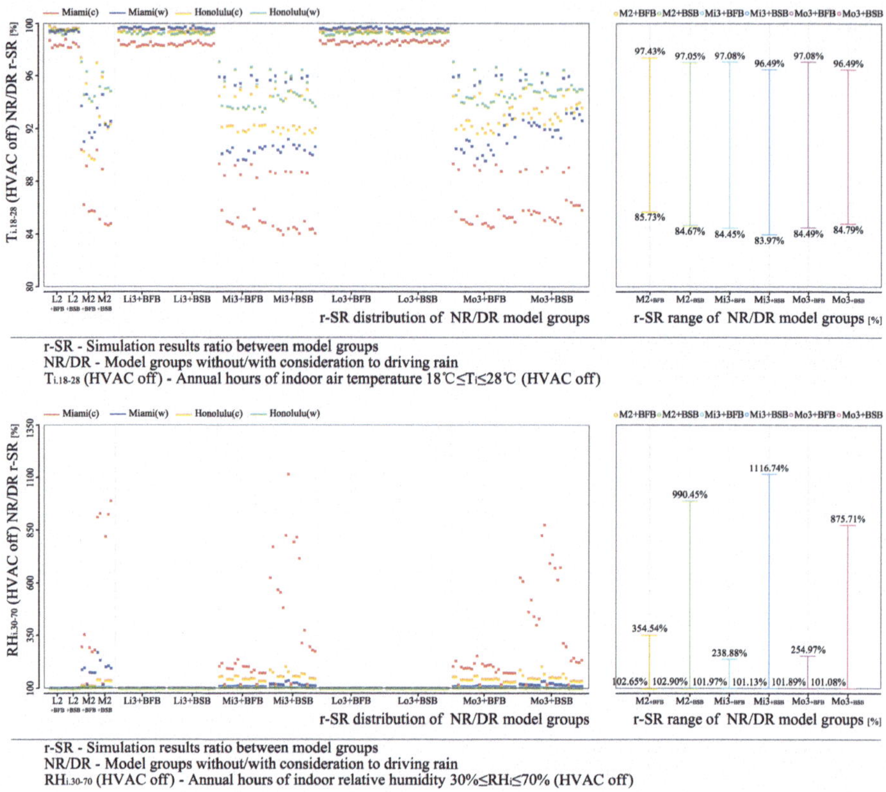

Fig. 5.34 Simulation results' comparison between model groups without (NR)/with (DR) consideration to the driving rain on exterior board: annual hours of air temperature 18 °C $\leq T_i \leq$ 28 °C (up) and hours of relative humidity 30% $\leq \mathrm{RH_i} \leq$ 70% (down)

5.3.5.3 HVAC Demand

In the HVAC on conditions, setting DR model groups as 100%, the *P* value of NR groups is in the range 90.43–110.03% and 98.79–112.20%, respectively, for the groups with BFB and BSB as interior board, from which it can be seen that M-type construction with BSB as interior board effectively reduces the P value when the LB layer is in contact with the rainfall. Correspondingly, the *H* value is 25.86–93.81% and 14.62–93.06%, which shows the significant decrease of dehumidification demand by the avoidance of driving rain (Fig. 5.35).

The arrangement of vapor barrier on the inside of the LB board can effectively reduce the P value without excessive demand for dehumidification. For the model groups with 0.2 mm PE arranged on the inner side of the LB layer, the *H* ratio between NR and DR model groups is narrowed down to the range 67.67–93.81% and 61.29–93.06%, respectively, for the groups with BFB and BSB as interior board.

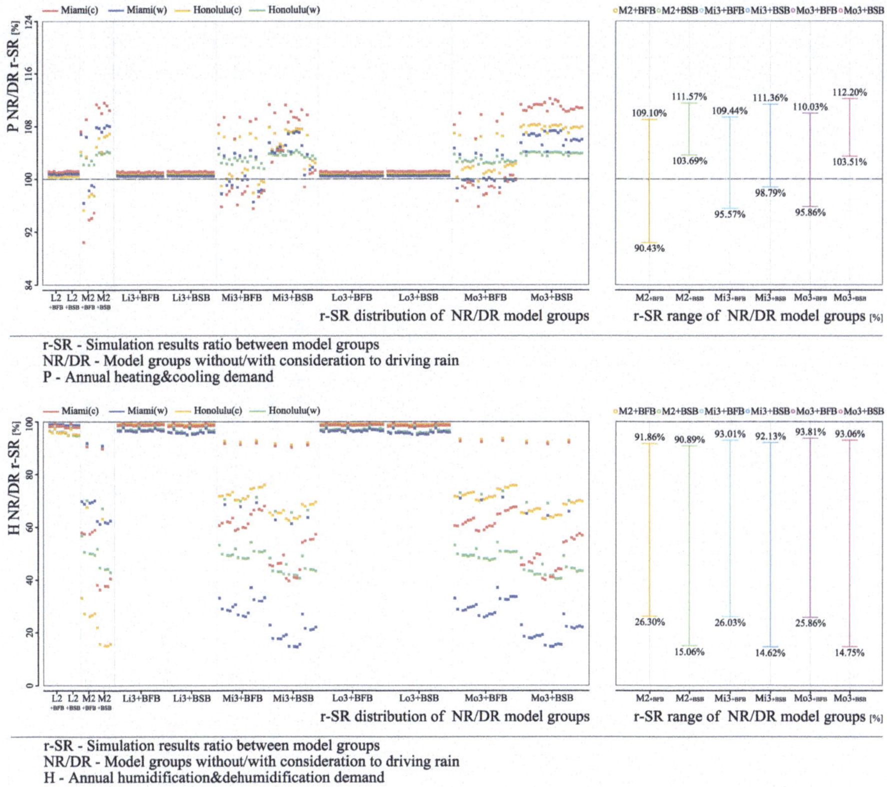

Fig. 5.35 Simulation results' comparison between model groups without (NR)/with (DR) consideration to the driving rain on exterior board: annual heating and cooling demand *P* (up) and humidification and dehumidification demand *H* (down)

5.3.5.4 Discussion

Taking the two-layer construction groups with LB and BSB correspondingly as the exterior and interior boards as the condition cases, correlation analyses are carried out between the simulation results and the hygric climate parameters, which include the annual outdoor air relative humidity RH_e and driving rain DR. Results show that, among the indicators referring to the hygric process, both the mean value and amplitude of H display a significant correlation with RH_e at 0.01 level (2-tailed), meanwhile the mean value and amplitude of dehumidification demand has a significant correlation with DR at 0.05 level (2-tailed). There is a significant correlation at 0.01 level (2-tailed) and at 0.05 level (2-tailed) correspondingly between $RH_{i.mean}$ and $RH_{e.mean}$, and between $RH_{i.mean}$ and DR, but $RH_{i.amp}$ has only a significant correlation at 0.01 level (2-tailed) with DR_{max} The mean value of M_{flow} shows a significant correlation with both $RH_{e.mean}$ and DR, but the amplitude of M_{flow} has only a significant correlation at 0.05 level (2-tailed) with $RH_{e.mean}$. Both the mean value and amplitude of $M_{i.flux}$ and *w*-BF have only a significant correlation with DR at 0.01 level (2-tailed). The mean value of $M_{e.flux}$ shows a significant correlation with $RH_{e.mean}$ and DR,

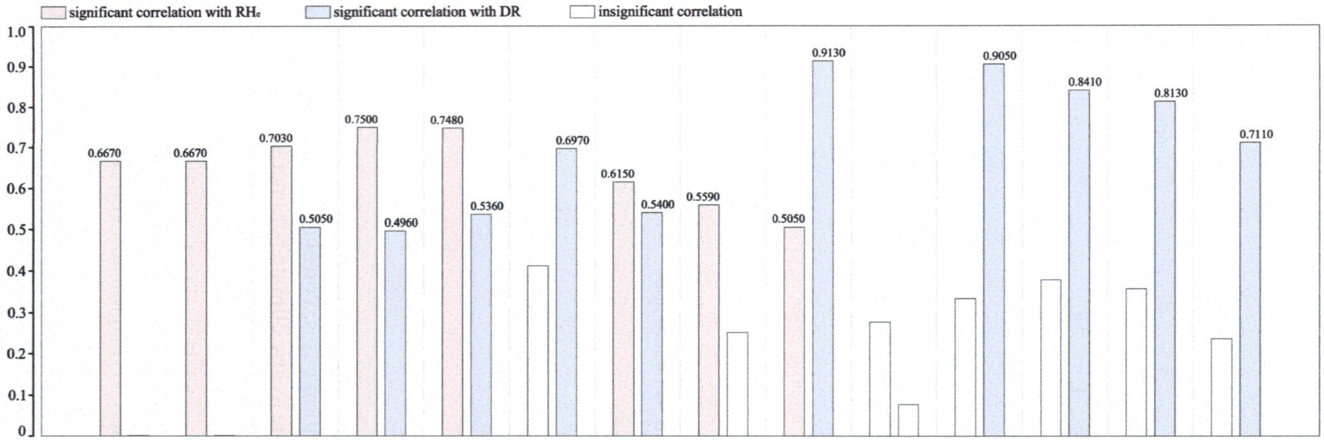

		H_{humid}	$H_{humid.max}$	$H_{dehumid}$	$H_{dehumid.max}$	$RH_{i.mean}$	$RH_{i.amp}$	M_{flow}	$M_{flow.max}$	M_{eflux}	$M_{eflux.max}$	M_{iflux}	$M_{iflux.max}$	$w\text{-}BF_{mean}$	$w\text{-}BF_{amp}$
RH_e	r:	0.667	0.667	0.703	0.750	0.748	-	0.615	0.559	0.505	-	-	-	-	-
	sig.:	0.002482	0.002482	0.001133	0.000342	0.000362	-	0.006558	0.015972	0.032589	-	-	-	-	-
	N:	18	18	18	18	18	-	18	18	18	-	-	-	-	-
DR	r:	-	-	0.505	0.496	0.536	0.697	0.540	-	0.913	-	0.905	0.841	0.813	0.711
	sig.:	-	-	0.032694	0.036140	0.021973	0.001297	0.020783	-	1.218E-07	-	2.456E-07	0.000012	0.000042	0.000939
	N:	-	-	18	18	18	18	18	-	18	-	18	18	18	18

Correlation indicators

RH_e - Relative humidity of the outdoor air; DR - Driving rain

r - Correlation coefficient; sig.- Statistical significance (2-tailed); N - Number of the sample

Fig. 5.36 Correlation analyses between hygric climate parameters (annual exterior air relative humidity RH_e, annual driving rain DR) and the simulation results (Conditions of two-layer construction with LB and BSB correspondingly as exterior and interior boards are shown as cases)

respectively, at 0.05 level (2-tailed) and 0.01 level (2-tailed), while the amplitude of $M_{e.flux}$ shows no significant correlation with any hygric climate parameters.

For a long period, the meteorological and material parameters related to liquid water process, such as the driving rain, liquid water transport coefficient, do not get equal attention as the parameters related to the gaseous moisture process, such as the relative humidity, water vapor transfer coefficient. In these cases, those simulation results that characterize the hygric performance of the construction show that the correlation with driving rain is generally stronger than the correlation with outdoor air relative humidity, which means that the effect of liquid water on the simulation results cannot be neglected, and the study accumulation for meteorological and material parameters related to the liquid water process is of necessity (Fig. 5.36).

The heavy rainfall and high-temperature happen normally at the same period in Hot-Humid climate region. For M-type construction, the evaporative cooling effect caused by rainfall on the facade can improve the indoor thermal environment and reduce the heating and cooling demand to a certain extent. In contrast, for L-type construction, due to the large moisture transport resistance of the exterior board, the influence of rainfall on the hygrothermal performance is ignorable.

The driving rain has a significant effect on the moisture content of the construction layers. As an organic material, bamboo is easy to suffer from mold growth caused by the high moisture content, and stress deformation or cracking due to the repeated moisture adsorption and desorption process, therefore, the avoidance of direct rainfall for L-type construction is recommended. On the contrary, for M-type construction, the evaporative cooling effect generated by rainfall on the exterior board (LB) of M-type construction can be utilized, while vapor barrier is suggested to be arranged inside the LB layer, and BSB selected as the interior board, so as to block the liquid water at the exterior layer, and facilitate the exhausting of the moisture inside the construction.

5.4 Summary

With four groups of climate data from two representative cities in the Hot-Humid climate region of North America as external conditions, exterior walls are constructed in WUFI Plus and the coupled heat and moisture process simulation is performed. Annual exterior walls' hygrothermal performance, indoor hygrothermal environment, and HVAC demand are chosen as indicators, on which factor impact analysis is carried out to investigate the effect of the partition boards, air layer, thermal insulation and moisture-control infill layers, and the façade rainfall treatment on the hygrothermal performance of the building envelope. Climate-responsive design suggestions in Hot-Humid climate region are generated from the analysis results, in terms of material and construction parameters' optimization.

(1) L-type construction (with BFB as exterior board) has moisture resistance advantage that is beneficial for blocking the influence of external hygric environment on the inside of the construction and the indoor hygric environment, and reducing the humidification and dehumidification demand, while M-type construction

(with LB as exterior board) has great heat storage advantage that is conducive to the indoor thermal environment and lowering the heating and cooling demand, while the moisture transport weakness can be significantly improved by arranging vapor barrier on its inner side. Choosing BFB as interior board for two-layer construction of L-type, while BSB for M-type and three-layer construction of L-type can have better comprehensive performance. BPB performs the best as interlayer board, but requires vapor barrier on its outer side in M-type construction to avoid the moisture transport resistance shortcomings.

(2) The improvement effect of independent closed air layer is close to the closed air layer within insulation cavity; therefore, the latter is recommended. Multi-layer air layers contribute better effect than single air layer of the same total thickness.

(3) The insulation layer arranged in the inner side of the insulation cavity is beneficial for weakening the external hygric environmental impact, reducing heat flow through exterior walls and the heating and cooling demand. The practical application of BF in Hot-Humid climate region requires either moisture resistance treatment or the control of hygric environment.

(4) The moisture-control BC arranged on the outer side of BF layer helps to reduce the moisture content of BF in M-type construction significantly, simultaneously the heat and moisture flow through exterior walls, as well as the HVAC demand, and improve the indoor hygric environment. Unlike the moisture-control interior finishes that directly interact with the indoor air, the mechanism of BC as construction infill is to control the moisture process inside the construction, then affect the moisture exchange between indoor and outdoor, and finally realize the regulation of the indoor hygric environment.

(5) As for the rainfall on building façade, L-type construction is suggested to avoid rain; on the contrary, the evaporative cooling effect generated by rainfall on the exterior board (LB) of M-type construction can be utilized, while vapor barrier is suggested to be arranged inside the LB layer, and BSB selected as the interior board, so as to block the liquid water at the exterior layer, and facilitate the exhausting of the moisture inside the construction.

References

1. Kaufmann H, Krötsch S, Winter S (2017) Atlas—Mehrgeschossiger Holzbau. Editor Detail
2. Götz K-H, Hoor D, Möhler K, et al (1980) Holzbau Atlas. J.P. Himmer GmbH & Co. KG, Reproduktionen: graphik-kunstklischee, Augsburg
3. Fritzen K, Metzger P, Krämer F-J (2014) Holzrahmenbau: Bewährtes Hausbau-System 5. Auflage. Bruderverlag Albert Bruder GmbH & Co. KG, Köln
4. Steiger L (2011) Timber construction. 中国建筑工业出版社, 2011. Steiger L (2011) Timber construction. China Architecture & Building Press
5. Holcroft N, Shea A (2013) Moisture buffering and latent heat effects in natural fibre insulation materials. Portugal SB13-contribution of sustainable building to meet the EU, 2013, pp 221–228

Conclusion

Bamboo has much potential in the construction industry in applications such as high-value utilization of bamboo forest resources and improvement of building performance. The research project 'Application of Bamboo in Building Envelope' defines typical industrial bamboo products and the building envelope, respectively, as the research subjects at the material and building levels. There are deficiencies in terms of the comprehensiveness of the materials objects, completeness of the material parameters, and systematicness of the research hierarchies among previous studies. In particular, those studies mainly focus on raw bamboo, which can not represent the wide variety of industrial bamboo products; These studies lack material parameters to support an entire heat and moisture process model; They focus only on material properties, rather than the performance of the building system. On the other hand, the development of theoretical models and computer programmes for the heat-air-moisture (HAM) transfer model greatly improves the computing capacity for the performance of the building envelope, and this makes it possible to simulate the coupled heat and moisture process with the HAM model instead of a simple evaluation by the Glaser model.

The motivation of this study is to gain systematically the material parameters of typical industrial bamboo products and then carry out a performance evaluation and optimization for building components and enclosed spaces, with comprehensive consideration of the representative conditions in terms of the external climate, building function, construction type, and HVAC. These are expected to serve as a reference for the application of bamboo in building envelope.

For the above goals, the study has completed four main tasks.

(1) Tests of hygrothermal properties are performed for raw bamboo and bamboo-based panels as construction partition boards, natural bamboo fiber as a thermal insulation infill, and bamboo charcoal as a moisture-control infill.
(2) The performance of bamboo's application in the building envelope is assessed through comparison with reference timber units at progressive levels of 'material-building component-enclosed space'.

Z. Huang, *Application of Bamboo in Building Envelope*,
Green Energy and Technology, https://doi.org/10.1007/978-3-030-12032-0

(3) Material and construction parameter optimization for layered construction is investigated through correlation analysis and factor impact analysis of a large sample of the coupled heat and moisture simulation results in Hot-Humid climate region.
(4) The HAM theoretical model and computer programme based on the Künzel equations are discussed and systematically performed to suggest improvements to the related studies on meteorologic data, material parameters, and evaluation method.

Main Conclusion

1 Hygrothermal property test on bamboo

Six construction partition materials, including flattened bamboo panel (FB), bamboo laminated lumber (BSB), plybamboo (BMB), bamboo scrimber (BFB), bamboo particleboard (BPB), and bamboo oriented strand board (BOSB), and two construction infill materials, namely the natural bamboo fiber and bamboo charcoal, are systematically tested for material properties, based on the material parameter requirements of the HAM model. Test items include bulk density test and vacuum saturation test for basic properties; sorption tests for moisture storage properties; capillary absorption, water vapor transmission, and drying tests for moisture transport properties; thermal analysis for heat storage properties; thermal conductivity test, surface light and thermal property test for heat transport properties. The test results initially form a properties database that could provide support for the application in the building envelope.

(1) Construction partition group

Correlation analysis between hygrothermal properties and basic properties shows that macroscopically gaseous moisture and heat storage properties have insignificant correlation with basic properties. Moisture and heat transport properties show stronger correlation with open porosity Φ than with dry bulk density ρ. Taking the correlation coefficient (r) and the statistical significance (sig) as indicators, the correlations between the hygrothermal properties and Φ are, respectively, 107.23–140.00% and 1.44–11.94% of those between the hygrothermal properties and ρ. Curve fitting between moisture transport properties and the moisture content shows that there is an exponential function relation between water vapor transfer coefficient δ and mass rate moisture content u that can be defined as $\delta = a \times \mathrm{EXP}(b \times u)$, and a logarithmic function relation between drying rate U and u that can be defined as $U = a \times \ln(u) + b$. For heat transport properties, a linear function $\lambda = a_w \times u + \mathrm{b}$ is generally used to approximate the relation between thermal conductivity λ and u, resulting in the thermal conductivity supplement, a_w [(W/m K)/u(–)] values, which are in the range 0.2137–0.6890. Spearman correlation analysis shows that the impact of moisture content in the gaseous moisture region on the surface light and thermal transport properties can be ignored, but further study is required in the liquid water region (*a*, *b*, and *c* in the formulae are fitting parameters).

The comparison between the five bamboo-based panels (BBPs, including BSB, BMB, BFB, BPB, and BOSB) and the FB (raw bamboo) shows that the BBP process can improve homogeneity and broaden the spectrum of material properties. In particular, the dry bulk density inhomogeneity of the BBPs are narrowed down from ca. 20% to ca. 10%. At FB 100%, the dry bulk density range increases to 84.60–166.39%; open porosity is 33.21–120.92%; hygric properties, including moisture storage and moisture transport, are 7.50–604.62%; and thermal properties, including heat storage and heat transport, are 50.88–159.28%. During the BBP process, the change magnitude of hygric and transport properties is greater than that of the thermal and storage properties, respectively.

(2) Construction infill group

For natural bamboo fiber (BF), results show that there are exponential relations between equilibrium moisture content u, moisture adsorption and desorption rate U, thermal conductivity λ, and the ambient relative humidity φ that can generally be defined with exponential function $Y = a \times EXP(b \times \varphi) + c$, of which the fitting degree of λ–φ ($R^2 = 0.9955$, sig = 0.000008) performs better than that of the linear function λ–u ($R^2 = 0.8976$, sig = 0.004073). Within the range of the test, the u–φ curve and U–φ curve grow rapidly when $\varphi = 85.4$–96.3% and $\varphi = 75$–95%, respectively, and there is a hysteresis effect of 0.52–3.50% in the range $\varphi = 11.2$–85.4%. The dry λ values of BF with bulk density $\rho = 70$–170 kg/m^3 are 0.0423–0465 W/(m K), which is comparable to the general vegetable fiber insulation materials. The wet λ values increase significantly with the rise of φ, and the growth rate is in positive correlation with the ρ values. The practical application of BF requires either moisture resistance treatment by reducing the U value or the control of the hygric environment (Y in the formula represents u, U, and λ, whereas a, b, and c in the formulae are fitting parameters.).

For bamboo charcoal (BC), compared with the five uncarbonized bamboo boards, the moisture-control capacity of BC is expanded when $\varphi = 33.4$–85.4%, and there is a hysteresis effect of 0.25–1.75% in the range $\varphi = 11.2$–85.4%. The water vapor transmission coefficient is enlarged to 9.99–213.2 times, which is 9.32E^{-11} kg/(m s Pa). The improvement of the moisture-control capacity and transport rate contributes better moisture-control ability to BC.

2 Hygrothermal performance assessment of bamboo building envelope

With the meteorological data of the 10 representative cities from nine typical climate zones of North America as the external conditions, timber and timber units are set as reference models and, accordingly bamboo and bamboo units of the same construction and space size as evaluation models. The performance on levels of material, building component, and space unit are compared, with comprehensive consideration of the representative conditions in terms of the external climate, building function, construction type, and HVAC. The reference timber units for comparison are set as benchmarks to provide performance assessment on the application of bamboo in building envelope.

(1) Timber property parameters from various databases are collected as the reference timber (RT) for the comparison with bamboo, of which the maximum and minimum values are adopted to define the RT range. The relative position (RP) of the bamboo parameters in RT is used to characterize the relative relation between bamboo and the corresponding timber products. Comparison results show that the basic properties, including the bulk density (ρ) and open porosity (Φ) of raw bamboo, are closer to hardwood (RP_ρ = 18.20%, RP_Φ = 46.59%) and far different from softwood (RP_ρ = 175.77%, RP_Φ = −122.12%) that is normally the raw material for wood-based panels. Except for BPB, the other BBPs have higher bulk density (RP_ρ = 122.15–182.14%) and lower porosity (RP_Φ = −28.92–2.44%), lower moisture storage and transport properties (RP_{hygric} = −20.79–49.02%), and higher heat storage and transport properties than RT ($RP_{thermal}$ = 77.61–232.14%, and S_{24h} is used as the indicator to characterize the thermal storage properties). The BBP process strengthens the distinction between bamboo and timber.

(2) A dynamic test in a wind tunnel with the meteorologic data of a typical Guangzhou summer meteorological day is performed to examine the moisture adsorption and desorption rate difference between bamboo scrimber (BFB) and corresponding antiseptic hardwood (HW). The results show that BFB has a lower moisture adsorption and desorption rate (U) than the reference HW, that the mean value and maximum value of U of the BFB are, respectively, 64.35 and 66.02% of the HW. The maximum adsorption and desorption rates of BFB are, respectively, $2.91E^{-05}$ kg/(m^2 s) and $4.45E^{-05}$ kg/(m^2 s), which are far greater than the $7.74E^{-07}$–$12.58E^{-07}$ kg/(m^2 s) tested in an indoor steady-state temperature and humidity environment (T = 23 °C, RH = 50%). The significant magnitude difference between the steady and dynamic test results show the necessity of taking practical meteorological conditions as an external environment for comprehensive evaluation.

(3) The annual simulation of the coupled heat and moisture process for building components and enclosed space units with the same construction and space sizes in WUFI Plus shows that bamboo units has a heat storage advantage and heat transport disadvantage that can vary with the operating conditions, including the external climate, building function, construction type, and the HVAC conditions. Judging from the annual hours of $-1.0 \leq \text{PMV} \leq 1.0$ in HVAC-off conditions ($PMV_{1.0}$) and the heating and cooling demand in HVAC-on conditions (P), bamboo units surpasses timber units in L-type construction (lightweight) in hot climate regions. The $PMV_{1.0}$ and P values of the bamboo units are, respectively, in the ranges 100.65–107.69% and 98.35–99.73% while setting the timber units as 100%. The comparison result is just the opposite in cold climate regions, where the corresponding values are 93.79–101.36% and 100.20–103.14%. The difference between bamboo and timber units is narrowed in M-type construction (heavyweight). Comparison

results show that 'substitute timber with bamboo' in tropical and subtropical regions, where bamboo forest is widely distributed, is conducive to using the strengths of the bamboo heat storage properties, and this is beneficial for utilizing the local forest resources and improving the building's physical performance.

(4) The larger moisture transport resistance of bamboo at the material level shows dialectical features when it comes to the building component and enclosed space levels. On the one hand, it can help to weaken the influence of the external hygric environment on the inside of the construction and the vapor transport. The annual moisture flow through exterior walls (M_{flow}) of bamboo units is 87.39–96.07% compared with the timber units. On the other hand, it might also cause water retardation on the boundaries, slow down the condensed water and indoor moisture from exhausting outward, and weaken the moisture buffering effect of the building envelope. The annual moisture exchange between indoor air and the inner surface of the exterior walls ($M_{i.flux}$) of the bamboo units is 77.67–86.53% compared with the timber units. It ultimately results in hygric performance weakness of the bamboo units; taking the annual hours of relative humidity $30\% \leq RH_i \leq 70\%$ in HVAC-off conditions ($RH_{i.30–70}$) and the humidification and dehumidification demand in HVAC-on conditions (H) as indicators, the $RH_{i.30–70}$ and H values of the bamboo units are, respectively, in the range 94.06–101.22% and 96.80–104.23% while setting the timber units as 100%.

3 Hygrothermal performance optimization on bamboo building envelope

With four groups of meteorological data from two representative cities in the Hot-Humid climate region of North America as external conditions, exterior walls are constructed in WUFI Plus, and the coupled heat and moisture process simulation is performed. Annual exterior walls' hygrothermal performance, the indoor hygrothermal environment, and HVAC demand are chosen as indicators, for which factors impact analysis is carried out to investigate the effect of the partition boards, air layer, thermal insulation infill, and moisture-control infill layers, as well as the façade rainfall treatment, on the hygrothermal performance of the building envelope. Climate-responsive design suggestions in Hot-Humid climate region are generated from the analysis results, in terms of material and construction parameter optimization.

(1) Partition boards

BFB (L-type construction) and LB as exterior board (M-type construction) are compared. L-type construction has a moisture resistance advantage that is beneficial for blocking the influence of the external hygric environment on the inside of the construction and the indoor hygric environment, as well as reducing humidification and dehumidification demand. M-type construction has a great heat storage advantage that is conducive to the indoor thermal environment and lowering the heating and cooling demand. The moisture transport weakness of M-type construction can be significantly improved by arranging a vapor barrier on its inner side. With this treatment, the P and H values in HVAC for M-type construction are, respectively, 80.63–102.84% and 102.17–133.33% while setting L-type construction as 100%, and correspondingly the $T_{i.18–28}$ and $RH_{i.30–70}$ values are 99.49–117.62% and 84.06–98.98%.

The comparison between BFB and BSB as interior board shows that the thermal capacity advantage of BFB is highlighted in L-type construction, whereas the BSB is beneficial for playing the moisture-control role of the building envelope and exhausting the inner moisture of the construction. Choosing BFB as the interior board for 2-layer construction of L-type is recommended, while BSB for M-type and 3-layer construction of the L-type can have better comprehensive performance. The annual BF infill moisture content mean value (w-BF_{mean}) and its amplitude (w-BF_{amp}) of the construction groups with BSB as interior boards are, respectively, in the range 39.15–99.61 and 63.93–109.45% while setting BFB groups as 100%. Correspondingly, the P value is 99.89–101.10% and 81.82–100.29%, respectively, for L-type and M-type construction.

The comparison among BMB, BPB, and BOSB as interlayer boards shows that, because the contribution proportion of the interlayer board to the hygrothermal performance of the whole component is relatively small, the difference among the three boards is slight. BPB performs overall the best as an interlayer board, but it requires a vapor barrier on its outer side in M-type construction to avoid the weakness of hygric performance.

(2) Air layer

In HVAC-on conditions, the improvement effect of an independent closed air layer is close to the closed air layer within the insulation cavity; therefore, the latter is recommended. Multiple air layers of smaller thickness ($d \leq 50$ mm) have a better effect than a single air layer of the same total thickness. Compared with the 2-layer (1-cavity) construction groups with a single air layer (thickness: 20–140 mm), the P value of the 3-layer (2-cavity) construction groups with two air layers (each thickness: 10–70 mm) is reduced to 82.59–95.59%. However, in HVAC-off conditions, the air layer (thickness: 10–70 mm) added to each cavity of the 3-layer and 2-layer construction groups reduces the $T_{i.18–28}$ value respectively to 86.37–97.81 and 93.52–99.31%, which indicates that the closed air layer is unsuitable for the building envelope projects that are mostly in HVAC-off conditions.

(3) Thermal insulation infill layer

The insulation layer arranged in the inner side of the insulation cavity is beneficial for weakening the external hygric environmental impact, reducing heat flow through exterior walls, and reducing the heating and cooling demand. Setting 30 mm BF on the outer side of the 30 mm air layer, compared with the opposite arrangement, results in an increase to 96.40–134.04, 98.87–112.35, and 99.06–111.82%, respectively, for the w-BF_{mean}, H_{flow}, and P values. Correspondingly, for the 3-layer construction, the setting of BF in the outer cavity brings about an increase to 97.82–227.12, 98.77–113.43, and 99.20–111.26%. The practical application of BF in Hot-Humid regions requires either moisture resistance treatment or the control of the hygric environment. The projection results show that the thermal performance of BF can be effectively improved by reducing its moisture adsorption and desorption rate.

(4) Moisture-control infill layer

The moisture-control BC arranged on the outer side of the BF layer helps to reduce the moisture content of BF in M-type construction significantly. Simultaneously, the heat and moisture flow through exterior walls and the HVAC demand are reduced, and the indoor hygric environment improved. Unlike the moisture-control interior finishes that directly interact with the indoor air, the mechanism of BC as a construction infill is to control the moisture process inside the construction then affect the moisture exchange between indoor and outdoor, and finally realize the regulation of the indoor hygric environment. When 30 mm BC is arranged on the outer side of the 30 mm BF layer, the w-BF_{mean}, M_{flow}, and H are reduced to 80.51–100.42%, while the $RH_{i.30–70}$ is improved to 97.67–164.58%.

(5) Rainfall treatment

As for the rainfall on the building façade, L-type construction is suggested to avoid rain. However, the evaporative cooling effect generated by rainfall on the exterior board (LB) of M-type construction can be utilized, while a vapor barrier should be arranged inside the LB layer and BSB to be used as the interior board, to block the liquid water at the exterior layer and facilitate the exhausting of moisture inside the construction. With this arrangement, the P value in HVAC under these conditions is reduced to 89.13–101.22%, whereas the $T_{i.18–28}$ value in HVAC-off conditions is improved to 103.04–119.35%.

4 Heat and moisture process model study on bamboo building envelope

With 20 cities from 15 representative climate regions of North America as external conditions, 15 groups of exterior walls are constructed in WUFI Plus and simulated for annual hygrothermal performance of the exterior walls, indoor hygrothermal environment, and HVAC demand of the enclosed space units. The HAM model based on Künzel equations is analyzed and compared with the widely used Glaser model. Suggestions are given for the studies on similar materials and their application in the building envelope, in terms of the meteorological parameters, material parameters, and evaluation methods.

(1) Correlation analyses are carried out between the simulation results and the meteorological parameters and show that the outdoor solar radiation, air temperature, relative humidity, and driving rain have complementary significant correlation with the corresponding results. This indicates the necessity of setting complete meteorological parameters as the external conditions. Among the simulation results, the annual humidification maximum value and, exterior wall moisture flow mean and maximum values are only significantly

correlated with the driving rain, of which the correlation coefficients are, respectively, 0.721–0.773 (at 0.01 level, 2-tailed), 0.498–0.514 (at 0.05 level, 2-tailed) and 0.523–0.555 (at 0.05 level, 2-tailed) The simulation without consideration of driving rain results in significant underestimation of moisture content, heat, and moisture flow, as well as error estimating HVAC demand, of which the error range is 92.84–100.88 and 31.93–108.19%, respectively, for the L-type and M-type construction. For construction with core layer exposed, the influence of liquid water is far greater than that of gaseous water.

(2) The moisture content-dependent thermal conductivity has a significant impact on the heat flow through the building component, indoor thermal environment, and heating and cooling demand. Taking the H_{flow}, $T_{i.amp}$, and P values as indicators, the assignment of a constant value lead to an error range of 78.10–117.38%. Correpondingly, the moisture content-dependent water vapor transfer coefficient significantly affects the moisture flow, hygric environment, and humidification and dehumidification demand. Taking the M_{flow}, $RH_{i.amp}$, and H values as indicators, the assignment of constant value lead to an error range of 59.79–143.71%. The deviation caused by the constant valuing shows greater uncertainty in the simulation results of the moisture process.

(3) The Glaser model is still widely used for simplified steady assessment of the hygric performance of certain constructions. However, affected by such factors as climate conditions and construction types, the annual dynamic simulation results shows different laws compared with the steady assessment, which means a long-period dynamic evaluation of the building envelope under practical conditions is necessary. Studies of the building envelope in many areas such as China has a strong bias toward heat process, whereas the moisture process is simply evaluated by the Glaser model. The hygric environmental parameters, such as driving rain, are not available, and the existing parameters for most building materials can not support the full HAM model. Therefore, the promotion of the coupled heat and moisture process model makes it necessary to carry out long-term and extensive accumulation of meteorologic and material data.

Main Innovations

(1) For the first time, the typically utilized industrial bamboo products are systematically tested for the properties required for application in the building envelope, and the property parameters obtained can support the complete HAM model. The correlation among the bamboo-based panel process, basic properties, and hygrothermal properties is quantified, and this provides reference for the hygrothermal property-oriented industrial production and application of bamboo.

(2) A performance comparison is made between bamboo and timber building components and enclosed space units, and the strengths and weaknesses

between bamboo and timber units under different conditions of external climate and construction type are clarified. A performance optimization study of the bamboo building envelope is conducted. It gives a systematic factor impact analysis of the material and construction parameters and provides guidance for the hygrothermal performance-oriented design for bamboo layered construction.

(3) The correlation between coupled heat and moisture process simulation results and the outdoor meteorological parameters is elucidated. The simulation results' error range is quantified, resulting from the lack of driving rain as a meteorological parameter, the absence of liquid water-related material parameters, and the constant valuing of the water vapor transfer coefficient and thermal conductivity, which provides a reference for the establishment of a model for the heat and moisture process of the bamboo building envelope.

Deficiencies in this Study

Insufficient test verification. The performance studies on building components and enclosed space units are carried out by means of a computer programme, without verification by measurement. Compared with pure thermal testing, the measurement of coupled heat and moisture process in the building envelope is far more difficult to perform efficiently. The moisture process is quite slow, and it might take several years to reach a state that can show certain regularities. Because of the durability of bamboo, problems such as construction layer aging, corrosion, shedding, and so on might occur during the long monitoring period and lead to unexpected errors in the measured results. However, further work to obtain verification and feedback for the simulation results and corresponding conclusions is necessary, through either experimental houses or practical projects in the future.

Inadequacy in the WUFI Plus model design. The WUFI Plus model design takes comprehensive consideration of external climate, indoor heat and moisture loads, construction types, and HVAC conditions. As the model input, the setting of meteorological parameters and construction types in this study is relatively sufficient. However, for the indoor heat and moisture loads and HVAC conditions, the default values of standard office in the computer programme, ideal heating and cooling, humidification and dehumidification, and mechanical ventilation values are adopted. The enclosed space units are simply set with a closed, independent, and dark room that is far away from practical conditions. In future work, the setting of space units, internal loads, and HVAC conditions should be refined to make the work more relevant to practical application.

The defects of the WUFI Plus computer programme. The heat and moisture process simulation is mainly aimed at the interior of the material that can not describe and fully quantify the exchange characteristics between the material surface and the air and between the adjacent construction layers; therefore, default empirical values are partially used. The ventilated air layer can not be simulated, so the ventilated construction that has the potential for hygrothermal performance optimization can not be discussed. Only one-dimensional simulation of the boundary condition is possible at the enclosed space unit level, while the problems caused by the thermal bridges can not be described. A study of thermal bridges must be carried out with the aid of WUFI 2D as a supporting analysis at the building component level.

Prospects for Future Work

Considering the achievements and deficiencies in this study, further work should be undertaken in the future.

(1) In terms of material. The weaknesses of bamboo and their prevention should be examined, for example, the hygrothermal environmental requirement for the typical bamboo mold growth should be tested and integrated with the hygrothermal environment simulation results from the HAM model to provide suggestions about the material treatment and environment control. The strengths of bamboo and their utilization should be studied, for example, combining the knowledge of bamboo-fiber-reinforced composites with the correlation analysis conclusions concerning hygrothermal properties and basic properties, and developing hygrothermal property-oriented building envelope materials with high performance.
(2) In terms of building component and enclosed space. Through detailed setting in terms of the internal load and HVAC conditions of the model, the performance' simulation should be more relevant to practical applications. Through experimental houses or practical projects, verification of and feedback for the simulation sections in this study should be obtained. Through complementary materials and composite construction methods, bamboo building envelope products of high performance should be developed.
(3) In terms of HAM model. Based on the HAM model systematic material property testing should be done for common building materials in building envelope. Furthermore, construction hygrothermal performance should be assessed and optimized, to gradually accumulate a research foundation for the application of the HAM model to the building envelope.

Appendix

Main data collection of the demonstration section (Chaps. 2–5)

Model study (Chap. 2)

Hygrothermal performance projection of building component in hygric environment
Moisture storage—Areal moisture content ξ/A [kg/m²]: $\xi/A = \sum_{i=1}^{n} u_i \cdot \rho_i \cdot d_i$ Moisture transport—vapor diffusion thickness s_d [m]: $s_d = \sum_{i=1}^{n} \mu_i \cdot d_i$ Heat storage—thermal inertia index D [–]: $D = \sum_{i=1}^{n} R_i \cdot S_i$ Heat transport—thermal resistance R [m² K/W]: $R = \sum_{i=1}^{n} d_i/\lambda_i$
The ratio of the max. value to the min. value in the range φ = 40–95%, which is divided into two groups—BF (with BF infill layer) and NF (without BF infill layer): $\xi/A_{RH=95\%}$—$\xi/A_{RH=40\%}$ ratio, BF: 453.36–557.17%, NF: 453.16–557.02% $s_{dRH=40\%}$—$s_{dRH=95\%}$ ratio, BF: 3903.08–8261.54%, NF: 4488.35–9920.29% $D_{RH=40\%}$—$D_{RH=95\%}$ ratio, BF: 105.25–108.98%, NF: 101.26–106.38% $R_{RH=40\%}$—$R_{RH=95\%}$ ratio, BF: 154.38–174.31%, NF: 106.86–114.16%
Simulation and analysis with HAM model for building envelope in North American typical climate zones
In WUFI Plus, comparative model groups are constructed by controlling the material parameters: Group HAM—complete material parameter Group HAM_0—based on the HAM above, the liquid water related material parameters are set as 0 Group $Glaser_{0/50/75}$—constant material parameter corresponding to φ = 0/50/75% External condition: 20 cities from 15 representative climate regions of North America: Groups: U0.15(zone 8); U0.2(zone 6A-8); U0.4(zone 4A-5B); U0.6(zone 1A-3C); U1.0 (zone 1A-2B) Meteorological parameters of driving rain: on/off Internal condition: $(3\ m)^3$ units, with indoor heat and moisture loads of standard office: convective heat: 33.3 W radiant heat: 25.2 W moisture: 17.55 g/h

(continued)

Z. Huang, *Application of Bamboo in Building Envelope*,
Green Energy and Technology, https://doi.org/10.1007/978-3-030-12032-0

(continued)

CO_2: 20.79 g/h human activity: 1.2 met Boundary condition: U_c group: 0.15/0.2/0.4/0.6/1.0 W/(m^2 K) Construction type: L-type: exterior board + infill layer + interior board M-type: masonry layer + infill layer + interior board H-type: masonry layer + interior board HVAC condition: Off: simulate the indoor hygrothermal environment On: maintaine the indoor air temperature within T_i = 20–26 °C, RH_i = 40–60%, simulate the exterior wall hygrothermal performance and the HVAC demand Operation period is set from 2015-01-01 to 2017-01-01, in which the hourly data for the second complete year are collected for further analysis
Meteorologic parameters: Correlation analyses with the simulation results (HAM model group) Group1 annual total/mean values: Annual outdoor air temperature mean value $T_{e.mean}$: H_{flow} (0.05 level, 0.541–0.734); $T_{i.mean}$ (0.01 level, 0.996–0.997); P (0.01 level, 0.683–0.710) Annual solar radiation SR: H_{flow} (0.01 level, 0.654–0.812); $T_{i.mean}$ (0.01 level, 0.928–0.929); P (0.01 level, 0.579–0.623) Annual outdoor relative humidity mean value $RH_{e.mean}$: $RH_{i.mean}$ (0.01 level, 0.653–0.680) Annual driving rain DR: H_{flow} (0.01 level, 0.586–0.630); M_{flow} (0.05 level, 0.498–0.514); $T_{i.mean}$ (0.05 level, 0.535–0.537) Group 2 annual amplitude/max. values: Annual outdoor air temperature max. value $T_{e.max}$: $T_{i.amp}$ (0.01 level, 0.860–0.966); $RH_{i.amp}$ (0.01 level, 0.741–0.825); $P_{cooling.max}$ (0.05 level, 0.477–0.545) Annual outdoor air temperature min. value $T_{e.min}$: $H_{flow.max}$ (0.01 level, 0.701–0.769); $T_{i.amp}$ (0.01 level, 0.860–0.966); $RH_{i.amp}$ (0.01 level, 0.741–0.825); $H_{humid.max}$ (0.01 level, 0.922–0.944) Annual outdoor relative humidity min. value $RH_{e.min}$: $RH_{i.amp}$ (0.01 level, 0.568–0.645) Annual driving rain max. value DR_{max}: $M_{flow.max}$ (0.05 level, 0.523–0.555); $RH_{i.amp}$ (0.05 level, 0.539–0.578); $H_{dehumid.max}$ (0.01 level, 0.721–0.773) Material parameters: liquid water (Simulation results ratio of the HAM_0/HAM model groups) w-BF_{mean}: at lowest 92.84/31.93/5.86% respectively for L/M/H-type construction For L-type construction, the lowest, in the range 97.68–100.88%; For M-type construction, between L-type and H-type, in the range 59.78–108.19%; For H-type construction: Exterior walls hygrothermal performance: H_{flow}: 94.95–100.07% M_{flow}: 92.08–100.00% Indoor hygrothermal environment: $T_{i.mean}$: 100.00–104.92%, $T_{i.amp}$: 96.43–102.75% $RH_{i.mean}$: 94.01–100.00%, $RH_{i.amp}$: 91.61–106.57% HVAC demand:

(continued)

(continued)

P: 78.07–100.00%, $P_{heating.max}$: 78.77–100.00%, $P_{cooling.max}$: 95.02–125.95%
H: 94.94–103.51%, $H_{humid.max}$: 99.90–112.65%, $H_{dehumid.max}$: 51.91–106.83%

Material parameters: moisture content-dependent water vapor transfer coefficient (Simulation results ratio of the Glaser/HAM_0 model groups)
Moisture process simulation results:
M_{flow}: 81.72–143.71%
$RH_{i.mean}$: 92.78–100.45%, $RH_{i.amp}$: 76.86–115.70%
H: 59.79–121.91%, $H_{humid.max}$: 24.61–199.37%, $H_{dehumid.max}$: 0.00–115.78%
Material parameters: moisture content-dependent thermal conductivity (Simulation results ratio of the Glaser/HAM_0 model groups)
Heat process simulation results:
H_{flow}: 97.06–117.38%
$T_{i.mean}$: 100.00–162.50%, $T_{i.amp}$: 78.52–102.06%
P: 78.10–116.35%, $P_{heating.max}$: 44.02–106.75%, $P_{cooling.max}$: 52.73–146.66%

Material property study (Chap. 3)

Properties test

Basic properties:
Bulk density test: Dry bulk density ρ_d [kg/m^3]
Vacuum saturation test: Open porosity Φ [%]
Hygric properties:
Sorption test: (Mass-volume rate) equilibrium moisture content w [kg/m^3]
Capillary absorption test: Capillary saturation moisture content w_{cap} [kg/m^3]
Vacuum saturation test: Vacuum saturation moisture content w_{vac} [kg/m^3]
Calculated with w_{cap} and A_{cap}: Liquid transport coefficient D_l [m^2/s]
Capillary absorption test: Water absorption coefficient A_{cap} [kg/(m^2 s$^{0.5}$)]
Water vapor transmission test: Water vapor transfer coefficient δ [kg/(m s Pa)]
Drying test: Drying rate/Adsorption and desorption rate U [kg/(m^2 s)]
Thermal properties:
Thermal analysis: Specific heat capacity c [J/(kg K)]
Calculated with ρ_d, c and λ: Thermal storage coefficient S_{24h} [W/(m^2 K)]
Thermal conductivity test: Thermal conductivity λ [W/(m K)]
Surface light and thermal properties' test—Hemispherical emissivity detector: Hemispherical emissivity ε [–]
Surface light and thermal properties' test—UV, VIS, NIR spectrometer: Light reflectivity (380–780 nm) ρ_v [%]; Solar direct reflectivity (200–2600 nm) ρ_e [%]; Solar direct absorptivity (200–2600 nm) α_e [%]

Test results:
Properties parameter database

BP—Bamboo panel:
Comparison between BBPs (bamboo-based panel) and FB (flattened bamboo panel):
ρ_d inhomogeneity are narrowed down from 21.58% to 6.97–11.38%
material properties spectrum is broadened, in which the BBP/FB ratio:
dry bulk density: 84.60–166.39%
open porosity: 33.21–120.92%
Hygric properties: 7.50–604.62%, of which

(continued)

(continued)

moisture storage: 33.21–159.90%
moisture transport: 7.50–604.62%
Thermal properties: 50.88–159.28%, of which
heat storage: 86.30–112.47%
heat transport: 50.88–159.28%
Correlation between basic properties and hygrothermal properties:
Between ρ_d–Φ: r = –0.9429, sig = 4.7743E^{-09}, significant negative correlation
Liquid water storage properties, both moisture and heat transport properties show stronger correlation with open porosity Φ than with dry bulk density ρ_d, Taking the r and sig value as indicators, for the ratio of the correlation with Φ and that with ρ_d, there are:
Correlation with w_{cap}: r–Φ/ρ_d is 125.72%, sig–Φ/ρ_d is 9.89%
Correlation with δ: r–Φ/ρ_d is 112.15%, sig–Φ/ρ_d is 2.45%
Correlation with A_{cap}: r–Φ/ρ_d is 109.13%, sig–Φ/ρ_d is 1.44%
Correlation with U: r–Φ/ρ_d is 107.23%, sig–Φ/ρ_d is 3.42%
Correlation with λ: r–Φ/ρ_d is 140.00%, sig–Φ/ρ_d is 11.94%
Gaseous moisture storage properties (u value) and heat storage properties (c value) show no significant correlation with basic properties
Curve fitting for the impact of moisture content on heat and moisture transport properties:
δ–u relation: can be defined as $\delta = a \times \exp(b \times u)$, ($R^2$ = 0.9083–0.9972, sig < 0.01)
U–u relation: can be defined as $\delta = a \times \ln(u) + b$, ($R^2$ = 0.9379–0.9904, sig < 0.01)
λ–u relation: can be defined approximatively as $\lambda = a_w \times u + b$, resulting in a_w [(W/m K)/u(–)]: FB—0.2587, BSB—0.2137, BMB—0.2185, BFB—0.3289, BPB—0.4583, BOSB—0.6890
ε, ρ_v, ρ_e, α_e–u relation: for ε–φ, r = 0.1951 and sig = 0.1575; for ρ_e–φ, r = –0.0598 and sig = 0.6674, show no significant correlation
BF—Bamboo fiber:
Dry thermal conductivity λ_d values of BF with bulk density ρ = 70–170 kg/m^3 are 0.0423–0.0465 W/(m K)
Relations of u, U and λ respectively with φ can generally be defined with exponential function:
u–φ relation: $u = 0.0088 \times \exp(3.3637 \times \varphi)$, of which R^2 = 0.9495, sig = 0.000009; In the range φ = 11.2–85.4%, there is a hysteresis effect of 0.52–3.50%
U–φ relation: U = 2.7610E^{-10} × exp(0.1110 × φ), of which R^2 = 0.9960, sig = 9.0054E^{-10}
λ–φ relation: $\lambda = 0.000457 \times \exp(4.3974 \times \varphi) + 0.0437$, of which R^2 = 0.9955, sig = 0.000008; In comparison, when linear function is used for relation fitting, there is $\lambda = 0.0421 - 0.016870 \times u$, of which R^2 = 0.8976, sig = 0.004073
BF moisture resistance treatment—When the U value is reduced to 1/2, the S and D values are calculated according to the fitting functions:
λ—24 h mean value: decreases from 0.0536 to 0.0518 W/(m K)
λ—24 h amplitude: decreases from 16.60 to 0.80%
S—24 h amplitude: decreases from 12.12 to 0.59%
D—24 h amplitude: decreases from 4.28 to 0.21%

BC—Bamboo charcoal:
u values: Performed the best moisture-control capacity when φ = 33.4–59.7% and $\varphi \geq$ 85.4%; In the range φ = 11.2–85.4%, there is a hysteresis effect of 0.25–1.75%
U–φ relation: U = 6.0040E^{-09} × exp(0.0731 × φ), of which R^2 = 0.987587, sig = 6.2315E^{-08}
Δ values: dry cup 1.01E^{-10} kg/(m s Pa), wet cup 8.58E^{-11} kg/(m s Pa), mean value is enlarged to 9.99–213.2 times compared with uncarbonized bamboo boards

(continued)

(continued)

Hygrothermal performance assessment on building component and enclosed space units in North American typical climate regions (Chap. 4)
Properties comparison between corresponding material varieties of bamboo and timber
Related parameters from various databases are collected to form the reference timber (RT) , the RT range is defined with the available maximum and minimum values, and the relative position (RP) of the bamboo parameters in RT is used to characterize the relative relation between bamboo and the corresponding timber products: Group 1 Raw material: FB—softwood and hardwood Group 2-1 Exterior board: BFB—Hardwood, Plywood high, Wood fiberboard hard Group 2-2 Interior board: BSB—Laminated board Group 2-3 Interlayer board: BMB—Laminated veneer lumber (LVL), Plywood; BPB—Particleboard/Chipboard, Fiberboard (MDF); BOSB—Wood oriented strand board (OSB)
RP: $$RP = \frac{\text{Bamboo value–RT.min value}}{\text{RT.max value–RT.min value}} \times 100\%$$ Raw material: FB versus hardwood: RP_{ρ} = 18.20%, RP_{Φ} = 46.59% FB versus softwood: RP_{ρ} = 175.77%, RP_{Φ} = −122.12% Bamboo/Timber based panels: (here used S_{24h} as indicator for heat storage property) BFB: RP_{ρ} = 126.79%, RP_{Φ} = –28.92%; RP_{hygric} = –20.79–30.65%, $RP_{thermal}$ = 84.38–112.54% BSB: RP_{ρ} = 163.81%, RP_{Φ} = –5.97%; RP_{hygric} = 4.69–35.54%, $RP_{thermal}$ = 113.89–143.65% BMB: RP_{ρ} = 122.15%, RP_{Φ} = –0.91%; RP_{hygric} = 4.23–46.60%, $RP_{thermal}$ = 169.84–232.14% BPB: RP_{ρ} = 24.66%, RP_{Φ} = 43.90%; RP_{hygric} = 10.15–172.00%, $RP_{thermal}$ = –3.28–17.64% BOSB: RP_{ρ} = 182.14%, RP_{Φ} = 2.44%; RP_{hygric} = 14.35–49.02%, $RP_{thermal}$ = 77.61–94.30%
Wind tunnel test on bamboo scrimber (BFB) and hardwood (HW)
Solar radiation: sunrise—06:00; sunset—18:00; max. value—539.0 W/m^2 Air temperature: mean value—28.7 °C; amplitude—2.6 °C; peak time—15:00 Relative humidity: mean value—80.3%; amplitude—12.5%; peak time—05:00 Wind speed: constant as 0.6 m/s
Comparison between BFB and HW: $U_{ab.max}$: BFB is 4.45E^{-05} kg/(m^2 s); HW is 6.74E^{-05} kg/(m^2 s); the BFB/HW ratio is 66.02% $U_{ab.mean}$: BFB is 1.49E^{-05} kg/(m^2 s); HW is 2.32E^{-05} kg/(m^2 s); the BFB/HW ratio is 64.35% Comparison between the results from the wind tunnel test and the standard test of BFB: The U values of BFB are respectively up to 50.22E^{-06} kg/(m^2 s) and 67.41E^{-06} kg/(m^2 s), much higher than the values 7.74E^{-07}–12.58E^{-07} kg/(m^2 s) resulted from the static drying test (T = 23 °C, RH = 50%)

(continued)

(continued)

Performance comparison between bamboo and timber building envelope
Simulation model design: External conditions: U0.4 group: from Zone 5-cool to Zone 4-mixed: Chicago (Ch), Salt Lake City (Sl), Seattle (Se) U0.6 group: Zone 3-warm: Atlanta (At), Los Angeles (Lo), San Francisco (Sa) U1.0 and U1.5 group: from Zone 2-hot to Zone 1-very hot: Tampa (Ta), Jacksonville (Ja), Honolulu (Ho), Miami (Mi) Internal conditions: 3 × 6 × 3 m^3 units, with standard office indoor heat and moisture load: convective heat: 66.6 W radiant heat: 50.4 W moisture: 35.10 g/h CO_2: 41.58 g/h human activity: 2.4 met Boundary conditions: Group of U_c: 0.4/0.6/1.0/1.5 W/(m^2 K) Construction type: L-type: exterior board + infill + interior board (full bamboo/timber) M-type: masonry + infill + interior board (bamboo/timber as internal finishes) Material group: Exterior board: BFB—Beech Interlayer board: BOSB—wood OSB Interior board: BSB—Spruce HVAC conditions: Off: export the indoor hygrothermal environment On: maintain the indoor air as T_i = 20–26 °C, RH_i = 40–60% with ideal HVAC device, export the exterior walls hygrothermal performance and the HVAC demand The operation period is set from 2015-01-01 to 2017-01-01, in which the hourly data for the second complete year are collected
Comparison of the simulation results between bamboo (B) and timber (T) model groups, for B/T ratio: Building component units: (HVAC on) Hygric performance: w-BF_{mean}: L-type construction, U0.4 group, 108.24–125.50%, U1.0 group, 94.33–97.72%; M-type construction, 100.06–102.91% (except Sl) w-BF_{amp}: L-type construction, 39.11–92.01% (except Se and Sa), of which U1.0 group, 9.11–59.81%; M-type construction, 80.23–130.24% M_{flow}: L-type construction, 87.39–96.00%; M-type construction, 90.48–96.07% $M_{e.flux}$: L-type construction, 37.73–53.59% $M_{i.flux}$: 77.67–86.53% Thermal performance: $T_{is.amp}$: 96.21–99.10% H_{flow}: U0.4 and U0.6 groups, 101.66–106.06%, U1.0 and U1.5 groups, 98.74–101.62% Enclosed space units: HVAC off: $PMV_{1.0}$: L-type construction, U0.4 and U0.6 groups, 93.79–101.36%, U1.0 and U1.5 groups, 100.65–107.69%; M-type construction, 97.23–102.98% $T_{i.amp}$: 96.91–99.25%

(continued)

(continued)

$T_{i.18–28}$: L-type construction, U0.4 and U0.6 groups, 96.34–100.37%, U1.0 and U1.5 groups, 101.78–106.11%; M-type construction, 98.82–103.19%
$RH_{i.amp}$: L-type construction, 101.88–114.72%; M-type construction, 101.11–108.67%
$RH_{i.30–70}$: L-type construction, 94.06–99.49%; M-type construction, 94.18–101.22%

HVAC on:
P: L-type construction, U0.4 and U0.6 groups, 100.20–103.14%, U1.0 and U1.5 groups, 98.35–99.73%; M-type construction, 99.94–102.46%
H: Except for individual conditions of Sa, L-type construction, 96.80–102.94%; M-type construction, 98.35–104.23%

Hygrothermal performance optimization on bamboo building envelope layered construction in North American Hot-Humid climate region (Chap. 5)

Building envelope coupled heat and moisture process simulation

Simulation model design:
External conditions:
Meteorological parameters of the 2 cities, Miami and Honolulu, from the North American 1A (very hot and humid) climate region, of which both include cool year and warm year
Internal conditions:
$(3\ m)^3$ units, with standard office indoor heat and moisture load:
convective heat: 33.3 W
radiant heat: 25.2 W
moisture: 17.55 g/h
CO_2: 20.79 g/h
human activity: 1.2 met
Boundary conditions:
Construction type:
L-type: exterior board + infill layer + interior board (18 mm BFB as exterior board)
M-type: masonry layer + infill layer + interior board (120 mm LB as exterior board)
Both included 2-layer group (2 partition boards with 1 cavity) and 3-layer group (3 partition boards with 2 cavities), for the U_c value control corresponding to RH = 0%:
2-layer: 0.9745–1.1375 $W/(m^2\ K)$
3-layer: 0.7654–0.9270 $W/(m^2\ K)$, add '12 mm thick interlayer board + 30 mm thick air layer' to the inside of the cavity to form the Li3/Mi3 groups; Correspondingly, add the interlayer board and air layer to the outside of the cavity to form the Lo3/Mo3 groups
6 conditions are arranged within the core cavity, including:
a: 30 mm air layer
b: 30 mm BF thermal insulation layer
c: inside 30 mm BF thermal insulation layer + outside 0.2 mm PE foil
d: inside 30 mm BF thermal insulation layer + outside 30 mm air layer
e: inside 30 mm air layer + outside 30 mm BF thermal insulation layer
f: inside 30 mm BF thermal insulation layer + outside 30 mm BC moisture-control layer
HVAC conditions:
Off: export the indoor hygrothermal environment
On: maintain the indoor air as T_i = 20–26 °C, RH_i = 40–60% with ideal HVAC device, export the exterior walls hygrothermal performance and the HVAC demand
The operation period is set from 2015-01-01 to 2017-01-01, in which the hourly data for the second complete year are collected

(continued)

(continued)

Layered construction parameters impact analysis
Partition board: Exterior board: BFB for L-type construction, LB for M-type construction, for the simulation results ratio between LB/BFB model groups: Exterior walls hygrothermal performance: w-BF_{mean}: for BFB as interior board, 106.30–352.92%; for BSB as interior board, 103.03–275.12%; of which for the model groups with 0.2 mm PE foil, 106.30–113.20%, 103.03–110.23% w-BF_{amp}: for BFB as interior board, 63.03–3126.80%; for BSB as interior board, 78.81–1590.98% M_{flow}: for BFB as interior board, 80.89–188.33%; for BSB as interior board, 73.36–260.98% H_{flow}: for BFB as interior board, 81.56–106.55%; for BSB as interior board, 80.35–100.62% Indoor hygrothermal environment: $T_{is.amp}$: 82.69–96.18% $T_{i.18–28}$: 99.49–117.62% $RH_{i.30–70}$: for BFB as interior board, 29.24–98.43%; for BSB as interior board, 9.42–98.98%; of which for the model groups with 0.2 mm PE foil, 84.06–98.43%, 84.42–98.98% HVAC demand: P: 80.63–102.84% H: 102.17–500.39%; of which for the model groups with 0.2 mm PE foil, for BFB as interior board, 104.00–129.07%; for BSB as interior board, 102.17–133.33%
Interior board: simulation results ratio between the model groups with BSB and BFB as interior board: Exterior walls hygrothermal performance: w-BF_{mean}: L-type construction, 96.41–99.61%; M-type construction, 39.15–99.40% M_{flow}: L-type construction, 120.10–151.48%; M-type construction, 114.76–212.00% H_{flow}: L-type construction, 99.62–101.10%; M-type construction, 81.43–99.69% Indoor hygrothermal environment: $T_{is.amp}$: L-type construction, 101.10–103.63%; M-type construction, 97.50–103.06% $T_{i.amp}$: L-type construction, 100.68–104.55%; M-type construction, 97.64–103.16% $RH_{i.amp}$: L-type construction, 75.07–87.08%; M-type construction, 63.93–109.45% HVAC demand: P: L-type construction, 99.89–101.10%; M-type construction, 81.82–100.29% H: 3-layer groups of the L-type construction, 84.19–100.00%; M-type construction, 87.61–155.40%, of which for the model groups with 0.2 mm PE foil, 87.61–99.17% Interlayer board: model groups with BMB/BPB/BOSB as interlayer boards, for the simulation results ratio between BPB/BMB and between BOSB/BMB model groups: Exterior walls hygrothermal performance: M_{flow}: L-type construction, BPB/BMB 99.43–102.26%, BOSB/BMB 99.28–99.95%; M-type construction, BPB/BMB 100.39–122.65%, BOSB/BMB 72.83–99.90% H_{flow}: L-type construction, BPB/BMB 96.14–99.98%, BOSB/BMB 98.99–99.68%; M-type construction, BPB/BMB 92.17–100.03%, BOSB/BMB 97.44–103.23% HVAC demand: P: L-type construction, BPB/BMB 96.47–99.47%, BOSB/BMB 99.13–99.76%; M-type construction, BPB/BMB 92.63–99.63%, BOSB/BMB 98.24–103.40% H: L-type construction, BPB/BMB 98.39–101.06%, BOSB/BMB 98.71–101.24%; M-type construction, BPB/BMB 99.05–117.15%, BOSB/BMB 82.57–100.24%

(continued)

(continued)

Air layer:
Independent closed air layer: Simulation results comparison between model groups with independent closed air layer (CA) and without air layer (NA):

Simulation results ratio between CA/NA model groups, show the impact of 30 mm independent closed air layer: (the following improvement of thermal performance include the contribution from the additional interlayer board)

$T_{i.amp}$: L-type construction, 86.49–93.72%; M-type construction, 88.28–96.14%
$T_{is.amp}$: L-type construction, 84.12–93.04%; M-type construction, 87.66–95.33%
H_{flow}: L-type construction, 66.42–83.40%; M-type construction, 69.35–92.99%
P: L-type construction, 68.28–87.70%; M-type construction, 71.17–94.95%

Of which the two comparison groups of Li3/L2 and Mi3/M2 characterize the impact of arranging independent closed air layer on the inner side of the construction, while the groups of Lo3/L2 and Mo3/M2 show the impact of the air layer arranged on the outer side, see 'Thermal insulation construction infill—Insulation position within double cavity'

Closed air layer within insulation cavity: Simulation results comparison between model groups with closed air layer (CA) and without air layer (NA) :

Simulation results ratio between CA/NA model groups, showed the impact of 30 mm closed air layer within insulation cavity:

$T_{i.amp}$: L-type construction, 95.77–98.29%; M-type construction, 96.04–99.12%
$T_{is.amp}$: L-type construction, 95.42–97.86%; M-type construction, 95.87–98.64%
H_{flow}: L-type construction, 84.99–90.12%; M-type construction, 85.96–91.82%
P: L-type construction, 88.55–92.79%; M-type construction, 89.86–94.72%
$T_{i.18\text{–}28}$: L2 and M2 construction groups, 95.10–98.68%; L3 and M3 construction groups, 94.35–98.78%

Air layer thickness:
Arrange air layer with thickness of 10/20/30/40/50/60/70 mm for the 3-layer construction groups (3 partition boards with 2 cavities), correspondingly, 20/40/60/80/100/120/140 mm for the 2-layer construction groups (2 partition boards with 1 cavity):

P: For the construction conditions with BF layer, 2-layer groups, 85.63–91.30%; 3-layer groups, 77.13–86.55%; For the conditions without BF layer, 2-layer groups, 67.01–76.03%; 3-layer groups, 56.61–68.53%; When the single 20–140 mm thick air layer is evenly divided into double 10–70 mm thick air layer, P is reduced further to 82.59–95.59%

$T_{i.18\text{–}28}$: L2 and M2 construction groups, 93.52–99.31%; L3 and M3 construction groups, 86.37–97.81%

Thermal insulation construction infill:
Insulation position within single cavity: Simulation results comparison between model groups of BF arranged on exterior (EBF) and interior side (IBF) in the cavity:

For L2 and M2 construction groups, Simulation results comparison between EBF/IBF model groups, show the impact of the position relation of the 30 mm BF layer and the 30 mm air layer within the cavity:

L2-type construction, w-BF_{mean} is 98.23–103.56%; H_{flow} is 99.97–101.47%; P is 99.98–101.39%

M2-type construction, w-BF_{mean} is 97.52–134.04%; H_{flow} is 98.87–112.35%; P is 99.06–111.82%

Insulation position within double cavity: Simulation results comparison between model groups of BF arranged on exterior (EBF) and interior side (IBF) in the cavity:

For L3 and M3 construction groups, Simulation results comparison between EBF/IBF model groups, show the impact of the position relation of the 30 mm BF layer and the 30 mm air layer in inner or outer cavity:

(continued)

(continued)

Li3, Lo3, Mi3 and Mo3 construction groups, w-BF_{mean} is 96.40–107.42%; H_{flow} is 99.15–100.86%; P is 99.24–100.67%

Simulation results comparison between model groups with 30 mm BF in the inner cavity (Lo3 and Mo3 construction groups) and the groups with the BF layer in the outer cavity (Li3 and Mi3 construction groups):

L-type construction, w-BF_{mean} is 97.59–102.23%; H_{flow} is 97.83–101.25%; P is 99.06–100.81%

M-type construction, w-BF_{mean} is 44.03–100.24%; H_{flow} is 88.16–100.71%; P is 89.88–100.61%

Moisture-control construction infill:
Simulation results comparison between model groups with (BC)/without (NC) BC layer arranged on the outer side of BF layer:
Simulation results ratio between BC/NC model groups, show the impact of 30 mm BC arranged on the outer side of BF layer:

Exterior walls hygrothermal performance:

w-BF_{mean}: L-type construction, 99.95–100.66%; M-type construction, 80.51–100.42%
M_{flow}: L-type construction, 88.13–96.19%; M-type construction, 85.77–97.97%
H_{flow}: L-type construction, 86.95–91.20%; M-type construction, 84.03–91.72%

Indoor hygrothermal environment:

$T_{i.amp}$: L-type construction, 92.86–96.70%; M-type construction, 93.86–97.17%
$T_{is.amp}$: L-type construction, 92.43–96.02%; M-type construction, 93.57–96.46%
$M_{i.exchange}$: L-type construction, 82.02–91.13%; M-type construction, 86.00–96.60%
$RH_{i.30–70}$: L-type construction, 100.29–103.95%; M-type construction, 97.67–164.58%

HVAC demand:

P: L-type construction, 88.18–93.37%; M-type construction, 85.12–94.34%
H: L-type construction, 83.47%–98.91%; M-type construction, 87.47–98.13%

Treatment against rainfall on facade:
Simulation results comparison between model groups without (NR)/with (DR) consideration to the driving rain on exterior board. Due to the large moisture resistance of BFB, the impact of fainfall on L-type construction can be ignored, while for the M-type construction:

Exterior walls hygrothermal performance:

w-BF_{mean}: M2 construction groups, 12.27–95.83%; Mi3 construction groups, 30.03–94.79%; Mo3 construction groups, 27.09–96.76%; of which for the model groups with 0.2 mm PE foil, 90.12–95.83%, 86.94–94.79%, 91.78–96.76%

M_{flow}: for BFB as interior board, 50.70–105.28%; for BSB as interior board, 33.08–105.40%; of which for the model groups with 0.2 mm PE foil, 100.85–105.28%, 99.80–105.40%

H_{flow}: for BFB as interior board, 85.72–107.80%; for BSB as interior board, 93.52–108.96%; of which for the model groups with 0.2 mm PE foil, 101.59–107.80%, 102.06–108.38%

Indoor hygrothermal environment:

$T_{i.18–28}$: for BFB as interior board, 84.45–97.43%; for BSB as interior board, 83.79–97.05%
$RH_{i.30–70}$: for BFB as interior board, 101.89–354.54%; for BSB as interior board, 101.08–1116.74%; of which for the model groups with 0.2 mm PE foil, 101.89–121.02%, 101.08–120.99%

HVAC demand:

P: for BFB as interior board, 90.43–110.03%; for BSB as interior board, 98.79–112.20%
H: for BFB as interior board, 25.86–93.81%; for BSB as interior board, 14.62–93.06%; of which for the model groups with 0.2 mm PE foil, 67.67–93.81%, 61.29–93.06%

Index

Z. Huang, *Application of Bamboo in Building Envelope*,
Green Energy and Technology, https://doi.org/10.1007/978-3-030-12032-0

The manufacturer's authorised representative in the EU is Springer Nature Customer Service Centre GmbH, Europaplatz 3, 69115 Heidelberg, Germany. If you have any concerns regarding our products, please contact ProductSafety@springernature.com

Printed and bound by CPI Group (UK) Ltd, Croydon, CR0 4YY
01/10/2026
02226185-0006